T0138181

Smart Innovation, Systems and Technologies

Volume 63

Series editors

Robert James Howlett, KES International, Shoreham-by-sea, UK
e-mail: rjhowlett@kesinternational.org

Lakhmi C. Jain, University of Canberra, Canberra, Australia;
Bournemouth University, UK;
KES International, UK
e-mails: jainlc2002@yahoo.co.uk; Lakhmi.Jain@canberra.edu.au

About this Series

The Smart Innovation, Systems and Technologies book series encompasses the topics of knowledge, intelligence, innovation and sustainability. The aim of the series is to make available a platform for the publication of books on all aspects of single and multi-disciplinary research on these themes in order to make the latest results available in a readily-accessible form. Volumes on interdisciplinary research combining two or more of these areas is particularly sought.

The series covers systems and paradigms that employ knowledge and intelligence in a broad sense. Its scope is systems having embedded knowledge and intelligence, which may be applied to the solution of world problems in industry, the environment and the community. It also focusses on the knowledge-transfer methodologies and innovation strategies employed to make this happen effectively. The combination of intelligent systems tools and a broad range of applications introduces a need for a synergy of disciplines from science, technology, business and the humanities. The series will include conference proceedings, edited collections, monographs, handbooks, reference books, and other relevant types of book in areas of science and technology where smart systems and technologies can offer innovative solutions.

High quality content is an essential feature for all book proposals accepted for the series. It is expected that editors of all accepted volumes will ensure that contributions are subjected to an appropriate level of reviewing process and adhere to KES quality principles.

More information about this series at http://www.springer.com/series/8767

Jeng-Shyang Pan · Pei-Wei Tsai
Hsiang-Cheh Huang
Editors

Advances in Intelligent Information Hiding and Multimedia Signal Processing

Proceeding of the Twelfth International Conference on Intelligent Information Hiding and Multimedia Signal Processing, November, 21–23, 2016, Kaohsiung, Taiwan, Volume 1

Springer

Editors
Jeng-Shyang Pan
Fujian Provincial Key Laboratory
   of Big Data Mining and Applications
Fujian University of Technology
Fujian
China

Hsiang-Cheh Huang
National University of Kaohsiung
Kaohsiung
Taiwan

Pei-Wei Tsai
Fujian University of Technology
Fujian
China

ISSN 2190-3018   ISSN 2190-3026   (electronic)
Smart Innovation, Systems and Technologies
ISBN 978-3-319-84346-9   ISBN 978-3-319-50209-0   (eBook)
DOI 10.1007/978-3-319-50209-0

Printed on acid-free paper

This Springer imprint is published by Springer Nature
The registered company is Springer International Publishing AG
The registered company address is: Gewerbestrasse 11, 6330 Cham, Switzerland

Preface

Welcome to the 12th International Conference on Intelligent Information Hiding and Multimedia Signal Processing (IIH-MSP 2016) held in Kaohsiung, Taiwan on November 21–23, 2016. IIH-MSP 2016 is hosted by National Kaohsiung University of Applied Sciences and technically co-sponsored by Tainan Chapter of IEEE Signal Processing Society, Fujian University of Technology, Chaoyang University of Technology, Taiwan Association for Web Intelligence Consortium, Fujian Provincial Key Laboratory of Big Data Mining and Applications (Fujian University of Technology), and Harbin Institute of Technology Shenzhen Graduate School. It aims to bring together researchers, engineers, and policymakers to discuss the related techniques, to exchange research ideas, and to make friends.

We received a total of 268 papers and finally 84 papers are accepted after the review process. Keynote speeches were kindly provided by Prof. Chin-Chen Chang (Feng Chia University, Taiwan) on "Some Steganographic Methods for Delivering Secret Messages Using Cover Media," and Prof. Shyi-Ming Chen (National Taiwan University of Science and Technology, Taiwan) on "Fuzzy Forecasting Based on High-Order Fuzzy Time Series and Genetic Algorithms." All the above speakers are leading experts in related research area.

We would like to thank the authors for their tremendous contributions. We would also express our sincere appreciation to the reviewers, Program Committee members, and the Local Committee members for making this conference successful. Finally, we would like to express special thanks for Tainan Chapter of

IEEE Signal Processing Society, Fujian University of Technology, Chaoyang University of Technology, National Kaohsiung University of Applied Sciences, and Harbin Institute of Technology Shenzhen Graduate School for their generous support in making IIH-MSP 2016 possible.

Fujian, China   Jeng-Shyang Pan
Fujian, China   Pei-Wei Tsai
Kaohsiung, Taiwan   Hsiang-Cheh Huang
November 2016

Conference Organization

Honorary Chairs

- Lakhmi C. Jain, University of Canberra, Australia and Bournemouth University, UK
- Chin-Chen Chang, Feng Chia University, Taiwan
- Xin-Hua Jiang, Fujian University of Technology, China
- Ching-Yu Yang, National Kaohsiung University of Applied Sciences, Taiwan

Advisory Committee

- Yoiti Suzuki, Tohoku University, Japan
- Bin-Yih Liao, National Kaohsiung University of Applied Sciences, Taiwan
- Kebin Jia, Beijing University of Technology, China
- Yao Zhao, Beijing Jiaotong University, China
- Ioannis Pitas, Aristotle University of Thessaloniki, Greece

General Chairs

- Jeng-Shyang Pan, Fujian University of Technology, China
- Chia-Chen Lin, Providence University, Taiwan
- Akinori Ito, Tohoku University, Japan

Program Chairs

- Mong-Fong Horng, National Kaohsiung University of Applied Sciences, Taiwan
- Isao Echizen, National Institute of Informatics, Japan
- Ivan Lee, University of South Australia, Australia
- Wu-Chih Hu, National Penghu University of Science and Technology, Taiwan

Invited Session Chairs

- Chin-Shiuh Shieh, National Kaohsiung University of Applied Sciences, Taiwan
- Shen Wang, Harbin Institute of Technology, China
- Ching-Yu Yang, National Penghu University of Science and Technology, Taiwan
- Chin-Feng Lee, Chaoyang University of Technology, Taiwan
- Ivan Lee, University of South Australia, Australia

Publication Chairs

- Hsiang-Cheh Huang, National University of Kaohsiung, Taiwan
- Chien-Ming Chen, Harbin Institute of Technology Shenzhen Graduate School, China
- Pei-Wei Tsai, Fujian University of Technology, China

Finance Chairs

- Jui-Fang Chang, National Kaohsiung University of Applied Sciences, Taiwan

Local Organization Chairs

- Jung-Fang Chen, National Kaohsiung University of Applied Sciences, Taiwan
- Shi-Huang Chen, Shu-Te University, Taiwan

Invited Session Organizers

- Feng-Cheng Chang, Tamkang University, Taiwan
- Yueh-Hong Chen, Far East University, Taiwan
- Hsiang-Cheh Huang, National University of Kaohsiung, Taiwan
- Ching-Yu Yang, National Penghu University of Science and Technology, Taiwan
- Wen-Fong Wang, National Yunlin University of Science and Technology, Taiwan
- Chiou-Yng Lee, Lunghwa University of Science and Technology, Taiwan
- Jim-Min Lin, Feng Chia University, Taiwan
- Chih-Feng Wu, Fujian University of Technology, China
- Wen-Kai Tsai, National Formosa University, Taiwan
- Tzu-Chuen Lu, Chaoyang University of Technology, Taiwan
- Yung-Chen Chou, Asia University, Taiwan
- Chin-Feng Lee, Chaoyang University of Technology, Taiwan
- Masashi Unoki, Japan Advanced Institute of Science and Technology, Japan.

- Kotaro Sonoda, Nagasaki University, Japan
- Shian-Shyong Tseng, Asia University, Taiwan
- Hui-Kai Su, National Formosa University, Taiwan
- Shuichi Sakamoto, Tohoku University, Japan
- Ryouichi Nishimura, National Institute of Information and Communications Technology, Japan
- Kazuhiro Kondo, Yamagata University, Japan
- Shu-Chuan Chu, Flinders University, Australia
- Pei-Wei Tsai, Fujian University of Technology, China
- Mao-Hsiung Hung, Fujian University of Technology, China
- Chia-Hung Wang, Fujian University of Technology, China
- Sheng-Hui Meng, Fujian University of Technology, China
- Hai-Han Lu, National Taipei University of Technology, Taiwan
- Zhi-yuan Su, Chia Nan University of Pharmacy and Science, Taiwan
- I-Hsien Liu, National Cheng Kung University, Taiwan
- Tien-Wen Sung, Fujian University of Technology, China
- Yao Zhao, Beijing Jiaotong University, China
- Rongrong Ni, Beijing Jiaotong University, China
- Ming-Yuan Cho, National Kaohsiung University of Applied Sciences, Taiwan
- Yen-Ming Tseng, Fujian University of Technology, China

Contents

Part III   Communication Protocols, Techniques, and Methods

Part I
Information Hiding and Secret Sharing

Demonstration Experiment of Data Hiding into OOXML Document for Suppression of Plagiarism

Akinori Ito[1]

Graduate School of Engineering, Tohoku University
`aito@spcom.ecei.tohoku.ac.jp`

Abstract. When a teacher gathers the students' assignment electronically, one big problem is plagiarism of report from documents in a Web site or other learner's report. This paper proposes a framework using data hiding technology to suppress plagiarism. In this framework, a teacher embeds ID of a student into a template file and sends the template file to the student. The student writes a report using the template file and submits it. The teacher extracts the ID from the report file to validate the file's originality. The Open Office XML (OOXML) format was chosen as the format of the template file because of its popularity. In the experiment, two methods were examined. The first method inserts small images with the ID into the template file. The second method embeds the ID into the fonts of the heading. According to the results of the experiments, the method using images was fragile against format conversion into PDF, and the method of font switching was more robust while the amount of embedded information was small.

1 Introduction

With the popularization of information technology (IT) in education, many activities in a class are now conducted on-line, such as presenting and retrieving reports or assignments. Such digitization of class activities is considered to be useful for digitization of class management [13, 7].

Plagiarism, i.e. copying descriptions from other books, the same report of other student or websites, is a problem recognized from a long time ago. When preparing a report digitally, so-called "copy and paste" activity become drastically easier than the case when writing the report with a pen on paper. Since the digital copy is easy to make, some students just copy the report from another student without reading its content other than the student's name.

To suppress students' plagiarism, many methods and systems have been developed for measuring the similarity between multiple report files and documents on the Internet. If a report has high similarity to other documents, the system points out that as a possibility of plagiarism [5, 17].

Even with these systems, when a teacher finds the contents of several reports very similar, it is difficult to determine which one is the original and which one is a copy.

© Springer International Publishing AG 2017
J.-S. Pan et al. (eds.), *Advances in Intelligent Information Hiding and Multimedia Signal Processing*, Smart Innovation, Systems and Technologies 63,
DOI 10.1007/978-3-319-50209-0_1

This paper proposes a framework in which data hiding technique is used for keeping copying history in a file. There are two kinds of plagiarism: one is copying contents of other documents in a book or a website, and the other one is copying the content of other student's report. The former one is rather easier to find. We can decide that a student plagiarized the content of a website when there is a high similarity between the report and the website. Thus, this paper treats the latter problem.

2 Problem Statement

In this paper, we assume the following framework. First, a teacher presents a problem to students. The teacher deals template files customized to each student out to the students, and the teacher instructs the students to write the report using that templates. A unique ID is embedded into the template before dealing it to the student. The student writes a report using the template and submits it to the teacher. The teacher extracts the ID from the document file and compares the ID with the original one. If they coincide, the report file is considered to be original; if they are different, there is a possibility that the report is copied from the other template.

There are several issues to be considered. The first one is the format of the template. In this work, Open Office XML (OOXML) format [11] is chosen, which is the standard format of Microsoft Word 2007 or later. One reason for choosing OOXML is its popularity, and another reason is its easiness of manipulation because OOXML format is a set of XML documents.

The second issue is to determine desired behavior of embedded ID when writing a report using a template. Those two properties are desirable:

1. When copying contents of the other report into the current template, the ID should be destroyed or some trace of copy is stored in the ID.
2. The ID should not be removed or destroyed by ordinary writing behaviors of a report.

In general, a writer has complete control to the document file, and thus it is impossible to completely satisfy the above two requirements. Therefore the requirements are relaxed as follows.

1. When a writer obtains the other one's document file and rewrites the minimum part such as student ID and name, or when a writer copies all of other one's report into the template, it affects the embedded ID.
2. Other *ordinary* writing behaviors do not harm the embedded ID.

These requirements mean that we cannot detect a partial plagiarism. Detection of such case is a problem for future works.

Figure 1 shows an example of a template file. The template file is not blank, having a header, forms of filling student ID and name, and numbers of questions etc.

Fig. 1. An example of a template file

3 Related Works

In this section, I review data hiding methods into documents (especially structured documents such as Word documents) and discusses relation to this work.

The main purpose of data hiding into a text-based document is steganography. There are two types of data hiding methods for documents: methods embedding information into text image [18, 4] and methods embedding information into character sequences or document format. The character-sequence-based hiding uses number of space characters between words [14], inflection of verbs [9], and selection of homonyms [16] for making payload in the stego data. The document-format-based methods include inter-character distances in a PDF document [19] and the methods that embed data into XML format [12].

OOXML format is a ZIP file packing multiple folders and XML files. It is easy to manipulate because most of its contents are text files. Liu and Tsai proposed a method for embedding data into OOXML using revision history as payload [8]. This method uses the existence of revision history in a text block as hidden data. Park et al. proposed a method to insert undefined XML tags as hidden data [12]. Castiglione et al. investigated various hiding methods into an OOXML document [3]. Their methods include a method to embed information into compression algorithm of ZIP, a method using Office macro, a method that inserts images with zero size (zero dimension image) into the document, and a method that manipulates the Revision Identifier (RI) in the XML. Sarsoh et al. proposed a method to embed information into number of space characters in a paragraph [15], which did not explicitly use the structure of OOXML. Bhaya et al. developed a method using similar font pairs [1]. They prepared several similar-looking font pairs, and information bits are embedded by altering these fonts. Mohamed et al. proposed a method to encrypt a part of an OOXML document and embed it into the document using zero-dimension images [10].

Most of these methods assume that the data are embedded into an existing document. However, the assumption of this work is to embed information into a template (which is almost blank) and the writer appends contents to the template. Therefore, we cannot use methods that assume a certain length of contents, such as embedding using space characters, homonyms, inflections, revision history etc. The methods we can use are the methods to embed information into hidden XML tags, zero dimension images or font-based methods.

4 OOXML Format

Open Office XML (OOXML) is a document format proposed by Microsoft Corp. as a format of Microsoft Office 2007 [11]. It is standardized by Ecma International in 2006 [6]. An OOXML document is a ZIP-compressed file of a folder containing multiple XML documents and multimedia files such as image, video and audio files. Those files are arranged in multiple folders and compressed into one file using ZIP compression. The OOXML standard contains various files such as Word document, Excel worksheet, and PowerPoint presentation file. A Word document includes an XML file containing the main body of the document (document.xml), XML files of headers, footers, footnotes, and styles.

5 Examined Methods

5.1 A method based on images

In this section, the following two methods are examined that satisfy the above-mentioned requirements. The first method is similar to the one using zero dimension images [3]. First, a white PNG image [2] with just one pixel is prepared, and the ID is stored in the comment field if the image. Since the image is small enough, the image can be inserted into the header, title, and headings. When copying the contents of other student's report, the writer may copy all the contents including title and headings. In this case, a teacher can know the original ID of the report. If we embed different IDs to different images and insert them into different headings, we can detect plagiarism section by section.

5.2 A method based on similar fonts

This method is similar to the one proposed by Bhaya et al. [1]. When using the Japanese version of Windows, following four fonts are provided as the standard fonts: MS-Gothic, MS P-Gothic, MS-Mincho and MS P-Mincho. Among them, MS-Gothic and MS P-Gothic, MS-Mincho and MS P-Mincho have the same shapes of characters with different spacing, respectively (P means the proportional spacing, and the fonts without P have fixed character width). Therefore, when we alter the font between them within a sentence, the difference is difficult to perceive. Figure 2 shows an example of font switching. In this figure, (a) uses MS-Gothic font, (b) changes MS Gothic and MS P-Gothic character by character and (c) uses MS P-Gothic font. From this figure, we can confirm that the character shapes the two fonts are identical. By assigning bit 0 to MS-Gothic font and 1 to MS P-Gothic font, we can embed information into a string. Since

(a) レポート提出用ファイル

(b) レポート提出用ファイル

(c) レポート提出用ファイル

Fig. 2. Mixed use of MS Gothic and MS P Gothic fonts. (a): MS Gothic, (b): Mixed, (c) MS P Gothic

this method embeds only one bit per character, this method can be applied to a heading with a certain number of characters. This method cannot be applied to short headings, such as "Q1" or "Q2". Since this method is robust against format conversion such as OOXML to PDF, we can expect that the embedded information can be extracted from the converted file.

6 Experiments

The above-mentioned data hiding methods are actually examined in classes of Tohoku University. Template files with individual student IDs were generated and dealt to the participants of the class. The students wrote reports using the templates and submitted to the author. The IDs were extracted from the reports and compared with the original student IDs. The reports were marked regardless of the result of the ID extraction. No explicit plagiarism was detected in the experiment.

Two experiments were conducted. The method with white images was examined in the first experiment, and the method is font switching was used in the second experiment for embedding IDs into the templates.

6.1 Experiment 1

This experiment was conducted in August 2015. In this experiment, an image with individual student ID (seven characters) was inserted into the document header and headings of each question. The image was one-pixel white PNG image and the ID was embedded into the comment field. In addition, the same ID was embedded into the OOXML metadata of the creator of the document.

An assignment was presented to 25 students and 24 of them submitted reports. Table 1 shows the document formats of offered documents. In addition to OOXML format, some students offered documents of DOC, PDF, and ODF. A student who offered report with ODF format might use OpenOffice instead of Microsoft Office. Five students offered reports with PDF. By analyzing the metadata of the documents, it was found that four of them used Microsoft Office and one used LibreOffice. The embedded image was extracted from ODF and PDF documents by opening the documents using Microsoft Word 2013 and converting into the OOXML format.

Table 1. Data formats of the submitted report files (experiment 1)

Format	OOXML(.docx)	.doc	ODF(.odt)	PDF
Population	18	0	1	5

Table 2. Result of ID detection (experiment 1)

Result	OOXML	ODF	PDF
Creator and image (two kinds)	10	0	0
Creator and image (one kind)	6	1	0
Creator only	0	0	4
No detection	2	0	1

Table 2 shows the detection results. This result suggests that the creator information was well preserved. However, tracing ID using *only* creator field is not suitable for this purpose. Since the creator information can be easily manipulated by the writer, the writer might change its creator field into right ID after copying the document. We need to combine creator field and other embedded information. In this experiment, two images were embedded into a template (into the header and the headings); two images were extracted from only 10 of 24 documents. For the case when only one image detected, the other image was not deleted from the document but merged with the other image because the two images were identical. No images were detected from PDF documents because the OOXML to PDF conversion filter of Microsoft Word converts all images in the document into JPEG format. In this process, all comments and metadata in the images are removed. From the experiment, it is found that a data hiding using PNG image cannot be used when we permit PDF conversion.

6.2 Experiment 2

The second experiment was conducted in October 2015. In this experiment, IDs were embedded into the title of the template using font switching. MS-Gothic and MS P-Gothic were used as the fonts. In this method, it was difficult to embed seven-character ID (56 bits) into the title, so an ID was compressed into 20 bits and embedded into the title. In this experiment, the teacher (the author) announced the participants that there was some mechanism that detects plagiarism if a student copied the contents of a report.

An assignment was presented to 67 participants and 57 reports were submitted. Among them, since two reports were offered as paper reports, they are excluded from the results. Table 3 shows the document formats of 55 reports. One out of 55 was a PDF document, two were DOC documents and the other 52 documents were in OOXML format. The documents in PDF and DOC were converted into OOXML format using Microsoft Word 2013.

Table 3. Data formats of the submitted report files (experiment 2)

Format	OOXML(.docx)	.doc	ODF(.odt)	PDF
Population	52	2	0	1

Table 4. Result of ID detection (experiment 2)

	OOXML	doc	PDF
Correct detection	48	2	0
No detection	4	0	1

Table 4 shows the results of ID detection. The student IDs could not be extracted from four documents out of 52 OOXML documents. The documents were written without using the specified template file in all of those cases. Two documents in DOC format used the specified templates, and the IDs could be extracted correctly. The ID could not be extracted from the PDF document; this was because the specified template file was not used for preparing the document. In the preliminary experiment, I confirmed the ID could be extracted after the OOXML–PDF–OOXML format conversion if the proper template file was used.

7 Conclusion

This paper investigated data hiding methods to embed IDs into OOXML format document to suppress plagiarism by students when offering reports digitally. Two methods were examined; the first one was based on small white images and the second one was based on font switching. According to the experimental results, it was found that the former method could embed a large amount of information but fragile against PDF conversion, and the latter method was robust against format conversion but the capacity was small.

In future works, it is desired to develop tools for embedding/extracting information into OOXML documents.

References

1. Bhaya, W., Rahma, A.M., AL-Nasrawi, D.: Text steganography based on font type in MS-Word documents. J. of Computer Science 9(7), 898–904 (2013)
2. Boutell, T. et al.: PNG (Portable Network Graphics) specification version 1.0. RFC2083 (1997)
3. Castiglione, A., D'Allessio, B., Santis, A.D., Palmieri, F.: Hiding information into OOXML documents: New steganographic perspectives. Journal of Wireless Mobile Networks, Ubiquitous Computing, and Dependable Applications 2(4), 59–83 (2011)

4. Chen, M., Wong, E.K., Memon, N.D., Adams, S.F.: Recent developments in document image watermarking and data hiding. In: Proc. SPIE. vol. 4518, pp. 166–176 (2001), http://dx.doi.org/10.1117/12.448201
5. Chuda, D., Navrat, P.: Support for checking plagiarism in e-learning. Procedia - Social and Behavioral Sciences 2(2), 3140 – 3144 (2010), http://www.sciencedirect.com/science/article/pii/S1877042810005185, innovation and Creativity in Education
6. Ecma International: Office Open XML File Formats (4th edition). ECMA-376 (2012)
7. Heinrich, E., Milne, J., Moore, M.: An investigation into E-Tool use for formative assignment assessment — status and recommendations. J. of Educational Technology & Society 12(4), 176–192 (2009)
8. Liu, T.Y., Tsai, W.H.: A new steganographic method for data hiding in Microsoft Word documents by a change tracking technique. IEEE Trans. Information Forensics and Security 2(1), 24–30 (2007)
9. Meral, H.M., Sankur, B., Özsoy, A.S., Güngör, T., Sevinç, E.: Natural language watermarking via morphosyntactic alterations. Computer Speech & Language 23(1), 107–125 (2009), http://www.sciencedirect.com/science/article/pii/S0885230808000284
10. Mohamed, M.A., Altrafi, O.G., Ismail, M.O., Elobied, M.O.: A novel method to protect content of Microsoft Word document using cryptography and steganography. Int. J. of Computer Theory and Engineering 7(4), 292–296 (2015)
11. Organization, I.S.: Information technology – document description and processing languages – office open xml file formats. ISO/IEC-29500:2008 (2008)
12. Park, B., J.Park, Lee, S.: Data concealment and detection in Microsoft Office 2007 files. Digital Investigation 5, 104–114 (2009)
13. Petrus, K., Sankey, M.: Comparing Writely and Moodle online assignment submission and assessment. In: Proc. 3rd Int. Conf. on Pedagogies and Learning (2007)
14. Por, L.Y., Delina, B.: Information hiding: A new approach in text steganography. In: Proc. 7th WSEAS Int. Conf. on Applied Computer & Applied Computational Science (ACACOS '08), Hangzhou (2008)
15. Sarsoh, J.T., Hashem, K.M., Hendi, H.I.: An effective method for hidding data in microsoft word documents. Global Journal of Computer Science and Technology 12(12) (2012)
16. Shirali-Shahreza, M.H., Shirali-Shahreza, M.: A new synonym text steganography. In: Proc. Int. Conf. Intelligent Information Hiding and Multimedia Signal Processing. pp. 1524–1526. IEEE Computer Society, Los Alamitos, CA, USA (2008)
17. Tresnawati, D., R, A.S., Kuspriyanto: Plagiarism detection system design for programming assignment in virtual classroom based on Moodle. Procedia - Social and Behavioral Sciences 67, 114 – 122 (2012), http://www.sciencedirect.com/science/article/pii/S1877042812052986
18. Villán, R., Voloshynovskiy, S., Koval, O., Vila, J., Topak, E., Deguillaume, F., Rytsar, Y., Pun, T.: Text data-hiding for digital and printed documents: theoretical and practical considerations. In: Proc. SPIE. vol. 6072, p. 607212 (2006), http://dx.doi.org/10.1117/12.641957
19. Zhong, S., Cheng, X., Chen, T.: Data hiding in a kind of PDF texts for secret communication. International Journal of Network Security 4 (2007)

A Revisit to LSB Substitution Based Data Hiding for Embedding More Information

Yanjun Liu[1,*], Chin-Chen Chang[1], and Tzu-Yi Chien[2]

[1]Department of Information Engineering and Computer Science,
Feng Chia University, Taichung 40724, Taiwan, R.O.C
yjliu104@gmail.com, alan3c@gmail.com
[2]Department of Information Engineering and Computer Science,
National Chung Cheng University, Chiayi 62102, Taiwan, R.O.C.
m80429@yahoo.com.tw

Abstract. Steganography is a widely used approach to embed tremendous amount of secret message while maintaining satisfactory visual quality. Least-significant-bit (LSB) substitution is one of the famous techniques applied in steganography, which makes modifications to the cover image by simply substituting secret bits for the LSBs of the cover pixel. This paper presents a novel data hiding scheme based on LSB substitution in which binary secret data can directly be concealed into the cover image. To enhance the embedding capacity, as much as 2 secret bits can be embedded into each cover pixel by modifying 3 LSBs with the guidance of a reference table. Experimental results confirm that the proposed scheme outperforms the related schemes in terms of embedding capacity and visual quality.

Keywords: Steganography, least-significant-bit (LSB), embedding capacity, visual quality

1 Introduction

Nowadays, steganography (also called data hiding) has played an important role in information security due to the advantage that it not only embeds a massive amount of secret message into cover media, but also maintains satisfactory image quality without visual perception. Therefore, more and more researchers concentrate on data hiding in which digital images, videos and audios are commonly used as cover media. Different data hiding methods such as LSBs [1–8], pixel value difference (PVD) [9, 10] and Gray code [11–14] have been developed.

Data hiding technique can be traced back to the method proposed by Petitcolas et al. [1] in 1999 in which only the communicational entities know that the secret message was embedded into the multimedia. To increase the security, EL-Emam [2] proposed a method that embedded different numbers of secret bits using LSB in colorful images by three channels, i.e. red channel, green channel and blue channel. Although this method can strengthen the security during the transmission, the embedding capacity is much lower by some restrictions. Thus,

© Springer International Publishing AG 2017
J.-S. Pan et al. (eds.), *Advances in Intelligent Information Hiding and Multimedia Signal Processing*, Smart Innovation, Systems and Technologies 63,
DOI 10.1007/978-3-319-50209-0_2

to enhance the embedding capacity, Ioannidoua et al. [3] proposed a scheme based on the method of EL-Emam [2] to embed one extra secret bit in the edge area of the image.

Least-significant-bit (LSB) substitution [1–8] is one of the famous techniques applied in data hiding, which makes modifications to the cover image by simply substituting secret bits for the LSBs of the cover pixel. LSBs can achieve high embedding capacity while keeping good image quality. However, it cannot resist statistical analysis on the stego-image. In order to overcome the disadvantage of LSBs, Chen and Chang [11] proposed a method which followed the rule of Gray codes and divided the Gray codes into odd Gray codes and even Gray codes. As a result, the secret data can be embedded in the cover image by odd Gray codes or even Gray codes using 4-LSB. However, this method can neither accommodate a massive amount of secret data and nor maintain great visual quality of stego-image due to the fact that it modified four bits to only embed one secret bit.

On the other hand, Mielikainen [8] developed a method which exploited pairs of LSB matching to improve the embedding efficiency. Shen and Huang [9] utilized this advantage to propose a method that adopted Hilbert curve to segment the cover image into several pixel pairs. The difference of the two pixels in each pair is calculated and simple function (i.e., modular function) is used to embed secret data in each pixel pair. However, this method may cause overflow or underflow problems. In 2015, Jung and Yoo [10] proposed another method based on pixel value difference (PVD). Unfortunately, the visual quality of stego-image decreased rapidly when the cover image accommodated a massive amount of secret data.

In this paper, we propose a novel data hiding scheme based on LSB substitution in which binary secret data can directly be concealed into the cover image. According to a constructed reference table, 2 secret bits can be embedded into each cover pixel by modifying the 3 LSBs at most by 2. The proposed scheme can achieve high embedding capacity while maintaining great visual quality of the stego-image.

The rest of the paper is organized as follows. In Section 2, we briefly review Chen and Chang's method [11]. Our proposed scheme is described in detail in Section 3. Finally, our experimental results and conclusions are presented in Sections 4 and 5, respectively.

2 Review of Chen and Chang's scheme

Chen and Chang's data hiding scheme followed the rule of Gray code [12] to embed secret message. The Gray code is an approach to encode an integer in the 2^n-ary numeral system as an n-bit binary code in such a manner that two consecutive integers have only one bit different. The method of reflected Gray code is used to create n-bit Gray codes in Chen and Chang's scheme. Let G_n be an n-bit Gray code sequence, then G_1, G_2 and G_3 are shown as follows:

$$G_1 = \begin{cases} 0 = 0 \\ 1 = 1 \end{cases}, G_2 = \begin{cases} 00 = 0 \\ 01 = 1 \\ 11 = 2 \\ 10 = 3 \end{cases} \text{ and } G_3 = \begin{cases} 000 = 0 \\ 001 = 1 \\ 011 = 2 \\ 010 = 3 \\ 110 = 4 \\ 111 = 5 \\ 101 = 6 \\ 100 = 7 \end{cases}.$$

Denote $g = g_n g_{n-1} \cdots g_1$ as a Gray code in G_n, where g_i is one binary bit. Then the Gray code function $\text{Gray}(g)$ is denoted as the corresponding 2^n-ary value of g. In Chen and Chang's scheme, g is regarded as the n-LSB string of a cover pixel. One secret bit is embedded into a cover pixel in such a manner that just the right-most bit (i.e., g_1) of g in each cover pixel is flipped according to the characteristic of Gray code. Therefore, this scheme significantly reduces the distortion by modifying each pixel at most by 1. The embedding algorithm is illustrated below:

Embedding algorithm

Input: Cover image CI, binary secret data stream S, n-bit Gray code sequence G_n

Output: Stego image SI

Step 1. Obtain a pixel p from CI.

Step 2. Extract the n LSBs of p and denote the n-LSB string as g.

Step 3. Read one bit m from S. If $m = 1$, then go to Step 4; otherwise, go to Step 5.

Step 4. If $\text{Gray}(g)$ is odd, then remain p unchanged; otherwise, flip the right-most bit of p. Go to Step 6.

Step 5. If $\text{Gray}(g)$ is even, then remain p unchanged; otherwise, flip the right-most bit of p.

Step 6. Go to Step 1 until all secret bits have been embedded.

To clearly understand Chen and Chang's scheme, we give an example to demonstrate the embedding algorithm under $n = 4$ as follows. Assume that a cover pixel is $214_{10} (= 11010110_2 2)$ and the secret bit to be embedded is 1. Obviously, the last four bits of the cover pixel is $g = 0110$. We obtain $\text{Gray}(g) = \text{Gray}(0110) = 4$, which is an even value. Consequently, the right-most bit of the cover pixel is flipped and the cover pixel is modified as $215_{10} (= 11010111_2)$.

3   Proposed scheme

Chen and Chang's scheme can significantly reduce the distortion by modifying each pixel at most by 1 to the characteristic of Gray code. However, their scheme has a disadvantage that the embedding capacity is very low because each pixel can accommodate only one secret bit in spite of the value of n. Consequently, in order to enhance the capacity, we propose a novel data hiding scheme based on LSB substitution such that 2 secret bits can be embedded into each cover pixel.

In our proposed scheme, the binary secret message is divided into a sequence of 2-bit segments and the value of each segment is represented as a 4-ary digit. To conceal a digit in to a cover pixel, four candidate reference tables are constructed to guide the embedding process. As shown in Tables 1 (a)-(d), each element in the first row of each table represents 3 LSBs of a cover pixel, which corresponds to a 4-ary digit in the second row. It should be noticed that each table can imply the embedding of only three digits that are included in it. For example, Table A is used to embed the digits 0, 1 and 2 but not the digit 3. Similarly, Table B, C and D cannot be used to embed 0, 1 and 2, respectively. To solve this problem, we can simply employ an indicator to identify the digit that does not occur in a specified reference table, and then transform this digit to one of the digits included in this table to perform embedding. In order to minimize the number of indicator bits so as to further increase the embedding capacity, we count the frequency of the secret digits 0, 1, 2 and 3, respectively, and select one reference table in which the digit with the lowest frequency does not occur. This digit is then transformed to the digit with the second lowest frequency and the embedding is conducted according to the selected reference table. For example, if the digit 3 has the lowest frequency in the secret message, we select reference table A to for embedding. If the digit to be embedded is 0, 1 or 2, it can be easily embedded according to table A. In particular, we set the indicator as 0 for the embedded digit 0 that has the second lowest frequency. If the digit to be embedded is 3, we set the indicator as 1 meanwhile transform the digit to 0 and embed it just as the way that 0 does.

Table 1. Candidate reference tables

(a) Reference table A in which digit 3 does not occur

3 LSBs of a cover pixel	000	001	010	011	100	101	110	111
Corresponding digit	0	1	2	0	1	2	0	1

(b) Reference table B in which digit 0 does not occur

3 LSBs of a cover pixel	000	001	010	011	100	101	110	111
Corresponding digit	1	2	3	1	2	3	1	2

(c) Reference table C in which digit 1 does not occur

3 LSBs of a cover pixel	000	001	010	011	100	101	110	111
Corresponding digit	2	3	0	2	3	0	2	3

(d) Reference table D in which digit 2 does not occur

3 LSBs of a cover pixel	000	001	010	011	100	101	110	111
Corresponding digit	3	0	1	3	0	1	3	0

The embedding algorithm of our proposed scheme is described as follows:

Embedding algorithm

Input: Cover image CI, secret image S

Output: Stego-image SI

Step 1. Covert S to a binary secret data stream S' and divide S' into a sequence of 2-bit segments. Represent the value of each segment as a 4-ary digit from 0-3.

Step 2. Count the frequency of the secret digits 0, 1, 2 and 3, respectively. Denote s_1 and s_2 as the digits with the lowest and second lowest frequency, respectively.

Step 3. Select the reference table in which s_1 does not occur.

Step 4. Obtain a pixel p from CI. Extract 3 LSBs of p and denote the 3-LSB string as p^*.

Step 5. Read a 4-ary digit s from S'. If $s = s_1$, set the indicator $flag = 1$ and let $s = s_2$; if $s = s_2$, set the indicator $flag = 0$.

Step 6. Find the corresponding digit d of p^* in the reference table.

Step 7. Choose the digit d' that is equal to s and has the shortest distance with d in the selected reference table.

Step 8. Find the first-row element p^{**} in reference table which corresponds to the digit d'.

Step 9. Modify p^* to p^{**}.

Step 10. Go to Step 4 until all secret data have been embedded. Output stego-image SI.

The above embedding algorithm indicates that the cover pixels are modified by +1 or -1 except for those of which the 3 LSBs are 000 and 111. Thus, the distortion of our proposed scheme can be very small with a high embedding capacity.

After the receiver obtains the stego-image, he/she can extract the secret image from the stego-image by inversing the data embedding process. The extraction algorithm of our proposed scheme is described as follows.

Extraction algorithm

Input: Stego-image SI

Output: Secret image S

Step 1. Obtain a pixel p' from SI. Extract 3 LSBs of p' and denote the 3-LSB string as p'^*.

Step 2. Find the corresponding digit d' of p'^* in the reference table.

Step 3. If $d' = s_2$ and $flag = 1$, s_1 is the hidden data; if $d' = s_2$ and $flag = 0$, $s_2 s_1$ is the hidden data; otherwise, d' is the hidden data.

Step 4. Go to Step 1 until all secret data have been extracted. Output the secret image S.

For clearer explanation, we give an instance to demonstrate our scheme. Assume that a cover pixel p is 155 and the secret bits to be embedded are $s = 01_2(1_4)$. Suppose 01_2 and 10_2 have the lowest and second lowest frequency in the binary secret stream S', respectively. Therefore, $s_1 = 01_2(1_4)$ and $s_2 = 10_2(2_4)$ and Table C is selected as the reference table. Because the extracted 3 LSBs of p are 011_2 and $s = s_1$, we set $flag = 1$ and let $s = s_2 = 10_2(2_4)$. Then, we find the corresponding digit $d = 2_4$ of 011_2 in the reference table C. Thus, 3 LSBs of p remain unchanged because the corresponding digit $d' = s = 2_4$ of 011_2 (3-LSB of p itself) has the shortest distance with d. Finally, the stego pixel is 155. In the extraction, the extracted 3 LSBs of stego pixel 155 is 011_2 and we find the corresponding digit $d' = 2_4$ of 011_2 in the reference table C. Since $d' = s_2$ and $flag = 1$, the hidden data is $s_1 = 1_4 = 01_2$.

4   Experimental results

Peak signal to noise ratio (PSNR) is used in our experiments to measure the similarity between the stego-image and the cover image. PSNR is defined as follows:

$$\text{PSNR} = 10\log_{10}\left(\frac{255^2}{\text{MSE}}\right), \tag{1}$$

where the mean square error (MSE) for a $W \times H$ grayscale image is defined as follows:

$$\text{MSE} = \frac{1}{W \times H}\sum_{i=1}^{W}\sum_{j=1}^{H}(I_{i,j} - S_{i,j})^2, \tag{2}$$

where $I_{i,j}$ and $S_{i,j}$ are the original pixel value and the stego pixel value, respectively. In the following, we employ four 512×512 standard grayscale images, Lena, Baboon, GoldHill and Peppers as test images in our experiments. As shown in Figure 1 (a), a 512×512 binary image Peppers is adopted as the secret image. Table C is selected as the reference table due to the fact that the bit string 01 has the lowest frequency occurring in the secret image Peppers. The experimental results are shown in Table 2 which indicates that our proposed scheme outperforms Chen and Chang's scheme in terms of the embedding capacity (EC) while great PSNR is still achieved. That is, PSNR in our scheme can reach 49 dB even if we embed 517,388 bits in the cover images. In particular, Figure 1 shows the stego-image for Lena (see Figure 1 (b)) and the extracted binary image Peppers (see Figure 1 (c)). It can be observed that it is impossible to distinguish between the cover image and stego-image by human eyes since the PSNR value is achieved up to 49.04 dB.

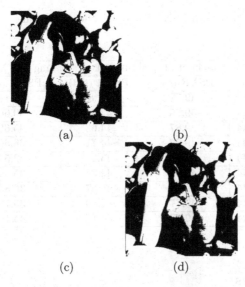

(a)   (b)

(c)   (d)

Fig. 1. Results of the proposed scheme (a) original secret image (b) original cover image (c) stego-image (PSNR = 49.04 dB) (d) extracted secret image

Table 2. Performance comparisons between Chen and Chang's scheme and our scheme

	Chen and Chang [11]		Our scheme	
Cover image	EC (bits)	PSNR (dB)	EC (bits)	PSNR (dB)
Lena	262,144	51.15	517,388	49.04
Baboon	262,144	51.13	517,388	49.13
GoldHill	262,144	51.13	517,388	49.13
Peppers	262,144	51.14	517,388	49.11

To further evaluate the performance of our proposed scheme, Table 3 give a comparison of PSNR values among different data hiding schemes under the same embedding capacity (262,144 bits). The results are tested on 512×512 binary secret image Peppers and eight 512×512 grayscale cover images Lena, Baboon, Airplane, Boat, Barbara, Peppers, Tiffany and Man. From Table 3, we can see that the average PSNR of our proposed scheme is greater than 52 dB, which is better than that of other schemes.

Table 3. PSNR values of different schemes

| | | Maleki et al. [15] | Shen et al. [9] | Jung and Yoo [10] | Our scheme |
Cover image	EC (bits)	PSNR (dB)	PSNR (dB)	PSNR (dB)	PSNR (dB)
Lena	262,144	49.73	44.8	37.19	52.03
Baboon	262,144	49.26	42.56	30.66	52.16
Airplane	262,144	49.87	44.63	35.66	52.17
Tiffany	262,144	49.72	44.71	36.56	52.13
Barbara	262,144	49.46	43.28	32.54	52.12
Man	262,144	49.52	44.34	35.56	52.18
Peppers	262,144	49.70	44.67	36.85	52.12
Boat	262,144	49.77	44.39	35.76	52.12
Average	262,144	49.63	44.17	35.10	52.13

5 Conclusions

In this paper, we proposed a novel data hiding scheme based on LSB substitution. With the guidance of the constructed reference table, each cover pixel can embed two secret bits. The distortion of the cover pixel is at most 2, which leads to a great image quality. The experimental results show that for a 512×512 image, our proposed scheme can embed at most 517,388 bits while the image quality can maintain above 49 dB. Comparisons demonstrate that the proposed scheme outperforms the related schemes in terms of embedding capacity and visual quality.

References

1. F. A. P. Petitcolas, R. J. Anderson and M. G. Kuhn.: Information hiding a survey. Proceedings of the IEEE, Vol. 87, No. 7, pp. 1062–1078 (1999)
2. N. N. EL-Emam.: Hiding a large amount of data with high security using steganography algorithm. Journal of Computer Science, Vol. 3, No. 4, pp. 223–232 (2007)
3. A. Ioannidoua, S. T. Halkidisb and G. Stephanidesb.: A novel technique for image steganography based on a high payload method and edge detection. Expert Systems with Application, Vol. 39, No. 14, pp. 11517–11524 (2012)
4. H. R. Kanan and B. Nazeri.: A novel image steganography scheme with high embedding capacity and tunable visual image quality based on a genetic algorithm. Expert Systems with Application, Vol. 39, No. 14, pp. 11517–11524 (2014)
5. K. H. Jung and K. Y. Yoo.: Steganographic method based on interpolation and LSB substitution of digital images. Multimedia Tools and Applications, Vol. 74, No. 6, pp. 2143–2155 (2015)
6. C. H. Yang.: Inverted pattern approach to improve image quality of information hiding by LSB substitution. Pattern Recognition, Vol. 41, No. 8, pp. 2674–2683 (2008)

7. X. Liao, Q. Y. Wen and J. Zhang.: A steganographic method for digital images with four-pixel differencing and modified LSB substitution. Journal of Visual Communication and Image Representation, Vol. 22, No. 1, pp. 1–8 (2011)
8. J. Mielikainen.: LSB matching revisited. IEEE Signal Processing Letters, Vol. 13, No. 5, pp. 285–287 (2006)
9. S. Y. Shen and L. H. Huang.: A data hiding scheme using pixel value differencing and improving exploiting modification directions. Computer and Security, Vol. 48, pp. 131–141 (2015)
10. K. H. Jung and K. Y. Yoo.: High-capacity index based data hiding method. Multimedia Tools and Applications, Vol. 74, No. 6, pp. 2179–2193 (2015)
11. C. C. Chen and C. C. Chang.: LSB-based steganography using reflected gray code. IEICE Transactions on Information and Systems, Vol. E91-D, No. 4, pp. 1110–1116 (2008)
12. M. Schwartz and T. Etzion.: The structure of single-track gray codes. IEEE Transactions on Information theory, Vol. 45, No. 7, pp. 2383–2396 (1999)
13. C. C. Chang, C. C. Lin and Y. H. Chen.: A secure data embedding scheme using gray-code computation and SMVQ encoding. Information Hiding and Applications, Vol. 227, pp. 63–74 (2009)
14. X. Y. Luo, F. L. Liu, C. F. Yang, S. G. Lian and Y. Zeng.: Steganalysis of adaptive image steganography in multiple gray code bit-planes. Multimedia Tools and Applications, Vol. 57, No. 3, pp. 651–667 (2012)
15. N. Maleki, M. Jalali and M. V. Jahan.: Adaptive and non-adaptive data hiding methods for grayscale images based on modulus function. Egyptian Informatics Journal, Vol. 15, No. 2, pp. 115–127 (2014)

Behavior Steganography in Social Network

Xinpeng Zhang

School of Communication and Information Engineering, Shanghai University, Shanghai,
200444, China
Email: xzhang@shu.edu.cn

Abstract. Most modern steganographic techniques embed secret data into digital multimedia by slight modifying the cover data. This work proposes a novel steganographic scheme converting the secret data into the behaviors of individuals in social network, not the multimedia data. In the scheme, a sender makes "love" marks on the news published by his friends with given rates for representing the secret data, and a receiver who is a friend of the sender can extract the secret data from a part of the sender's "love" marks although some "love" marks made by the sender may be invisible to the receiver.

Keywords: Steganography, behavior, social network

1   Introduction

The purpose of steganography is to convert secret message into another inconspicuous form to avoid unwanted attention. While there were a lot of traditional steganographic manners in past centuries, most modern steganographic techniques embed the secret data into digital multimedia, such as image, video/audio, by slight modifying the cover data [1, 2]. At the same time, a number of steganalytic techniques are also developed to detect the secret data in multimedia according to statistical abnormality of cover data [3, 4]. In recent years, more and more data in various types besides digital multimedia are generated and transmitted in the Internet. This work attempts to present a novel steganographic manner using behaviors in social network as camouflage of secret data. In other words, the secret data are hidden in the behaviors of individuals in social network, not the multimedia data.

2   System Description

As well known, a social network, such as Facebook and WeChat, includes a number of individuals and their interactions. An individual is able to see the news published by his friends, and comment or mark "love" on them. This work proposes a steganographic scheme carrying the secret data using the "love"-marking behaviors. In the scheme, a sender selects a part of news published by his friends and makes "love" marks on them for representing the secret data, and a receiver who is a friend of the sender extracts the secret data from the sender's "love" marks that he can see.

J.-S. Pan et al. (eds.), *Advances in Intelligent Information Hiding and Multimedia Signal Processing*, Smart Innovation, Systems and Technologies 63,
DOI 10.1007/978-3-319-50209-0_3

Denote the sender as S and his friend set as $F^{(S)}$, which contains N individuals F_1, F_2, ..., F_N, and the receiver R is one of the N individuals in $F^{(S)}$. Note that, in some social networks, such as WeChat, if one of the sender's friends is not a friend of the receiver, his published news, as well as the sender's "love" marks on the news, cannot be seen by the receiver. Furthermore, in some cases, the sender does not know whether one of his friends is also a friend of the receiver. That implies some "love" marks made by the sender may be invisible to the receiver, so that the receiver should be capable of retrieving the secret data from only a part of "love" marking behaviors. In other words, the receiver has to extract the secret data from the visible "love" marks.

Another requirement of the system is that the sender should make "love" marks on his friend's news with a normal rate to avoid arousing any warden's suspicion. To this end, the sender may assign a reasonable "love" rate for each friend before making the "love" marks. In practical application, the reasonable "love" rate, denoted as r_n ($1 \le n \le N$), may be empirically given according to the relationship between the sender and his each friend, or be assigned as the actual "love" rate in a normal period when the steganographic method is not implemented.

3   Steganographic method

The secret data to be transmitted can be viewed as a binary sequence $\mathbf{m} = [m_1, m_2, ..., m_L]$. For each friend F_n ($1 \le n \le N$), the sender pseudo-randomly generates a number of binary L-length vectors $\mathbf{v}_n^{(j)}$ ($j = 1, 2, 3, ...$), which is determined by a secret key, and calculates the binary inner products

$$d_n(j) = \mathbf{v}_n^{(j)} \cdot \mathbf{m}' \tag{1}$$

Then, the sender forms N bit-sequences $\mathbf{d}_n = [d_n(1), d_n(2), d_n(3), ...]$, and converts each bit-sequence $\mathbf{d}_n$ into another binary sequence $\mathbf{g}_n$ using inverse arithmetic coding with a rate r_n, which is given in advance as described in the previous section. Here, while both the probabilities of 0 and 1 in $\mathbf{d}_n$ are 1/2 since the vector $\mathbf{v}_n^{(j)}$ is pseudo-random, the probabilities of 0 and 1 in $\mathbf{g}_n$ are $(1-r_n)$ and r_n, respectively. So, the sender may begin the transmission of secret data. When seeing a news published by his friend F_n, the sender takes a bit from $\mathbf{g}_n$, and marks a "love" if the taken bit is 1 or does nothing if the taken bit is 0. This way, the bits in $\mathbf{g}_n$ are sent in an one-by-one manner until the receiver gives a feedback indicating that the secret data has been received successfully.

Considering the receiving side, if the sender's friend F_n is also a friend of the receiver or the receiver himself, the news published by him and the sender's "love" marks on the news is visible to the receiver. Then, the receiver firstly collects the 0 and 1 according to the absence and presence of the sender's "love" marks on the news that he can see, and forms a prefix of $\mathbf{g}_n$ made up of the collected bits. The more the visible news, the longer obtained prefix of $\mathbf{g}_n$ is. Assuming the number of sender's friends who are the friends of the receiver or the receiver himself is N_F, the receiver can obtain N_F prefixes, and these prefixes can be converted into N_F prefixes of $\mathbf{d}_n$ using arithmetic coding. In other words, a number of $d_n(j)$ have been obtained. Then, the receiver concatenates the obtained $d_n(j)$ to be a column vector $\mathbf{D}$ with a length N_D, and there must be

$$\mathbf{D} = \mathbf{V} \cdot \mathbf{m}' \tag{2}$$

where $\mathbf{V}$ is a $N_D \times L$ matrix and its rows are the pseudo-random vectors $\mathbf{v}_n^{(j)}$ corresponding to $d_n(j)$. Clearly, $\mathbf{D}$ and $\mathbf{V}$ are known to the receiver. In the case of $N_D \geq L$, as long as the rank of $\mathbf{V}$ is L, the receiver can solve the secret data $\mathbf{m}$. Since $\mathbf{V}$ is a pseudo-random binary matrix, the probability of rank$(\mathbf{V}) = L$ is

$$P = \left(1 - 2^{-N_D}\right) \cdot \left(1 - 2^{-N_D+1}\right) \cdots \left(1 - 2^{-N_D+L-1}\right) \tag{3}$$

With a given L, the probability in (3) approaches 1 exponentially with increasing N_D. Thus, when the number of obtained $d_n(j)$ is enough to solve the secret data, the receiver solves it and sends a feedback signal to the sender. This way, the secret data are extracted by the receiver, and the sender may stop the data transmission process.

4   Conclusion

This work proposes a steganographic manner converting the secret data into "love" marks in social network. The data sender can let the "love" marking rate on each friend be a normal given value to keep the covert communication confidential, and the data receiver can extract the secret data from only the visible "love" marks. With the proposed steganographic scheme, only a little network resource is consumed to create a covert channel for transmitting a small amount of secret data. The behavior steganopraphy suitable for a large amount of secret data deserves further investigation in the future.

Acknowledgement

This work was supported by the Natural Science Foundation of China (61525203 and 61472235), the Shanghai Dawn Scholar Plan (14SG36) and the Shanghai Excellent Academic Leader Plan (16XD1401200).

References

1. J. Fridrich, Steganography in digital media: principles, algorithms and applications. Cambridge, U.K.: Cambridge Univ. Press, 2010.
2. B. Li, M. Wang, X. Li, S. Tan, and J. Huang, A Strategy of Clustering Modification Directions in Spatial Image Steganography. *IEEE Trans. Information Forensics and Security*, 10(9), pp. 1905-1917, 2015.
3. R. Cogranne and J. Fridrich, Modeling and Extending the Ensemble Classifier for Steganalysis of Digital Images Using Hypothesis Testing Theory. *IEEE Trans. Information Forensics and Security*, 10(12), pp. 2627-2642, 2015.
4. J. Yu, F. Li, H. Cheng, and X. Zhang, Spatial Steganalysis Using Contrast of Residuals. *IEEE Signal Processing Letters*, 23(7), pp. 989-992, 2016.

Robust Steganography Using Texture Synthesis

Zhenxing Qian[1], Hang Zhou[2], Weiming Zhang[2], Xinpeng Zhang[1]

1. School of Communication and Information Engineering, Shanghai University, Shanghai, 200444, China
2. School of Information Science and Technology, University of Science and Technology of China, Hefei, 230026, China
Email: zxqian@shu.edu.cn

Abstract. This paper proposes a robust steganography based on texture synthesis. Different from the traditional steganography by modifying an existing image, we hide secret messages during the process of synthesizing a texture image. The generated stego texture is similar to the sample image, preserving a good local appearance. Large embedding capacities can be achieved proportional to the size of the synthesized texture image. This algorithm also ensures that the hidden message can be exactly extracted from the stego image. Most importantly, the proposed steganography approach provides a capability of countering JPEG compression.

Keywords: Steganography, information hiding, texture synthesis

1 Introduction

Steganography is a technology of covert communication, which hides secret information into a cover media so as to avoid the eavesdropper's suspicious [1]. Nowadays, secret messages are always embedded into the digital media such as digital image, video, audio, text, etc. Many steganography methods have been developed in the past few decades [7-9].

Traditionally, secret messages are hidden by overwriting the insignificant data of a chosen cover. Given a multimedia, the embedding capacity is determined by the allowed distortions. More distortions would result in more risk of defeat by an eavesdropper using steganalysis tools. Another emerging problem is that the transmission is not always lossless. For example, when uploading an image onto the social network, this image is always compressed by the service provider. Since the feature of robustness is not considered in traditional steganography methods, message extraction in most of the algorithms would fail when the stego is processed. Hence, both the large embedding capacity and the robustness are required in modern steganography.

To this end, this paper proposes a novel steganography method based on texture synthesis. Both the capacity and robustness can be achieved. With a small texture pattern, we construct a message-oriented texture image with proportional size to accommodate message with arbitrary length. The generated stego texture image is ro-

J.-S. Pan et al. (eds.), *Advances in Intelligent Information Hiding and Multimedia Signal Processing*, Smart Innovation, Systems and Technologies 63,
DOI 10.1007/978-3-319-50209-0_4

bust to JPEG compression, i.e., hidden massage can still be correctly extracted from the compressed stego image.

Until now, few steganography works based on texture synthesis have been done. Pioneering works were done by Otori and Kuriyama [2][3]. Secret messages are regularly arranged into colored dotted pattern using the colors picked from a texture sample with features corresponding to the embedded data. These dotted patterns are written onto a blank canvas, the blank regions of which are synthesized using the texture samples. Embedding capacity of the method is determined by the dotted patterns painted on the image. Another synthesis based steganography method was investigated by Wu and Wang [4]. A texture synthesis process re-samples a smaller texture image to construct a new texture image. Secret messages are concealed through the process of texture synthesis, in which sorted candidate patches are mapped to secret bits. During data extraction, they first recover the original texture image, and then extract the hidden message using the reordered candidate patches. This method can achieve high embedding rate than the works in [2] and [3].

In these steganography methods in [2]~[4], the constructed stego texture images have good appearance and the embedding rates are considerable. However, once the stego image is compressed by the tools like JPEG, many errors would happen during messages extraction. To overcome this problem, we propose a new synthesis based steganography approach, in which both large capacity and robustness can be realized.

2   Proposed Method

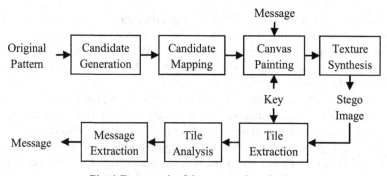

Fig. 1 Framework of the proposed method

Framework of the proposed method is shown in Fig. 1. To hide a secret message, a data hider uses a source texture pattern and divides this patter into overlapped candidate tiles. These candidates are mapped to several classifications by analyzing their texture complexities. With a secret key, the data hider pseudo-randomly distributes the selected candidate tiles to a blank canvas. Each tile is selected according to a segment of message. Other regions in blank are then synthesized by choosing the best tile from all candidates. On the receiver side, the secret key is used to identify the tiles containing secret bits. By analyzing texture of all the extracted tiles, several categories are reconstructed. Accordingly, the hidden bits can be extracted from each tile.

2.1   Data Hiding

Given a source texture pattern with the size of $S_r \times S_c$, we first divide it into a number of candidate tiles. The division is done by shifting each pixel following the raster-scan order, resulting in $(S_r - T_r + 1) \cdot (S_c - T_c + 1)$ overlapped tiles. Each tile is sized $T_r \times T_c$. We further divide each tile into the kernel area and boundary area, as shown in Fig. 2. Each kernel contains $K_r \times K_c$ pixels. The boundary depths in both directions are B_r and B_c, respectively. Therefore, there are $(S_r - T_r + 1) \cdot (S_c - T_c + 1)$ kernels corresponding to all the candidate tiles.

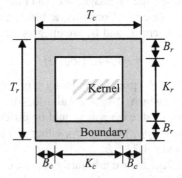

Fig. 2 A tile containing the kernel area and the boundary area

Denote all the tiles as $\{P_1, P_2, \ldots, P_N\}$ and corresponding kernels as $\{R_1, R_2, \ldots, R_N\}$, where $N = (S_r - T_r + 1) \cdot (S_c - T_c + 1)$. We evaluate the complexity degree of each kernel by standard deviation using

$$D_k = \left[\left\{ \frac{1}{K_r K_c - 1} \sum_{i=1}^{K_r} \sum_{j=1}^{K_c} \left(R_k(i,j) - A_k \right)^2 \right\}^{1/2} \right] \tag{1}$$

where $[\cdot]$ is the rounding operator, $k = 1, 2, \ldots, N$, and

$$A_k = \sum_{i=1}^{K_r} \sum_{j=1}^{K_c} R_k(i,j) \bigg/ K_r K_c \tag{2}$$

Denote that the degree values $\{D_1, D_2, \ldots, D_N\}$ range from $D_{\min}$ to $D_{\max}$. Accordingly, we compute M values $\{V_1, V_2, \ldots, V_M\}$, in which

$$V_i = \left[\frac{D_{\max} - D_{\min}}{M} (i - 1) + \frac{D_{\max} - D_{\min}}{2M} \right] \tag{3}$$

With these values, we construct M categories $\{\Pi_1, \Pi_2, \ldots, \Pi_M\}$ containing candidate tiles by

$$\Pi_i = \left\{ P_k \mid D_k \in (V_i - \delta, V_i + \delta) \right\} \tag{4}$$

where $i=1,2,...,M$, $k=1,2,...,N$, and δ is a smaller number satisfying $0\leq\delta<(D_{max}-D_{min})/2M$.

We use each category to represent several secret bits. Since there are M categories, each one stands for $\lfloor\log_2 M\rfloor$ bits, where $\lfloor\cdot\rfloor$ the rounding down operator. For example, if $M=4$, $\mathbf{\Pi}_1$, $\mathbf{\Pi}_2$, $\mathbf{\Pi}_3$, and $\mathbf{\Pi}_4$ stand for the secret bits of "00", "01", "10" and "11", respectively.

To hide a secret message, we turn the message into binary bits and divide these bits into segments with each containing $\lfloor\log_2 M\rfloor$ bits. Assuming L such segments $\{\mathbf{B}_1,\mathbf{B}_2,...,\mathbf{B}_L\}$ are included in the message, we calculate the decimal values E_i $(i=1,2,...,L)$ for all these segments. Construct a blank canvas with the size of $[(K_r+B_r)\cdot W_r]\times[(K_c+B_c)\cdot W_c]$, where W_r and W_c are integers satisfying

$$W_r\cdot W_c>4L/\lfloor\log_2 M\rfloor.$$

With a secret key, we pseudo-randomly generate L integer pairs $\{(p_1,q_1),(p_2,q_2),...,(p_L,q_L)\}$, in which $1\leq p_i\leq W_r$ and $1\leq q_i\leq W_c$. Two conditions are required during the generation. First, each pair should be different from any other one. Second, for arbitrary two pairs (p_i,q_i) and (p_j,q_j), either $p_i-p_j>1$ or $q_i-q_j>1$ should be satisfied, where $i=1,2,...,L$ and $j=1,2,...,L$.

For each segment $\mathbf{B}_i$ $(i=1,2,...,L)$, we arbitrarily choose one tile from the category $\mathbf{\Pi}_{Ei+1}$. Then we paint all $T_r\cdot T_c$ pixels in this tile onto the canvas from the pixel at $((K_r+B_r)\cdot(p_i-1)+1,\ (K_c+B_c)\cdot(q_i-1)+1)$ to the pixel at $((K_r+B_r)\cdot p_i+B_r,\ (K_c+B_c)\cdot q_i+B_c)$. Fig. 3(a) shows the diagram of painting one tile onto the canvas, in which the content surrounded by the blue square is the kernel of the candidate tile. Fig. 3(b) shows an example of painting 100 tiles containing 200 secret bits onto a blank canvas.

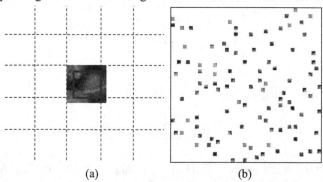

(a)   (b)

Fig. 3 Painting selected tiles onto a blank canvas. (a) Diagram of painting one tile onto a blank canvas. (b) An example of painting 100 tiles containing 200 secret bits onto a canvas.

The other blank regions on the canvas are then painted by texture synthesis. From all $(S_r-T_r+1)\cdot(S_c-T_c+1)$ candidates, proper tiles are identified. We slightly modify the "image quilting" algorithm proposed by Efros and Freeman in [5]. In [5], synthesis is realized by iteratively padding chosen identical sized candidates to a blank window. Since there are overlapped regions, errors between the chosen block and the existing blocks at the overlapped region are computed. Generally, a best tile that has the smallest mean square errors (MSE) of the overlapped parts is selected. A diagram is illustrated in Fig. 4(a). The regions in gray color stand for the synthesized contents. When synthesizing the content of "**C**", MSE of the overlapping regions between "**C**" and the *upper* tile "**A**", and MSE between "**C**" and the *left* tile "**B**" are calculated. One candidate that has the smallest MSE is chosen as the best. Then, the minimum cost path along the overlapped surface is computed to find the seam, see the red curves on the overlapped region, and the content is pasted onto the canvas along the seams. Details of the algorithm can be found in [5].

In the proposed method, the main difference is that many tiles containing secret bits have been painted on the canvas. We slightly modify the quilting algorithm in [5] to construct the synthesized image with good appearance. As aforementioned, the key controlled painting ensures that no blocks containing secret bits are adjacent. Instead of calculating MSE of *upper* and *left* overlapping parts, the *right* or *down* overlapping parts are also included to identify the best candidate. An example is shown in Fig. 4(b), in which "**D**" is the painted tile containing secret bits. We find the best candidate for "**C**" by computing MSE from three directions, and find the best seams for the overlapped surfaces, see the blue curves on overlapped regions.

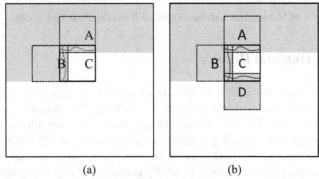

(a)   (b)

Fig. 4. Synthesizing the blank regions. (a) Diagram of the algorithm proposed by Efros and Freeman. (b) Diagram of the modified algorithm in the proposed method.

After painting all the blank regions of the canvas, a steganography image is finally generated. The embedding capacity C_e (bits) of the proposed algorithm is thus equal to

$$C_e = L \cdot \lfloor \log_2 M \rfloor \tag{5}$$

2.2   Data Extraction

On the recipient end, the secret key is used to identify the positions of tiles containing secret bits. The receiver extracts L tiles $\{Q_1, Q_2, ..., Q_L\}$ from the stego image. Correspondingly, L kernels $\{U_1, U_2, ..., U_L\}$ are separated from these tiles using the same way as illustrated in Fig. 2. Complexities of these kernels are calculated by (1) and (2). Denote these degree values as $\{G_1, G_2, ..., G_N\}$ range from G_{min} to G_{max}.

Since the stego image may be compressed by JPEG, the calculated degree values might not belong to the original degree set $\{V_1, V_2, \ldots, V_M\}$. The receiver reconstructs a new set $\{V_1', V_2', \ldots, V_M'\}$ by

$$V_i' = \left[\frac{G_{max} - G_{min}}{M}(i-1) + \frac{G_{max} - G_{min}}{2M} \right] \tag{6}$$

With these values, the tiles are re-classified into M new categories $\{\Lambda_1, \Lambda_2, \ldots, \Lambda_M\}$ containing candidate tiles by

$$\Lambda_i = \left\{ \mathbf{Q}_k \mid V_k' \in [V_i' - \Delta/2, V_i' + \Delta/2) \right\} \tag{7}$$

where $i=1,2,\ldots,M$, $k=1,2,\ldots,L$, and

$$\Delta = (G_{max} - G_{min})/M.$$

From the each extracted tile $\mathbf{Q}_k$ $(k=1,2,\ldots,L)$, $\lfloor \log_2 M \rfloor$ hidden bits can extracted. If $\mathbf{Q}_k \in \Lambda_i$ $(i=1,2,\ldots,M)$, the hidden bits would be the binary bits of $(i-1)$. This way, $L \cdot \lfloor \log_2 M \rfloor$ bits of hidden bits can be extracted from the received image.

3   Experimental Results

The proposed algorithm is carried out in many images. A group of image is shown in Fig. 5. The source patches sized 128×128 are shown in Fig. 5(a)~(h), containing several kinds of textures. Fig. 5(i)~(p) are the generated steganography images with the sizes of 653×653, corresponding to 2500 overlapped tiles, i.e., $W_r=50$ and $W_c=50$. All of the experiments use a fixed δ as 0. The main parameters are set as $M=16$ and $L=400$, while the tile size is 16×16 and the kernel size is 3×3. In each synthesized image, 1600 bits are hidden inside some selected tiles. The selection is controlled by a secret key. Results show that the stego textures preserve a good visual appearance.

(a)   (b)   (c)   (d)

(e)   (f)   (g)   (h)

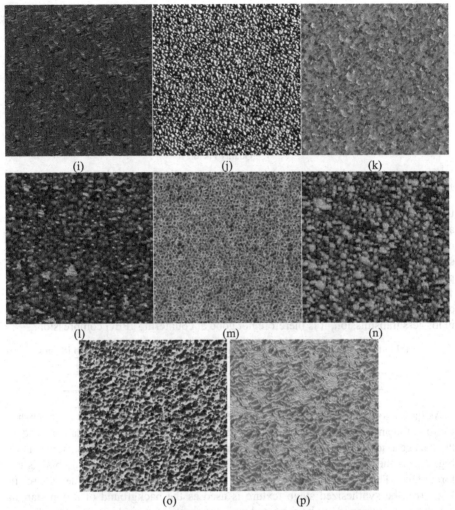

Fig. 5 Source patches and the stego textures. (a)~(h) are the source patches. (i)~(p) are the synthesized textures containing secret messages.

The proposed method is robust to JPEG compression. A group of results are shown in Fig. 6. We use the patterns listed in Fig. 5(a)~(h) to construct stego images. Size of the stego images is 653×653, in which the tile size is 16×16, the kernel size is 3×3, and L=400. Different values of M, from 2 to 16, are used to construct stegos, which are then compressed by JPEG using different quality factors from 10 to 90. After extracting hidden bits from the compressed stegos, the average error rates of the extracted bits in these images are calculated. Results in Fig. 6 show that the average error rate approaches 0 when the quality factor increases. Meanwhile, smaller M results in smaller error rate. When M=2, error rate is equal to zero even if the quality factor of compression is as poor as 10.

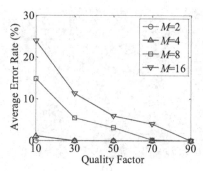

Fig. 6 Error rate of data extraction corresponding to quality factors of JPEG compression

We also use 80 source patches arbitrarily chosen from the database "Brodatz Textures" [5]. These images are rescaled to patterns with the size of 128×128. Parameters are selected to construct stego images with the embedding rate 0.5 bits per tile (non-overlapping) using both the proposed method and the method in [4]. We compress the stego images using different quality factors. Hidden data are extracted from the compressed images and the average error rates are calculated. The results are listed in Table I, indicating that the proposed steganography method has a good capability of countering JPEG compression. As the method in [4] was proposed for steganography in lossless transmission, it is therefore not good at countering JPEG compression.

Table I Average error rate of data extraction after compressing the stego images

Quality Factor	90	70	50
Proposed	0	5.6%	7.7%
[4]	14.5%	24.0%	26.3%

As the stego images are constructed by texture synthesis, sometimes it is somewhat weird to transmit a texture over internet. In real applications, the stegos can be used as the backgrounds of some images. Two examples are shown in Fig. 7. Background of Fig. 7(a) is the synthesized grass containing secret message, adding a football as the foreground. The secret message is hidden in the content outside the blue square. In Fig. 7(b), the synthesized stego texture is used as the background of a pop star, in which secret messages can be extracted from the regions squared by red rectangles.

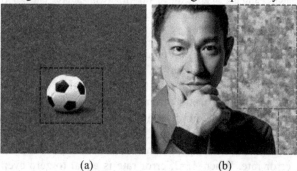

(a)   (b)

Fig. 7 Applications of texture synthesis based steganography. Backgrounds of both (a) and (b) are the synthesized stego textures.

4   Conclusions

This paper proposes a novel steganography method based on texture synthesis. Secret message is represented by different types of tiles captured from a source pattern with small size. Controlled by a secret key, tiles containing secret bits are painted onto the assigned positions in a blank canvas. With image quilting, this painted canvas is then filled with proper tiles to construct a texture image with good visual appearance. After capturing the tiles containing secret bits from the stego, hidden message can be extracted due to the complexities of the tiles. Different from traditional steganography methods, the proposed method provides an approach robustness and large payloads. Since complexities of the tiles containing secret bits slightly change after compression, data extraction is robust to JPEG compression. As a new camouflage way, stego images by texture synthesis can be used in many applications.

Acknowledgement

This work was supported by Natural Science Foundation of China (Grant U1536108, Grant 61572308, Grant 61572452 and Grant 61472235), 2015 Shanghai University Filmology Summit Research Grant, Shanghai Nature Science Foundation (Grant 14ZR1415900), and the Open Project Program of the State Key Lab of CAD&CG (A1502). Corresponding author: Zhenxing Qian, E-mail: zxqian@shu.edu.cn.

References

1. J. Fridrich, Steganography in digital media: principles, algorithms and applications. Cambridge, U.K.: Cambridge Univ. Press, 2009.
2. H. Otori and S. Kuriyama, Data-embeddable texture synthesis, in Proc. of the 8th International Symposium on Smart Graphics, Kyoto, Japan, 2007, pp. 146-157.
3. H. Otori and S. Kuriyama, Texture synthesis for mobile data communications, IEEE Comput. Graph. Appl., vol. 29, no. 6, pp. 74-81, 2009.
4. K-C Wu and C-M Wang, Steganography using reversible texture synthesis, IEEE Trans. Image Process., 24(1): 130-139, 2015.
5. A. A. Efros and W. T. Freeman, Image quilting for texture synthesis and transfer, Proc. 28th Annu. Conf. Comput. Graph. Interact. Techn., pp. 341-346, 2001.
6. Brodatz   Textures,   Texture   Image   Database   [Online].   Available: http://www.ux.uis.no/tranden/brodatz.html. 1997.
7. W. Zhang, X. Zhang, and S. Wang, Near-Optimal Codes for Information Embedding in Gray-Scale Signals, IEEE Trans. Information Theory, 56(3), pp. 1262–1270, 2010.
8. J. Fridrich and J. Kodovsky, Rich Models for Steganalysis of Digital Images, IEEE Trans. Information Forensics and Security, 7(3), pp. 868-882, 2012.
9. Y.-Q. Shi, C. Chen, and W. Chen, A Markov Process Based Approach to Effective Attacking JPEG Steganography, Proceedings of the 8th International Workshop of Lecture Notes in Computer Science, LNCS, 4437, pp. 249-264, 2007.

A Reversible Data Hiding Scheme in AMBTC Compression Code Using Joint Neighborhood Coding

Pei-Feng Shiu, Chia-Chen Lin*, Jinn-ke Jan, Cheng-Lin Hsieh, Yi-Ming Tang

Department of Computer Science and Information Management
Providence University, Taiwan
Department of Computer Science and Engineering
National Chung Hsing University, Taiwan.
pfshiu@gmail.com,mhlin3@pu.edu.tw,jkjan@mail.nchu.edu.tw
k7173968@gmail.com,tch01182003@gmail.com

Abstract. Steganography is a well-known technique used to protect important secret information by embedding the secret data into a cover multimedia file, such as video, still images and audio. This study proposes a reversible data hiding method, which uses a joint neighborhood coding (JNC) technique to hide secret data in AMBTC compression code. A reversible data-hiding scheme proposed by Wang was modified by this proposed method. We apply an exclusive OR operator and modified JNC algorithm to provide higher capacity. The experimental results demonstrated that the proposed method improved capacity by more than 3% over Wangs scheme in terms of embedded rate (ER).

Keywords: Reversible data hiding, AMBTC, JNC, Compression code.

1 Introduction

In recent years, data hiding techniques [1-8] have been widely discussed due to increased demands to protect sensitive data in an open network environment. Data hiding schemes are a technique to protect important information by embedding the secret information into a cover multimedia file. Some approaches focus on embedding secret data into compression codes [5-8]. Sun et al. [7] presented a reversible data-hiding scheme based in absolute moment block truncation coding (AMBTC). Sun embedded secret bits using joint neighborhood coding. There is a disadvantage in Sun et al.s method [7], as it requires one extra bit to mark whether or not the difference value is positive. Thus, Wang [8] proposed a reversible data hiding scheme to solve Sun et al.s problem [7] using an exclusive OR operator. The experimental results in Wangs scheme demonstrated that Sun's problem had been solved and the data hiding performance was improved.

To further improve performance in terms of capacity, we propose a reversible data-hiding scheme in AMBTC compression codes. Wangs scheme was modified to embed more secret messages. This paper defines a new value to replace the

© Springer International Publishing AG 2017

J.-S. Pan et al. (eds.), *Advances in Intelligent Information Hiding and Multimedia Signal Processing*, Smart Innovation, Systems and Technologies 63,
DOI 10.1007/978-3-319-50209-0_5

original neighborhood value to extend embedded secret data. Compared with other AMBTC based data hiding schemes, the proposed method provided the highest embedded rate.

The rest of the paper is organized as follows. Section 2 contains a review of AMBTC, and the JNC algorithm from Wangs scheme. Section 3 describes this proposed method in detail. Section 4 shows the experimental results of the proposed scheme. Finally, conclusions are given in section 5.

2   Related works

In this section, we briefly review the AMBTC technique, joint neighborhood coding (JNC) technique, and Wangs scheme.

2.1   Absolute Moment Block Truncation Coding

Absolute Moment Block Truncation Coding (AMBTC) is a widely known image compression technique proposed by Lema and Mitchell in 1984 [9], which has high compression and loss characteristic. The equations for AMBTC are given below.

$$\bar{x} = \frac{1}{n} \sum_{i=1}^{n} x_i \tag{1}$$

In equation (1), value $\bar{x}$ is the average of current block, value x_i means the i th pixel in the current block, and value n means the pixel number in block. When value $\bar{x}$ is obtained, equation (2) is used to record a bitmap of the block.

$$y_i = \begin{cases} 1, if x_i > \bar{x} \\ 0, if x_i \leq \bar{x} \end{cases} \tag{2}$$

Here y_i is the corresponding bit of the i th pixel in current block. If x_i is larger than the average value $\bar{x}$ then record y_i as '1', otherwise record y_i as '0'. When the bitmap of block is recorded, the value H_{AMBTC} and value L_{AMBTC} are calculated by equation (3) as given below.

$$H_{AMBTC} = \frac{1}{q} \sum_{x_i > \bar{x}}^{n} x_i \\ L_{AMBTC} = \frac{1}{p} \sum_{x_i \leq \bar{x}}^{n} x_i \tag{3}$$

In equation (3), value p and value q are the numbers of character '0' and character '1' in bitmap, respectively. When the high average value H_{AMBTC}, low average value L_{AMBTC} and bitmap of block are obtained, the AMBTC codes are obtained. Repeat equation (1) to equation (3) for all blocks in an image, to obtain the AMBTC codes of an image. At the decompression phase, AMBTC directs use of the high average value H_{AMBTC}, low average value L_{AMBTC} according to the bitmap to rebuild it per corresponding pixel block.

2.2 Joint Neighborhood Coding

The joint neighboring coding (JNC) algorithm was first proposed by Chang et al. [1]. In Chang et al.s approach JNC is used to encode VQ indices and hide secret data during the encoding process. Subsequently, many modified JNC based data hiding schemes were proposed, such as schemes by Sun et al. [7] and Wang [8]. In Wangs scheme [8], each value of the high average table and low average table is used to embed secret data, with the exception of the first row, first column and last column.

$H_{x-1,y-1}$	$H_{x-1,y}$	$H_{x-1,y+1}$
$(01, H_B)$	$(10, H_C)$	$(11, H_D)$
$H_{x,y-1}$	$H_{x,y}$	
$(00, H_A)$		

Fig. 1. The index around of current value $H_{x,y}$

Assume the current process high average value is called $H_{x,y}$. The neighborhood value according to the index around of current value $H_{x,y}$ in Figure 1. Here value x and y are the row and column in the current table, respectively. The secret bits are defined as $S = s_1, s_2, \ldots, s_r$, where $s_k = 0, 1$ and $0 \leq k \leq r$. And then Wang [8] presents a JNC algorithm and data hiding method described as below in the following steps.

Step 1. Extract and concatenate two secret bits, i.e., $s_k \| s_{k+1}$.

Step 2. Select value H_{sel} from the neighboring values of $H_{x,y}$ according to the corresponding the two bits $s_k \| s_{k+1}$.
The relationship between the value $H_{x,y}$ and the selected neighboring value is demonstrated in Figure 1.

Step 3. Calculate the difference D between the current value $H_{x,y}$ and the selected neighboring value H_{sel} follow equation (4).

$$D = H_{x,y} \oplus H_{sel} \tag{4}$$

Step 4. Classify the difference value D into four cases follow equation (5). Where the IDX represents the index of each case.

$$IDX = \begin{cases} 11, if 0 \leq D \leq 7 \\ 10, if 8 \leq D \leq 15 \\ 01, if 16 \leq D \leq 31 \\ 00, Otherwise \end{cases} \tag{5}$$

Step 5. According to IDX control the D and additional secret bits $s_k \| s_{k+1}$ to concatenate following the cases below.
Case 11 or 10, concatenated 3-LSBs of D and additional five secret bits. i.e., $H'_{x,y} = s_k \| s_{k+1} \| 11 \| D' \| s_{k+2} \| s_{k+3} \| s_{k+4} \| s_{k+5} \| s_{k+6}$. Where D' is the

3-LSBs of D.

Case 01, set stream length of D as four bits and additional four secret bits. i.e., $H'_{x,y} = s_k||s_{k+1}||01||D"||s_{k+2}||s_{k+3}||s_{k+4}||s_{k+5}$. Where $D"$ is the 4-LSBs of D.

Case 00, just concatenated $s_k||s_{k+1}$, index IDX, and difference value D. i.e., $H'_{x,y} = s_k||s_{k+1}||00||D$.

Step 6 Repeat step 1 to step 5 until all of values $H_{(x,y)}$ has been processed.

When user gets the compression code stream CS, the user can inverse the data hiding procedure to extract secret data and rebuild the AMBTC image. The data extraction and JNC decoding procedure is given as below.

Step 1. Extract the secret bits from the first and second bits in CS, to record it as $s'_k||s'_{k+1}$. And then set third, fourth bit as IDX'.

Step 2. According to IDX distinguish the difference value D and extract remainder secret bits in $H'_{x,y}$.

Case 11 or 10, extract three bits to recover the difference D and the remainder five bits as the secret bits. i.e., $s_{k+2}||s_{k+3}||s_{k+4}||s_{k+5}||s_{k+6}$.

Case 01, extract four bits and add 16 to recover difference D. And then the remainder four bits as the secret bits. i.e.,$s_{k+2}||s_{k+3}||s_{k+4}||s_{k+5}$. **Case** 00, extract eight bits to obtain the difference D directly.

Step 3. Recover the value $H_{x,y}$ of HT follow equation (6).

$$H_{x,y} = H_{sel} \oplus D \tag{6}$$

Step 4. Repeat Step 1 to step 3 until all of code stream CS has been processed.

In the JNC algorithm of Wang proposed method, the low average table use the process as same as the high average table, and the bitmap without any processing.

3   The proposed method

This paper modifies Wang proposed scheme to provide a higher capacity of data hiding method. The procedures for data embedding and data extraction in this proposed method are similar to Wangs scheme [8]. This approach define a reference value to replace the neighboring values in JNC algorithm. The reference values are given in Table 1. In Table 1, Sb is construct from four secret bits of embedding phase in proposed scheme. The reference value is calculated from the neighboring value of $H_{x,y}$ in Figure 1.

3.1   Data hiding phase

This is the same as Wangs scheme [8], We set a gray image I with size $M \times N$ as the cover image. The secret data is defined as $S = s_1, s_2, \ldots, s_r$, where $s_k = 0, 1$ and $0 \le k \le r$.

The proposed data embedding procedure is given as below.

Table 1. The index of reference values of $H_{x,y}$

Sb state	Reference value	Sb state	Reference value
0001	H_A	1001	$\lfloor (H_A + H_D)/2 \rfloor$
0010	H_B	1010	$\lfloor (H_B + H_D)/2 \rfloor$
0011	$\lfloor (H_A + H_B)/2 \rfloor$	1011	$\lfloor (H_A + H_B + H_D)/3 \rfloor$
0100	H_C	1100	$\lfloor (H_C + H_D)/2 \rfloor$
0101	$\lfloor (H_A + H_C)/2 \rfloor$	1101	$\lfloor (H_A + H_C + H_D)/3 \rfloor$
0110	$\lfloor (H_B + H_C)/2 \rfloor$	1110	$\lfloor (H_B + H_C + H_D)/3 \rfloor$
0111	$\lfloor (H_A + H_B + H_C)/3 \rfloor$	1111	$\lfloor (H_A + H_B + H_C + H_D)/4 \rfloor$
1000	H_D	0000	average of others case

Step 1. Generate AMBTC code from image I using AMBTC algorithm.

Step 2. Construct high average table HT, low average table LT and bitmap BM from AMBTC codes.

Step 3. Build the head of compression code CS by concatenate BM and the unchanged value such as the first row, first column and last column in the average table.

Step 4. Extract four secret bits as $s_k \| s_{k+1} \| s_{k+2} \| s_{k+3}$ called Sb. Where the symbol $\|$ is a concatenation symbol.

Step 5. Scan each value $H_{x,y}$ of the high average table HT , where $x = 2, 3, \ldots, M/n - 1$ and $y = 2, 3, \ldots, N/n$.

Step 6. Calculate the reference value H_{ref} according to Sb and the corresponding formula in Table 1. Where H_A, H_B, H_C, H_D in Table 1 is corresponding the neighboring value in Figure 1, respectively.

Step 7. Calculate the difference value D follow equation (4). The same as Wangs scheme, the symbol $\oplus$ in equation (4) is an exclusive OR operator. Here we replace the H_{sel} with H_{ref}.

Step 8. Obtain the IDX by classify difference value D follow equation (5).

Step 9. According to IDX control the D and additional secret bits to concatenate Sb following the cases below.

Case 11 or 10, concatenate 3-LSBs of D and additional three secret bits. i.e., $H'_{x,y} = Sb \| 11 \| D' \| s_{k+4} \| s_{k+5} \| s_{k+6}$. Where D' is the 3-LSBs of D.

Case 01, set stream length of D as four bits and additional two secret bits. i.e., $H'_{x,y} = s_k \| s_{k+1} \| 01 \| D" \| s_{k+4} \| s_{k+5}$. Where $D"$ is the 4-LSBs of D.

Case 00, just concatenate Sb, index IDX, and difference value D. i.e., $H'_{x,y} = Sb \| 00 \| D$.

Step 10. Repeat step 4 to step 9 until all average value $H'_{x,y}$ are computed, and then the modified high average table HT' is obtained.

Step 11. Obtain the modified low average table thru the same steps as Table.

Step 12. Concatenate the head of compression code CS and the HT' and LT', the compression code CS is obtained.

3.2 Data extraction phase

The proposed data extraction procedure is inverse the steps in data-hiding procedure. These data extraction steps are given in below.

Step 1. Extract the head of compression code stream CS and reconstruct BM and the unchanged values to recover remainder average values.

Step 2. Obtain the HT' and LT' from CS.

Step 3. Extract the first 4 bits of $H'_{x,y}$ as Sb.

Step 4. Calculate the reference value H_{ref} according to sb and the corresponding formula in Table 1.

Step 5. Extract the fifth bit and the sixth bit of $H'_{x,y}$ as index IDX.

Step 6. According to IDX divide the difference value D and extract remainder secret bits from $H'_{x,y}$.

Case 11 or 10, extract three bits to recover the difference D and the remainder three bits as the secret bits. i.e., $s_{k+4}||s_{k+5}||s_{k+6}$.

Case 01, extract four bits and add 16 to recover difference D. And then the remainder two bits as the secret bits. i.e., $s_{k+4}||s_{k+5}$.

Case 00, extract eight bits to obtain the difference D directly.

Step 7. Rebuild the value $H_{x,y}$ of HT follow equation (6). Here, we replace H_{sel} with H_{ref}.

Step 8. Repeat the step 3 to step 7 until all value $h_{x,y}$ are recovered. And then the HT is recovered and the secret bits of HT part are obtained.

Step 9. The data extraction procedure of LT' as same as HT', repeat the same steps to recover LT and extract the secret bits of LT part.

Step 10. Rebuild the AMBTC image using HT, LT and BM, the cover image I' could be obtained.

4 Experimental results

The experimental results of this proposed scheme in this section, the simulation platform is Microsoft Windows 7, Intel i5 CPU with 8.0 GB RAM, using the software MATLAB. The Standard images are six commonly used grayscale images with 512×512 pixels size shows in Figure 2. To examine the quality of the compressed image, peak signal to noise ratio (PSNR) was used to measure the similarity between the original image and the rebuilt AMBTC image. The PSNR value is defined in equation (7). Where the mean square error (MSE) for an $M \times N$ grayscale image is defined in Equation (8)

$$PSNR = 10 \times log_{10} \frac{255^2}{MSE} \tag{7}$$

$$MSE = \frac{1}{M \times N} \sum_{i=1}^{M} \sum_{j=1}^{N} (x_{i,j} - x'_{x,y}) \tag{8}$$

Here, $x_{i,j}$ and $x'_{i,j}$ are the pixel values of the original image and modified image, respectively. PSNR is a kind of image quality reference, a higher PSNR value

(a)Lena   (b) Peppers   (c) Baboon

(d) Boat   (e) Goldhill   (f) F16

Fig. 2. Original six standard test images

indicates that the original image and modified image are more similar.
To compare the performance of capacity, the embedded rate (ER) is calculated
by equation (9).

$$ER = \frac{Capacity}{Filesize} \tag{9}$$

where $Capacity$ is the number of total embedded secret bits and $Filesize$ present
the size of compressed file.

Table 2. Experimental results compare with other schemes

	Proposed scheme				Wangs scheme [8]			
	PSNR (dB)	File size(bits)	Capacity (bits)	ER(%)	PSNR (dB)	File size(bits)	Capacity (bits)	ER(%)
Lena	33.72	668,170	195,414	**29.24%**	33.72	652,304	175,145	26.85%
Peppers	34.10	668,654	195,062	**29.16%**	34.10	652,304	174,222	26.71%
Baboon	27.78	676,152	652,304	**26.79%**	27.78	652,304	151,439	23.22%
Boat	31.16	669,204	194,486	**29.06%**	31.16	652,304	166,786	25.57%
Goldhill	33.72	671,168	190,104	**28.32%**	33.72	652,304	168,185	25.78%
F16	33.29	668,964	196,162	**29.32%**	33.29	652,304	177,814	27.26%

Table 2 is the comparison table of the proposed method and other AMBTC based reversible data hiding schemes. For image quality, all schemes in Table 2 are the same. The reason of the same image quality is that each scheme in Table1 is a reversible data hiding scheme, and the AMBTC code could be recovered in the final. For the performance of capacity, the proposed method clearly provided the highest ER in Table 2. Compared with Wangs scheme, our method provided a 3% ER enhancement.

5 Conclusions

A reversible data-hiding scheme was presented. This method modified the joint neighborhood coding (JNC) algorithm for hiding secret data into AMBTC compression code. We extended the fixed secret bits in each code stream, and modified the coding mechanism in JNC to enhance the embedded rate. The experimental results demonstrated that the proposed method provided a 3% improvement in ER over the scheme by Wang.

6 Acknowledgement

This work is supported by MOST of Taiwan, No. 105-2410-H-126-005-MY3

References

1. C.C. Chang, T.D. Kieu,and W.C. Wu,: A lossless data embedding technique by joint neighboring coding. *Pattern Recognition*, vol. 42, no. 7, pp.1597–1603, (2009)
2. Z.M. Lu, J. X. Wang, B.B. Liu,: An improved lossless data hiding scheme based on image VQ-index residual value coding. *Journal of Systems and Software*, (2009)
3. X. Zhang,: Reversible data hiding in encrypted image. *IEEE Signal Processing Letters*, vol. 18, no. 4, pp.255–258, (2011)
4. W. Hong, T.S Chen, and H.Y Wu,: An improved reversible data hiding in encrypted image using side match. *IEEE Signal Processing Letters*, vol. 19, no. 4, pp.199–202, (2012)
5. I. C. Chang, Y. C. Hu, W. L. Chen, and C. C. Lo,: High capacity reversible data hiding scheme based on residual histogram shifting for block truncation coding. *Signal Processing*, vol. 7, no. 2, pp. 297–306, (2013)
6. C. C. Lin, X. L. Liu, W. L. Tai, and S. M. Yuan,: A novel reversible data hiding scheme based on AMBTC compression technique. *Multimedia Tools and Applications*, vol. 74, no. 11, pp. 3823–3842, (2013)
7. W. Sun, Z. M. Lu, Y. C. Wen, F. X. Yu, and R. J. Shen,: High performance reversible data hiding for block truncation coding compressed images. *Signal, Image and Video Processing*, vol. 7, no. 2, pp 297-306, (2013)
8. Y.K. Wang,: Hiding Message in Lossless Data Compression Codes, Condensed Images and Encrypted Images. Masters Thesis of Department of Computer Science and Information Engineering National Chung Cheng University, Taiwan. 74 pp. (2016)
9. M. Lema, O. Mitchell,: Absolute moment block truncation coding and its application to color images. *Communications, IEEE Transactions on*, vol. 32, issue. 10, 1148–1157 (1984)

A Quantization-Based Image Watermarking Scheme Using Vector Dot Product

*Yueh-Hong Chen[1] and Hsiang-Cheh Huang[2]

[1] Department of Computer Science and Information Engineering
Far East University
Tainan, Taiwan
yuehhong@google.com
[2] Department of Electrical Engineering
National Kaohsiung University
Kaohsiung, Taiwan
huang.hc@google.com

Abstract. In this paper, we proposed a quantization-based watermarking scheme for image data. To increase the robustness of the watermark, each bit of a binary watermark string is embedded into a vector composed of several wavelet coefficients selected from low and middle frequency domain of an image. The proposed scheme is based on quantization technique, so the space of the vector is divided into several subspaces. Then, the vector is modified such that it will fall into the subspace corresponding to the value of the watermark bit. Since subspaces divided by quantization operation are distributed over the whole space, the vector could be modified so that the square error is minimized. The experimental results illustrate that the proposed method can generate a more robust watermark while keeping image fidelity.

Keywords: Image Watermarking, Quantization, Steganography

1 Introduction

Digital watermarking is one of hot research topics in the multimedia area. By using this technology, one can embed copyright information in digital content to prevent some possible piracy attempts. Among all kings of digital contents, images have attracted much attention because digital cameras and smartphones have become more and more popular over the years. For the purpose of copyright protection, watermarking schemes need to meet few qualifications, such as imperceptibility, security, and robustness to common image processing methods. In this paper, we propose an image watermarking approach for ownership proving. This approach hides binary watermarks in vectors consisting of wavelet coefficients. Several watermarking methods have been proposed to hide a binary

---

* This research was supported in part by the Ministry of Science and Technology under Grant MOST 104-2622-E-269-012-CC3.

© Springer International Publishing AG 2017
J.-S. Pan et al. (eds.), *Advances in Intelligent Information Hiding and Multimedia Signal Processing*, Smart Innovation, Systems and Technologies 63,
DOI 10.1007/978-3-319-50209-0_6

43

value into more than one pixels or transform domain coefficients. A brief review of those research efforts is as follows.

A number of methods have been proposed to insert robust and invisible watermarks. Some operate directly in pixel domain [7], other in a transform domain, such as Fourier[8], DCT[5], or wavelet domain[2]. In [6], selected coefficients of an 8×8 DCT block in the image are grouped into ordered pairs. Each bit of the watermark is then encoded using one of the coefficient pairs. However, there was no experimental result showing the robustness and imperceptibility of the watermark in the paper. Hsu and Wu [4] proposed an approach using middle frequency coefficients chosen from one or more 8×8 DCT blocks to embed watermarks. Quantization operation is taken into consideration in this approach so that watermarks can survive the JPEG lossy compression. However, when extracting the watermark, the approach required the original image and the watermark, and they are unavailable in some applications.

In some watermarking approaches, vectors composed of more than two coefficients are used to embed the watermark. Typically, these approaches adequately modify the selected coefficients such that a pre-specified one among them is smallest or largest, according to the binary value to be embedded. In [9], three coefficients selected from an 8×8 DCT block are altered to meet the situations in which the third coefficient is largest or smallest, based on the watermark bit to be embedded. However, the extractor proposed in [9] ignores the three-coefficient sets in which the third coefficient is between other two. As a result, a failure of detecting one single bit may make the whole watermark undetectable. A similar approach in [3] selects six coefficients from a DCT block and then exchanges the first coefficient with the largest or smallest coefficient among them. Nevertheless, a significant degradation in image quality may be caused if the difference between two coefficients to be exchanged is large.

Although the aforementioned researches did not precisely called the set of selected coefficients a vector, these schemes adaptively divide the space into two parts and then modify the vector so that it falls in the specified subspace. In this paper, we proposed a quantization-based watermarking scheme for image data. To increase the robustness of the watermark, each bit of a binary watermark string is embedded into a vector composed of several wavelet coefficients, In general, these coefficients are randomly selected from low and middle frequency domain. Comparing to the aforementioned researches, the proposed scheme is based on quantization technique, so the space of the vector is divided into several subspaces. Then, the vector is modified such that it will fall into the subspace corresponding to the value of the watermark bit. Since subspaces divided by quantization operation are distributed over the whole space, the vector could be modified so that the square error is minimized. The experimental results shown in Section 3 illustrate the performance of the proposed approach.

The rest of the paper is organized as follows. Section 2 introduces the assumption in this paper and presents the proposed watermarking approach. The experimental results are shown in Section 3. Finally, Section 4 summarizes our approach and provides a brief concluding remarks.

2 The Proposed Method

In this paper, a watermark is a binary string, consisting of two symbols, 0 and 1. generated randomly. All bits of the watermark are separately embedded into an image with the same scheme. The method to embed a single bit is first proposed in Subsection 2.2. Then, the algorithm to embed the whole binary watermark is discussed in Subsection 2.3.

2.1 An overview

Before a watermark is embedded into an image, some parameters should be specified:

1. **Dimension of vectors** D: This is also the number of wavelet coefficients selected to embedded a binary value.
2. **Watermark strength** δ: This parameter are used to decided the size of quantization step. The larger δ is, the more robust the watermark is. However, increase δ would decrease quality of the image.
3. **Reference vector** V_r: This vector is used to calculate the inner product with vectors into which watermark bits are going to be embedded. Reference vector can be generated randomly or selected from wavelet coefficients of the image, and then normalized to a unit vector.

To embed a watermark, the image is firstly transformed into wavelet domain. For each bit of the watermark, D coefficients in pre-specified subband (e.g., $LH2$, $HL2$ or $HH2$) are then randomly chosen and modified. Finally, inverse wavelet transform is applied to obtain the watermarked image.

When one bit of the watermark is to be embedded, D coefficients are chosen randomly to form a vector, referred to as v. The vector v is then modified to obtain $\hat{v}$ such that the relationship between V_r and $\hat{v}$ is as Eq. (1):

$$\left\lfloor \frac{(V_r \cdot \hat{v})}{\delta} \right\rfloor \bmod 2 = w, \tag{1}$$

where $w \in \{0,1\}$ is the value of a specific bit of the watermark code, and δ is the quantization step, also serve as the strength parameter. In this method, the value of δ should be pre-specified by the user.

2.2 Single bit embedding algorithms

To embed a binary value w, a vector $\boldsymbol{\Delta v}$ is added to the vector v to generate $\hat{v}$ that meets Eq. (1). Precisely, $\hat{v} \cdot V_r$ would in the middle of a quantization step, as shown in Eq. (2) and Fig. 1:

$$\frac{(V_r \cdot \hat{v})}{\delta} = 2k + w + \frac{1}{2}, \tag{2}$$

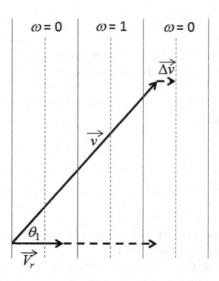

Fig. 1. An example of the proposed scheme, This example shows how a vector is modified to embed a binary value '0'.

where k is an integer. Clearly, the squared error between v and $\hat{v}$ is $|\boldsymbol{\Delta v}|^2$; thus, it is possible to find the optimal $\boldsymbol{\Delta v}$ such that the watermarked image have the best quality according to peak signal-to-noise ratio(PSNR), a commonly used quality metric. Supposed that the PSNR value is used as the quality metric; we can minimize mean square error (MSE) of the modified coefficients to maximize the PSNR value. In other words, minimizing all $|\boldsymbol{\Delta v}|$ values will minimize PSNR value of the image.

Since

$$V_r \cdot \hat{v} = V_r \cdot v + V_r \cdot \boldsymbol{\Delta v} = |v| \cos \theta_1 + |\boldsymbol{\Delta v}| \cos \theta_2 = \delta \left(2k + w + \frac{1}{2} \right), \quad (3)$$

where θ_1 is the angle between V_r and v, and θ_2 the angle between V_r and $\boldsymbol{\Delta v}$, therefore, assigning $\cos \theta_2$ to 1 or -1 will minimize $|\boldsymbol{\Delta v}|$.

Supposed that $|v| \cos \theta_1 < \delta \left(2k + w + \frac{1}{2} \right)$, minimum value of $|\boldsymbol{\Delta v}|$ can be obtain by assigning $\cos \theta_2$ to 1, and the $|\boldsymbol{\Delta v}|$ is

$$|\boldsymbol{\Delta v}| = \delta \left(2k + w + \frac{1}{2} \right) - |v| \cos \theta_1. \quad (4)$$

In other words, $\boldsymbol{\Delta v}$ is either in the same or opposite direction as V_r if $|\boldsymbol{\Delta v}|$ is minimized. Thus, Eq. (3) can be rewritten to

$$V_r \cdot v + V_r \cdot \boldsymbol{\Delta v} = V_r \cdot v + V_r \cdot (t \cdot V_r) = V_r \cdot v + t, \quad (5)$$

where t is a real number. Because

$$V_r \cdot v + t = |v| \cos \theta_1 + |\boldsymbol{\Delta v}| \cos \theta_2 = \delta \left(2k + w + \frac{1}{2} \right), \quad (6)$$

t can be calculated by Eq. (7):

$$t = \delta \left(2k + w + \frac{1}{2} \right) - \mathbf{V}_r \cdot \mathbf{v}. \tag{7}$$

The algorithm to find the optimal value t is as follows:

Algorithm 1
; Calculate weighted inner product between $\mathbf{V}_r$ and $\mathbf{v}$
$p = (\mathbf{V}_r \cdot \mathbf{v})/\delta$
; Calculate quantization index
$I = \lfloor p/\delta \rfloor$
If $I \bmod 2 = w$
    $t = I + 0.5 - p$
Else
    If $p > I + 0.5$
        $t = I + 1.5 - p$
    Else
        $t = I - 0.5 - p$
    End If
End If
$\mathbf{\Delta v} = t \cdot \mathbf{V}_r$

After the algorithm finishes, the optimal $\mathbf{\Delta v}$ can be found. $\mathbf{v}$ will then be modified to $\mathbf{v} + \mathbf{\Delta v}$,

The above algorithm is used to embed a watermark bit and optimize the PSNR value for a pre-specified strength parameter δ. However, keeping image quality acceptable is more meaningful and applicable. With some modifications, the proposed scheme can also calculate an appropriate δ such that δ is maximized while the squared error is just acceptable. Since $|\mathbf{\Delta v}|$ is squared error between $\hat{v}$ as well as v, $\mathbf{\Delta v} = t\mathbf{V}_r$ and $|\mathbf{V}_r| = 1$, the squared error equals to t. If an acceptable squared error of the vector v is E_s, δ can be evaluate with Eq. (8):

$$\delta = \frac{E_s + \mathbf{V}_r \cdot \mathbf{v}}{\left(2k + w + \frac{1}{2} \right)}. \tag{8}$$

If δ is adjusted with Eq. (8), each watermark bit may be embedded with different δ values (i.e., different quantization steps). Thus, these δ values should be preserved during watermark embedding process, and they will be required while watermark bits are decoded,

2.3 Image watermarking optimized for pre-specified PSNR

With the methods proposed in the previous subsection, a watermark bit can be embedded into the original image in accordance with a pre-specified strength parameter δ or squared error. Based on these methods, the scheme to embed a binary watermark string with L bits into the original image and optimize PSNR for a pre-defined strength parameter is proposed. The algorithm is as follows.

1. Apply discrete wavelet transform on the original image to obtain its wavelet domain coefficients.
2. Obtain L vectors by select L sets of wavelet coefficients from middle frequency bands of the original image. Each vector consists of D wavelet coefficients.
3. For vector v, use Algorithm 1 to find out the value of t and Δv according to δ pre-specified by the user.
4. For each vector v, embed the corresponding watermark bit by adding Δv to it.
5. Apply inverse discrete wavelet transform on the modified image to obtain the watermarked image.

3 Experimental Results

In this section, we present some experimental results to demonstrate the performance of the proposed method. A 512×512, 8 bit/pixel gray scale image, Lena, was used as the test image. An 1000-bit watermark was generated randomly and used throughout the experiments. To embed one bit of the watermark, twelve coefficients were chosen from $LH2$, $HL2$ or $HH2$ subband. In other words, D equals to 12 in our experiments. Three image processing operations: JPEG compression with quality factor $= 15$, Gaussian filtering and sharpening were applied on the watermarked images to demonstrate the robustness of the proposed method.

Table 1. The percentage of survival watermark bits after four image processing operations were applied.

Attack method	the proposed scheme	[1]
JPEG (Q=15)	71.5%	70.2%
Gaussian Filtering	89.1%	89.1%
Sharpening	99.6%	99.7%

In [1], the PSNR value of the watermarked image is 44.0. While keeping similar robustness, the proposed scheme can produce watermarked image with better quality. The obtained image is shown in Fig. 2, and its PSNR value is about 45.1. The results of applying three image processing operations on Fig. 2 is shown in Table 1. The experimental results revealed that the proposed watermarking scheme can embed into images the watermarks as robust as the approach proposed in [1]. This is because the proposed scheme changes the coefficient vectors to their nearest center of quantization step. Thus, to the three image processing methods adopted in our experiments, the proposed method can embed a more robust watermark into an image while keeping image quality the same with the scheme proposed in [1].

Fig. 2. The watermarked image with PSNR=45.1, obtained by the proposed watermarking scheme.

4 Conclusions

In this paper, we proposed a quantization-based watermarking scheme for image data. To increase the robustness of the watermark, each bit of a binary watermark string is embedded into a vector composed of several wavelet coefficients. In general, these coefficients are randomly selected from low and middle frequency domain of the image. Comparing to the aforementioned researches, the proposed scheme is based on quantization technique, so the space of the vector is divided into several subspaces. Then, the vector is modified such that it will fall into the subspace corresponding to the value of the watermark bit. Since subspaces divided by quantization operation are distributed over the whole space, the vector could be modified so that the square error is minimized. The experimental results shown in section 3 illustrate that the proposed method can embed a more robust watermark into an image while keeping image fidelity.

References

1. Chen, Y.H., Huang, H.C.: Adaptive image watermarking optimized for pre-specified fidelity. In: in International Conference on Innovative Computing, Information and Control. pp. 1305–1308 (December 2009)
2. Chen, Y.H., Su, J.M., Fu, H.C., Huang, H.C., Pao, H.T.: Adaptive watermarking using relationships between wavelet coefficients. In: 2005 IEEE International Symposium on Circuits and Systems (ISCAS 2005). vol. 5, pp. 4979– 4982 (May 2005)

3. Duan, F.Y., King, I., Chan, L.W.W., Xu, L.: Intra-block max-min algorithm for embedding robust digital watermark into images. Multimedia Information Analysis and Retrieval pp. 255–264 (1998)
4. Hsu, C.T., Wu, J.L.: Hidden digital watermarks in images. IEEE Trans. Image Processing 8(1), 58–68 (January 1999)
5. Huang, C.H., Wu, J.L.: A watermark optimization technique based on genetic algorithms. In: SPIE Electronic Imaging 2000. San Jose (January 2000)
6. Koch, E., Zhao, J.: Towards robust and hidden image copyright labeling. In: Proc. of 1995 IEEE Workshop on Nonlinear Signal and Image Processing. pp. 452–455. Halkidiki, Greece (June 1995)
7. Nikolaidis, N., Pitas, L.: Copyright protection of images using robust digital signatures. In: IEEE International Conference on Acoustics, Speech, and Signal Processing (ICASSP1996). vol. 4, pp. 2168–2171 (May 1996)
8. Ruanaidh, J.O., Dowling, W.J., Boland, F.: Phase watermarking of digital images. In: International Conference on Image Processing. vol. 3, pp. 239–242 (Septemter 1996)
9. Zhao, J., Koch, E.: Embedding robust labels into images for copyright protection. In: International Congress on Intellectual Property Rights for Specialised Information, Knowledge and New Technologies. Vienna, Austria (21–25 1995)

High-capacity Robust Watermarking Approach for Protecting Ownership Right

Chun-Yuan Hsiao[1], Ming-Feng Tsai[1], and Ching-Yu Yang[2] *

Dept. of Computer Science and Information Engineering
[1]National Kaoshiung University of Applied Science, Taiwan
cyhsiao@kuas.edu.tw, 1104308106@gm.kuas.edu.tw
Dept. of Computer Science and Information Engineering
[2]National Penghu University of Science and Technology, Taiwan
chingyu@gms.npu.edu.tw

Abstract. In this work, we proposed a high-capacity robust watermarking scheme for color images. Based on integer wavelet domain (IWT), the proposed scheme utilized the idea of computing the offset of two square-root-values, a number of data bits can be embedded in a host image. Simulations indicated that our method did provide a large hiding-storage while the perceived quality is good. Further, the proposed method is tolerant of various attacks such as brightness, cropping, edge sharpening, blurring, flip horizontal, inversion, and rotation. Additionally, the payload for the proposed method is larger than that for existing techniques. *abstract* environment.

Keywords: Data hiding, high-capacity robust watermarking, steganography.

1 Introduction

With the proliferation of Industry 4.0, or the fourth industrial revolution, the trend of automation and data exchange in manufacturing technologies is ubiquitous around the world. Namely, the organizations can effectively achieve their business goals by using the platform, which composed of internet of thing (IOT), intelligent robot (IR), cloud computing, and big data analytics. Consequently, the individuals and parties are easily to share their secret (or private) information from the Internet. However, data could be eavesdropped or tampered with during transmission. Data hiding can provide an economic ways to against the above issues. Generally, data hiding can be classified into two categories: steganography and digital watermarking [1, 2]. The steganographic methods provide a high payload with good perceived distortion, whereas robustness is the major goal of watermarking schemes. Recently, several researchers have presented their watermarking approaches for protecting copyright and ownership in color images [3–6]. However, either hiding capacity or robustness is not good enough. In this paper, we propose a high-capacity robust digital watermarking to achieve the goal.

---

* To whom correspondence should be addressed.

© Springer International Publishing AG 2017   51
J.-S. Pan et al. (eds.), *Advances in Intelligent Information Hiding and Multimedia Signal Processing*, Smart Innovation, Systems and Technologies 63,
DOI 10.1007/978-3-319-50209-0_7

2   Proposed Method

To achieve a high-capacity robust approach, we only embed data bits into the high-high (HH) subband of the level 1 (L1) of integer wavelet transform (IWT) domain. Namely, prior to bit embedment, an input image is decomposed to the IWT domain by using the following two formulas:

$$d_{j,k} = s_{j-1,2k+1} - s_{j-1,2k} \tag{1}$$

and

$$s_{j,k} = s_{j-1,2k} + \left\lfloor \frac{d_{j,k}}{2} \right\rfloor \tag{2}$$

where $s_{j,k}$ and $d_{j,k}$ are the k-th low-frequency and high-frequency wavelet coefficients at the j-th level, respectively [7].The $\lfloor x \rfloor$ is a floor function. The details of the scheme are described in the following sections.

A. Bit Embedding

Without loss of generality, let $W_j = \{(w_{rj}, w_{gj}, w_{bj})\}_{j=0}^{ab-1}$ be the j-th pixel derived from an input (scrambled) watermark of size $a \times b$. Also let $C = \{(c_{rj}, c_{gj}, c_{bj})\}_{j=0}^{ab-1}$ with $c_{rj} = \sqrt{w_{rj}} - round\left(\sqrt{w_{rj}}\right)$, $c_{gj} = \sqrt{w_{gj}} - round\left(\sqrt{w_{gj}}\right)$, and $c_{bj} = \sqrt{w_{bj}} - round\left(\sqrt{w_{bj}}\right)$ be the three deviation values of W_j.The main procedure of bit embedding is specified in the following algorithm.

Algorithm 1. Hiding data bits in an RGB color image.

Input: A host color image $S = \{(r_i, g_i, b_i)|i = 1, 2, ...MN\}$ and a scrambled watermark W.

Output: A marked image $\hat{S}$ with an auxiliary set of parameters C.

Method:

Step 0. Perform L1 IWT from host image S to obtain the HH-subband $H = \{(h_{rj}, h_{gj}, h_{bj})\}_{j=0}^{(MN/4)-1}$ of IWT coefficients.

Step 1. Input a set of coefficient H_j which derived from H . If the end of input is encountered, then proceed to Step 11.

Step 2. Assign a sign mark $s_{rj} = 1$ if $h_{rj} > 0$, $s_{gj} = 1$ if $h_{gj} > 0$, and $s_{bj} = 1$ if $h_{bj} > 0$, respectively, and go to Step 3. Otherwise, assign $s_{rj} = -1$ if $h_{rj} \leq 0$, $s_{gj} = -1$ if $h_{gj} \leq 0$, and $s_{bj} = -1$ if $h_{bj} \leq 0$ and get the absolute values $h_{rj} = |h_{rj}|$, $h_{gj} = |h_{gj}|$, and $h_{bj} = |h_{bj}|$, respectively.

Step 3. Round the values $d_{rj} = round\left(\sqrt{w_{rj}}\right)$, $d_{gj} = round\left(\sqrt{w_{gj}}\right)$, and $d_{bj} = round\left(\sqrt{w_{bj}}\right)$. In addition, calculate the deviation values $c_{rj} = \sqrt{w_{rj}} - d_{rj}$, $c_{gj} = \sqrt{w_{gj}} - d_{gj}$, and $c_{bj} = \sqrt{w_{bj}} - d_{bj}$ and save as an auxiliary information.

Step 4. Compute the values $h_{rj} = \frac{[h_{rj} - (h_{rj} mod 10)]}{10}$, $h_{gj} = \frac{[h_{gj} - (h_{gj} mod 10)]}{10}$, and $h_{bj} = \frac{[h_{bj} - (h_{bj} mod 10)]}{10}$.

Step 5. Set flag $\alpha = 1$.

Step 6. Assign $h_k = h_{rj}, d_k = d_{rj}, s_k = s_{rj}$ if $\alpha = 1$; $h_k = h_{gj}, d_k = d_{gj}, s_k = s_{gj}$ if $\alpha = 2$; and $h_k = h_{bj}, d_k = d_{bj}, s_k = s_{bj}$ if $\alpha = 3$, otherwise, if $\alpha > 3$ then go to Step 1.

Step 7. If $d_k \geq 10$, then do the following substeps:

Step 7a. If $(h_k mod 2) = 1$ then evaluate $h_k = \left[(h_k - 1) \times 10) + (\frac{d_k}{10}) + d_k mod 10\right] \times s_k$ else $h_k = \left[(h_k \times 10) + (\frac{d_k}{10}) + d_k mod 10\right] \times s_k$.

Step 7b. Compute $\alpha = \alpha + 1$ and go to Step 7.

Step 8. If $d_k < 10$, then do the following substeps:

Step 8a. If $(h_k mod 2) = 1$ then compute $h_k = (h_k \times 10 + d_k) \times s_k$, else $h_k = \left[(h_k + 1) \times 10 + d_k\right] \times s_k$.

Step 8b. Compute $\alpha = \alpha + 1$ and go to Step 6.

Step 9. Repeat Step 1 until all data bits have been processed.

Step 10. Perform inverse IWT from the (marked) IWT coefficients to obtain marked image $\hat{S}$.

Step 11. Stop.

B. Data Extraction

The primary procedure of the proposed bit extraction is described in the following algorithm.

Algorithm 2. Extracting hidden bits from a marked image.

Input: A marked image $\hat{S} = \left\{ (\hat{r}_i, \hat{g}_i, \hat{b}_i) | i = 1, 2, ..., MN \right\}$ and an auxiliary set of parameters C.

Output: An extracted watermark $W' = \left\{ \left(w'_{rj}, w'_{gj}, w'_{bj} \right) \right\}_{j=0}^{ab-1}$

Method:

Step 0. Perform L1 IWT from marked image $\hat{S}$ to obtain the HH-subband $\hat{H} = \left\{ \left(\hat{h}_{rj}, \hat{h}_{gj}, \hat{h}_{bj} \right) \right\}_{j=0}^{(MN/4)-1}$ of IWT coefficients.

Step 1. Input a set of coefficient $\hat{H}_j$, which derived from $\hat{H}$. If the end of input is encountered, then proceed to Step 9.

Step 2. Set the marks $s_{rj} = 1$ if $\hat{h}_{rj} > 0$, $s_{gj} = 1$ if $\hat{h}_{gj} > 0$, and $s_{bj} = 1$ if $\hat{h}_{bj} > 0$, respectively, and go to Step 3. Otherwise, set $s_{rj} = -1$ if $\hat{h}_{rj} \leq 0$, $s_{gj} = -1$ if $\hat{h}_{gj} \leq 0$, and $s_{bj} = -1$ if $\hat{h}_{bj} \leq 0$; besides, get the absolute values $\hat{h}_{rj} = \left| \hat{h}_{rj} \right|$, $\hat{h}_{gj} = \left| \hat{h}_{gj} \right|$, and $\hat{h}_{bj} = \left| \hat{h}_{bj} \right|$, respectively.

Step 3. Compute the values $w'_{rj} = \hat{h}_{rj} mod 10$, $w'_{gj} = \hat{h}_{gj} mod 10$, and $w'_{bj} = \hat{h}_{bj} mod 10$.

Step 4. Set flag $\beta = 1$.

Step 5. Obtain $h_k = \hat{h}_{rj}, w_k = w'_{rj}, s_k = s_{rj}, c_k = c_{rj}$ if $\beta = 1$; $h_k = \hat{h}_{gj}, w_k = w'_{gj}, s_k = s_{gj}, c_k = c_{gj}$ if $\beta = 2$, and $h_k = \hat{h}_{bj}, w_k = w'_{bj}, s_k = s_{bj}, c_k = c_{bj}$ if $\beta = 3$, otherwise, if $\beta > 3$ then go to Step 1.

Step 6. Compute the value $T = \frac{(h_k - w_k)}{10}$, if $T mod 2 = 1$ then do nothing, otherwise evaluate $w_k = w_k + 9$.

Step 7. Evaluate $w_k = (w_k + c_k)^2$, $h_k = h_k \times s_k$, and $\beta = \beta + 1$, respectively, go to Step 5.

Step 8. Repeat Step1 until all hidden bits have been extracted.

Step 9. Assemble, descrambled and form the watermark W'.

Step 10. Stop.

C. Analysis and Discussion

As specified previously, without the help of auxiliary information, the extraction of the hidden watermark would be unsuccessfully at the receiver. However, the adversaries (or the third parties) are incapable of extracting hidden message if they have no auxiliary information. The optimal payload for our method is $\frac{(4\times3\times256\times256)}{512\times512} = 3$ bpp(bit per pixel).The overhead $C = \{(c_{rj}, c_{gj}, c_{bj})\}_{j=0}^{ab-1}$ for the proposed scheme is associated with the size of the watermark. In addition, each non-integer value in the set of (c_{rj}, c_{gj}, c_{bj}) lies between -1.0 and +1.0. To reduce the transmission time and increase privacy, the overhead can be losslessly compressed by using either the run-length coding algorithm or JBIG2 [8]. The resultant coded data can then sent by an out-of-band transmission to the receiver.

3   Experimental Results

Four 512×512 color images, as shown in Fig. 1, were used as host images. Each RGB pixel of the host images is represented by 24 bits, 8 bits per component. The size of the test color watermark is 256×256, as depicted in Fig. 2. The marked images generated by the proposed method are depicted in Fig. 3. It can be seen from the figure that the perceived quality is good. No apparent color distortion appeared in the figures. Their average PSNR is 42.42 dB. The PSNR is defined by

$$PSNR = 10 \times \log_{10} \frac{255^2}{MSE} \tag{3}$$

with $MSE = \frac{(\sum_{i=1}^{MN}[(r_i-\hat{r}_i)^2+(g_i-\hat{g}_i)^2+(b_i-\hat{b}_i)^2])}{3MN}$. Here (r_i, g_i, b_i) and $(\hat{r}_i, \hat{g}_i, \hat{b}_i)$ denote the RGB pixel values of the host image and the marked image. Notice that an input watermark was fully embedded in the host images, namely, the bit rate for each marked images is $\frac{(2\times3\times256\times256)}{512\times512} = 2.17$ bpp. Tradeoff between PSNR and payload for our method in four test image were shown in Fig. 4. The figure indicated that the PSNR performance for the image Goldhill is the best among test images, while Jet has the least PSNR as payload was larger than 1 bpp. Generally, the average PSNR of the test images has significantly increased when payload being less than 1 bpp.

Fig. 1. The host images. (a) GoldHill, (b) Lena, (c) Jet, and (d) Baboon.

Fig. 2. The test watermark.

Fig. 3. The host images. The marked images generated by the proposed method. (a) GoldHill (PSNR=42.70 dB), (b) Lena (PSNR=42.53 dB), (c) Jet (PSNR=42.23 dB), and (d) Baboon (PSNR=42.35 dB).

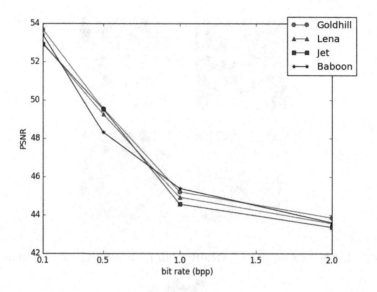

Fig. 4. Trade-off between PSNR and payload for the proposed method.

To demonstrate the robustness of the proposed method, examples of extracted watermarks after various manipulations of the image were given in Table 1. The normalized correlation (NC) value is also included. The NC_{RGB} is defined by

$$NC_{RGB} = \frac{NC_R + NC_G + NC_B}{3} \tag{4}$$

where $NC_R = \frac{\sum_i \sum_j w_R(i,j)w_R{}'(i,j)}{\sum_i \sum_j [w_R(i,j)]^2}$, $NC_G = \frac{\sum_i \sum_j w_G(i,j)w_G{}'(i,j)}{\sum_i \sum_j [w_G(i,j)]^2}$, and $NC_B = \frac{\sum_i \sum_j w_B(i,j)w_B{}'(i,j)}{\sum_i \sum_j [w_B(i,j)]^2}$. Here $w_R(i,j)$, $w_G(i,j)$, and $w_B(i,j)$ as well as $w_R{}'(i,j)$, $w_G{}'(i,j)$, and $w_B{}'(i,j)$ denote the RGB pixel values of the original watermark and the extracted one, respectively.

From Table 1 we can see that most of the extracted watermarks are recognized. Although the NC_{RGB} of the extracted watermark which attacked by cutting off 80% from the marked image, it is identifiable. Notice as well the NC_{RGB} of the survived watermark extracted from a marked image, which had undergone inversion attack, is still recognizable. In addition, the extracted watermarks are recognized when the marked images were rotated by 90 degrees. Similar performance can be found in the marked images which manipulated by brightness. From the above demonstration, we concluded that the marked images generated by the proposed method do resist from attacks including edge sharpening, cropping, rotation, brightness, and inversion.

Table 1. The survived watermarks extracted from the marked images which undergone various manipulations.

Attacks	Distorted Image		Survived Watermark	
Attack free NC_{RGB}=0.99662				
Cropping 80% NC_{RGB} = 0.46536				
Edge crispening NC_{RGB} = 0.74151				
Rotate 90 degrees NC_{RGB} = 0.78016				
Brightness -100 NC_{RGB} = 0.61653				
Negative NC_{RGB} = 0.99479				
Flip Horizontal NC_{RGB} = 0.77937				

Performance comparison between the proposed method and existing schemes: Yang's scheme [3], Yang and Wang[5], and Yang and Wang[6] is given in Table 2. It is obvious that the proposed method provides the largest PSNR and payload among these compared methods. Also notice that the average payload for the proposed method is approximated eleven times larger than that for the Yang and Wangs techniques [5].

Table 2. PSNR/payload (bit) comparison of various methods.

Images	Payload (bit)/ PSNR (dB)			
	Yang [3]+	Yang and Wang [5]	Yang and Wang [6]	Our method
Lena	17,874/39.59	16,042/43.54	57,600/39.07	177,068/46.00
Baboon	15,474/39.61	15,373/39.40	57,600/29.39	177,068/45.05
Jet	21,345/39.34	16,309/42.74	57,600/39.17	177,068/45.76
House	17,146/39.21	16,660/48.44	57,600/47.35	177,068/44.37
Tiffany	21,872/37.50	16,403/43.23	57,600/39.64	177,068/43.02
Average	18,742/39.05	16,157/43.47	57,600/38.92	177,068/44.84

+With the watermarking approach of Yang's technique.

4   Conclusion

By using the computation of the offset between two square-root-values, a large volume of secret bits can be successfully embedded in the host images by the proposed method. Experimental results confirmed that the perceptual quality of the marked images is good while the hiding-capacity is high. Moreover, the proposed method does resist several kinds of attacks such as brightness, cropping, edge sharpening, blurring, flip horizontal, inversion, and rotation. Additionally, the hiding-capacity of our proposed method is larger than that of existing techniques. Major applications of the proposed method can be found in the protection of copyright and ownership.

References

1. I.J. Cox, M.L. Miller, J.A. Bloom, J. Fridrich, and T. Kalker, *Digital Watermarking and Steganography*, 2nd Ed., Morgan Kaufmann., MA, USA, 2008.
2. E. Eielinska, W. Mazurczyk, and K. Szczypiorski, Trends in steganography, *Communications of the ACM* 57, 86-95, 2014.
3. C.Y. Yang, Robust watermarking scheme based on radius weight mean and feature-embedding technique, *ETRI Journal*, vol. 35, pp. 512-522, 2013.
4. C.C. Lin, C.C. Chang and Y.H. Chen, A novel SVD-based watermarking scheme for protecting rightful ownership of digital images, *Journal of Information Hiding and Multimedia Signal Processing*, vol. 5, pp. 124-143, 2014.
5. C.Y. Yang and W.F. Wang, Robust color image watermarking approach based on shape-specific points, *The 10th Int. Conf. on Intellig. Info. Hiding and Multim. Sig. Proc.*, Kitakyushu, Japan, Aug. 27-29, pp. 65-68, 2014.
6. C.Y. Yang and W.F. Wang, High-performance digital watermarking with L2-norm centroid for colour images, *The 11th Int. Conf. on Intellig. Info. Hiding and Multim. Sig. Proc.*, Adelaide, Australia, Sept. 23-25, pp. 29-31, 2015.
7. A. R. Calderbank, I. Daubechies, W. Sweldens, and B. L. Yeo, Wavelet transforms that map integers to integers, *Applied & Computational Harmonics Analysis*, vol. 5, no. 3, pp. 332-369, 1998.
8. P.G. Howard, F. Kossentini, B. Martins, S. Forchhammer, and W.J. Rucklidge, The emerging JBIG2 standard, *IEEE T. Circuits and Systems for Video Technology*, vol. 8, pp. 838-848, 1998.

High Payload Data Hiding Based on Just Noticeable Distortion Profile and LSB Substitution

Hui-Shih Leng, Hsien-Wen Tseng

Dept. of Mathematics
National Changhua Univ. of Education
Changhua, Taiwan
lenghs@cc.ncue.edu.tw
Dept. of Information Management
Chaoyang Univ. of Technology
hwtseng@cyut.edu.tw

Abstract. The growth of the Internet with the rapid advance of technology raises consumers' concerns regarding threats to their personal privacy. Data hiding is a technique that conceals data into a carrier for conveying the secret message confidentially and do not show any visible marks to draw the attention of the intruders. In this study, the grayscale image is divided into non-overlapping blocks. In each block, calculates the sum of the just noticeable distortion profile of each pixel. Based on this value, the block is classified into four levels, such as, lower, lower-middle, high-middle, and higher. According to these levels, 2, 3, 4, and 5 bits can be embedded by using the least significant bit (LSB) substitution method. The experimental results show that the proposed scheme achieves high payload with acceptable image quality according to the human visual system (HVS).

Keywords: data hiding; just distortion profile; least significant bit substitution

1 Introduction

The growth of the Internet with the rapid advance of technology raises consumers' concerns regarding threats to their personal privacy. Cryptography and data hiding are two kinds of techniques that are used to send information securely. Cryptography involves coding the secret message using an encryption key and sending it as cypher text. Different from the cryptography, data hiding involves conceals data into a carrier for conveying the secret message confidentially and do not show any visible marks to draw the attention of the intruders.

The least significant bit (LSB) substitution method is the simplest and well known data hiding scheme. The advantage of the LSB substitution method is that it can be implemented easily and provides high payload and low distortion.

© Springer International Publishing AG 2017

J.-S. Pan et al. (eds.), *Advances in Intelligent Information Hiding and Multimedia Signal Processing*, Smart Innovation, Systems and Technologies 63,
DOI 10.1007/978-3-319-50209-0_8

Chan and Cheng [1] proposed an optimal pixel adjustment process to the stego-image obtained by the LSB substitution method, the image quality of the stego-image can be greatly improved with low extra computational complexity. The disadvantage of the LSB substitution method is that it can be detected by RS-analysis [2] according to its asymmetry property. Swain [3] proposed an adaptive and modified LSB substitution method of categorizing the blocks into one of the four levels (lower, lower-middle, higher-middle, and higher) based on average of pixel value differences in 3-by-3 blocks. It can escape from RS-Steganalysis by hiding variable number of bits in different blocks.

In order to embed large amount of secret messages and maintain stego-image imperceptibility according to the human visual system (HVS), the just notice-able distortion (JND) profile is used to estimate the maximum payload. A variety of human visual system models have been proposed to deal with image process-ing problems and signal processing problems [4-9]. Chou and Li [4] proposed a perceptual model based on the background luminance and the change level in the background to estimate the JND profile of a grayscale image. The proposed scheme combines Swain's scheme and the concept of a JND profile to achieve high payload with acceptable image quality under HVS and the experimental results use PSNR and PNPNR(Peak Signal to Perceptible Noise Ratio) [10-13] to measure the image quality.

The rest of this paper is organized as follows. Section 2 introduces Swain's [3] data hiding scheme and Chou and Li's JND profile estimation method. In Section 3, our scheme is described. The experimental results are presented and discussed in Section 4. Conclusions are provided in Section 5.

2   Related Works

In this section, we first introduce Swain's data hiding scheme and Chou and Li's JND profile estimation method.

2.1   Swain's data hiding scheme

We describe the embedding procedure and extraction procedure as follows.

First, partition the grayscale image into 3-by-3 non-overlapping blocks in raster scan order as shown in Fig.1, where x_i $(i=0,1,2,,8)$ is the different pixel values.

x_0	x_1	x_2
x_3	x_4	x_5
x_6	x_7	x_8

Fig. 1. 3-by-3 sample block

Second, calculating the average difference value d, where x_{min} is the minimum value of this block.

$$d = \frac{1}{8} \sum_{i=0}^{8} |x_i - x_{min}| \tag{1}$$

The embedding bits depend on the value d: if $d \leq 7$, then this block belongs to lower level and 2 bits LSB substitution is applied; if $8 \leq d \leq 15$, then this block belongs to lower-middle level and 3 bits LSB substitution is applied; if $16 \leq d \leq 31$, then this block belongs to higher-middle level and 4 bits LSB substitution is applied; if $d \geq 32$, then this block belongs to higher level and 5 bits LSB substitution is applied. The last pixel in the block, $x_8 = b_1 b_2 b_3 b_4 b_5 b_6 b_7 b_8$, the last two bits $b_7 b_8$ are set to be 00, 01, 10, 11 if the block belongs to lower, lower-middle, higher-middle and higher level as shown in Table 1.

Table 1. Level Marks of the last two bits of x_8

Level	$b_7 b_8$ of x_8
lower	00
lower-middle	01
higher-middle	10
higher	11

In addition, if the block belongs to lower-middle, higher-middle and higher level, bits b_6, $b_5 b_6$, and $b_4 b_5 b_6$ of x_8 can embed extra bits respectively. Finally, we conclude the Swain's embedding procedure in Table 2.

Table 2. Swain's embedding procedure

Level	d	Embedding	Payload
lower	$d \leq 8$	Two LSBs for $x_0, \cdots, x_7$	16
lower-middle	$9 \leq d \leq 16$	Three LSBs for $x_0, \cdots, x_7$ and b_6 of x_8	25
higher-middle	$17 \leq d \leq 32$	Four LSBs for $x_0, \cdots, x_7$ and b_5, b_6 of x_8	34
higher	$d > 33$	Five LSBs for $x_0, \cdots, x_7$ and b_4, b_5, b_6 of x_8	43

The extraction procedure is very simple. Partition the stego-image into non-overlapping blocks in raster scan order as in Fig. 1. Let x'_0, x'_1, , and x'_8 are stego-pixels of x_0, x_1, , x_8, respectively. For each block, if bits $b_7 b_8$ of x'_8 is 00, then two LSBs from x'_0, x'_1,, and x'_7 are extracted; if bits $b_7 b_8$ of x'_8 is 01, then three LSBs from x'_0, x'_1,, and x'_7 are extracted and extract extra bit from bit b_6 of x'_8; if bits $b_7 b_8$ of x'_8 is 10, then four LSBs from x'_0, x'_1,, and x'_7 are extracted and extract extra bits from bit $b_5 b_6$ of x'_8; if bits $b_7 b_8$ of x'_8 is 11, then five LSBs from x'_0, x'_1,, and x'_7 are extracted and extract extra bits from bit $b_4 b_5 b_6$ of x'_8. Then, all secret messages can be extracting correctly. After embedding procedure, applied optimal pixel adjustment process (OPAP) to minimize the distortion.

2.2 Chou and Li's profile estimation method

Chou and Li proposed a perceptual model based on the background luminance and the change level in the background to estimate the JND profile of a grayscale image. The following eq. (2)-(6) are the details for the perceptual model for estimating the JND profile.

$$JND(x, y) = \max\{f_1(bg(x, y), mg(x, y)), f_2(bg(x, y))\} \tag{2}$$

$$f_1(bg(x, y), mg(x, y)) = mg(x, y)\alpha(bg(x, y)) + \beta(bg(x, y)) \tag{3}$$

$$f_2(bg(x, y)) = \begin{cases} T_0(1 - (bg(x, y)/127)^{1/2})) + 3 & \text{for} bg(x, y) \leq 127 \\ \gamma(bg(x, y) - 127) + 3 & \text{for} bg(x, y) > 127 \end{cases} \tag{4}$$

$$\alpha(bg(x, y)) = bg(x, y) \times 0.0001 + 0.115 \tag{5}$$

$$\beta(bg(x, y)) = \lambda - bg(x, y) \times 0.01 \text{for} 0 \leq x < H, 0 \leq y < W \tag{6}$$

where $bg(x, y)$ and $mg(x, y)$ are the average background luminance and the maximum weighted average of the luminance difference around the pixel at (x, y), respectively. H and W denote the height and weight of the image, respectively. $f_1(x, y)$ and $f_2(x, y)$ are the spatial masking effect and the visibility threshold due to background luminance. $\alpha(x, y)$ and $\beta(x, y)$ are the background luminance-dependent functions that specify the slope of the line and the intersection with the visibility threshold axis. In Chou and Li's experiment, T_0, γ , and λ are found to be 17, 3/128, and 1/2.

3   The Proposed Scheme

In our experiments, the minimal JND value is 3. According to Swain's scheme, we give a new definition to distinguish lower, lower-middle, higher-middle, and high levels and it satisfies each level has the same payload. We define the sum of JND value of each block D as follows:

$$D = \frac{1}{9} \sum_{i=0}^{8} (\lfloor JND(x_i) \rfloor - 1) \tag{7}$$

where $JND(x_i)$ is the JND value of x_i, i=0,1,2,,8. The embedding procedure is described in Table 3.

By the same way, the embedding bits depend on the value D: if $D \leq 16$, then this block belongs to lower level and 2 bits LSB substitution is applied;

Table 3. Proposed Embedding Procedure

Level	D	Embedding	Payload
lower	$D \leq 16$	Two LSBs for $x_0, \cdots, x_7$	16
lower-middle	$17 \leq D \leq 25$	Three LSBs for $x_0, \cdots, x_7$ and b_6 of x_8	25
higher-middle	$26 \leq D \leq 34$	Four LSBs for $x_0, \cdots, x_7$ and b_5, b_6 of x_8	34
higher	$D > 34$	Five LSBs for $x_0, \cdots, x_7$ and b_4, b_5, b_6 of x_8	43

if $17 \leq D \leq 25$, then this block belongs to lower-middle level and 3 bits LSB substitution is applied; if $26 \leq D \leq 34$, then this block belongs to higher-middle level and 4 bits LSB substitution is applied; if $D > 34$, then this block belongs to higher level and 5 bits LSB substitution is applied. The last pixel in the block, $x_8 = b_1b_2b_3b_4b_5b_6b_7b_8$, the last two bits b7b8 are set to be 00, 01, 10, 11 if the block belongs to lower, lower-middle, higher-middle and higher level as shown in Table 1.

4 Experimental Results

The most common tools to measure the image quality are Mean Square Error (MSE) and Peak Signal to Noise Ratio (PSNR).

$$MSE = \frac{1}{m \times n} \sum_{i=1}^{m} \sum_{j=1}^{n} \left(x_{ij} - x'_{ij} \right)^2 \tag{8}$$

$$PNSR = 10 \times \log_{10} \frac{255^2}{MSE} \tag{9}$$

where x_{ij} and x'_{ij} are the pixel values of the cover image and the stego-image in position (i, j). m and n represent the width and height of the image.

However, these measures use the same criterion in the smooth area and edge area without considering the human eye's property. Therefore, Chou and Li proposed another kind of image quality measure tool, called Peak Signal to Perceptible Noise Ratio (PSPNR).

$$MSE' = \frac{1}{m \times n} \sum_{i=1}^{m} \sum_{j=1}^{n} \left(|x_{ij} - x'_{ij}| - JND(x_{ij}) \right)^2 \cdot \delta_{ij} \tag{10}$$

$$PNPSR = 10 \times \log_{10} \frac{255^2}{MSE'}, \text{where} \delta_{ij} = \begin{cases} 1 \text{ if} |x_{ij} - x'_{ij}| > JND(x_{ij}) \\ 0 \text{ if} |x_{ij} - x'_{ij}| \leq JND(x_{ij}) \end{cases} \tag{11}$$

Obviously, the PSPNR is dependent upon the JND profile and satisfactorily match human's perception.

The proposed scheme is implemented using MATLAB and the secret messages are all randomly generated. Six standard test images 'Tiffany', 'Baboon',

Table 4. Six standard test images

'Lena', 'Jet', 'Scene' and 'Peppers' from SIPI image database are chosen as the cover image and shown in Fig. 2.

The experimental results are shown in Fig. 3.

Table 5. Six stego-images

It can be observed from Fig. 3. that the stego-images are of good image quality, and they are very similar to their respective cover images in Fig. 2. We compare PSNR and payload between Swain's scheme and the proposed scheme and show the results in Table 4. From Table 4, we find that the proposed enlarge the payload than Swain's method successfully.

Table 6. Compare of Swain's scheme with proposed scheme on PSNR and payload

	Swain's scheme		Proposed scheme	
	PSNR	Payload	PSNR	Payload
Tiffany	40.67	2.48	35.39	3.73
Baboon	35.86	3.66	33.49	3.51
Lena	40.21	2.57	33.44	3.35
Jet	40.04	2.55	35.11	3.53
Scene	38.18	2.99	31.14	4.24
Peppers	40.37	2.58	32.87	3.60

Moreover, we compare the stego-image quality by PSNR and PSPNR in Table 5.

Table 5 shows the proposed method achieves high payload with acceptable image quality. Though the swain's scheme use the d value to distinguish lower, lower-middle, higher-middle and higher levels. In fact, it only uses three levels,

Table 7. Compare of PSNR and PSPNR on the proposal scheme

	Proposed scheme		
	PSNR	PSPNR	Payload
Tiffany	35.39	37.03	3.73
Baboon	33.49	34.52	3.51
Lena	33.44	34.46	3.35
Jet	35.11	36.85	3.53
Scene	31.14	32.11	4.24
Peppers	32.87	34.20	3.60

and there is no block belongs to high level lead to low payload. We illustrate the distributions of swain's scheme on six standard test images as follows:

Table 8. The distributions of Swain's scheme on six standard test images

Levels	Tuffany	Baboon	Lena	Jet	Scene	Peppers
lower	19706	3685	18512	19475	11573	17350
lower-middle	5266	6564	5547	3993	8239	7400
higher-middle	3928	18651	4841	5432	9088	4150
higher	0	0	0	0	0	0

We also illustrate the distribution of the proposed scheme on six standard test images as follows:

Table 9. The distributions of the proposed scheme on six standard test images

Levels	Tuffany	Baboon	Lena	Jet	Scene	Peppers
lower	1083	7558	11294	1955	1577	7675
lower-middle	10404	9600	7146	16764	9068	8513
higher-middle	16952	4738	2770	6884	4755	4420
higher	461	7004	7690	3297	13500	8292

From Table 6 and Table 7, the proposed scheme can really distinguish the four different levels and embeds secret messages according to the HVS.

5 Conclusions

In this study, the grayscale image is divided into 3-by-3 non-overlapping blocks. If the width and height can't dividable by 3, we can keep the remainder section unchanged to raise the stego-image quality or embeds by LSB substitution to increase the payload.

We use Swain's scheme because it provides a good model to distinguish the complexity of each block, and it can escape from RS-Steganalysis for the reason that no asymmetry occurs. But the Swain's scheme can't really distinguish the four levels. The proposed scheme give a new definition to distinguish lower, lower-middle, higher-middle, and higher levels and it satisfies each level has the same payload. In addition, we use PSNR and PSPNR to evaluate the quality of the stego-image. The experimental results show that it achieves high payload with acceptable image quality according to the HVS.

References

1. C. K. Chan, L. M. Cheng: Hiding data in images by simple LSB substitution. Pattern recognition, 37(3), 469–474 (2004)
2. Fridrich, Jessica, Miroslav Goljan, Rui Du: Reliable detection of LSB steganography in color and grayscale images. Proceedings of the 2001 workshop on Multimedia and security: new challenges, ACM, (2001)
3. Swain, Gandharba: Digital image steganography using nine-pixel differencing and modified LSB substitution. Indian Journal of Science and Technology, 7(9), 1444–1450 (2014)
4. C. H. Chou, Y. C. Li: A perceptually tuned subband image coder based on the measure of just-noticeable-distortion profile. IEEE Transactions on Circuits and Systems for Video Technology, 5(6), 467–476 (1995)
5. Jayant, Nikil, James Johnston, Robert Safranek: Signal compression based on models of human perception. Proceedings of the IEEE, 81(10), 1385–1422 (1993)
6. Kutter, Martin, Stefan Winkler: A vision-based masking model for spread-spectrum image watermarking. IEEE Transactions on Image Processing, 11(1), 16–25 (2002)
7. Z. Lu, W. Lin, X. Yang, E. Ong, S. Yao: Modeling visual attention's modulatory aftereffects on visual sensitivity and quality evaluation. IEEE Transactions on Image Processing, 14(11), 1928-1942 (2005)
8. Z. H. Wei, P. Qin, Y. Q. Fu: Perceptual digital watermark of images using wavelet transform. IEEE Transactions on Consumer Electronic, 44(4), 1267-1272 (1998)
9. Peter H. W. Wong, Oscar C. Au: A capacity estimation technique for JPEG-to-JPEG image watermarking. IEEE Transactions on Circuits and Systems for Video Technology, 13(8), 746–752 (2003)
10. Yang, X. K., Ling, W. S., Lu, Z. K., Ong, E. P., Yao, S. S.: Just noticeable distortion model and its applications in video coding. Signal Processing: Image Communication, 20(7), 662–680 (2005)
11. Gao, Y., Xiu, X., Liang, J., Lin, W.: Perceptual multiview video coding using synthesized just noticeable distortion maps. IEEE International Symposium of Circuits and Systems, 2153–2156 (2011)
12. Chou, C. H., Li, Y. C.: A perceptually tuned subband image coder based on the measure of just-noticeable-distortion profile. IEEE Transactions on Circuits and Systems for Video Technology, 5(6), 467–476 (1995)
13. Yang, X., Lin, W., Lu, Z., Ong, E. P., Yao, S.: Perceptually adaptive hybrid video encoding based on just-noticeable-distortion profile. Visual Communications and Image Processing, 1448–1459 (2003)

An Improved Data Hiding Method of Five Pixel Pair Differencing and LSB Substitution Hiding Scheme

Tzu-Chuen Lu*, Yu-Ching Lu

Department of Information Management, Chaoyang University of Technology,
Taichung 41349, Taiwan

tclu@cyut.edu.tw
yuching1120@gmail.com

Abstract. In 2015, Gulve and Joshi proposed a Five Pixel Pair Differencing and least significant bit (LSB) substitution (FPPD) hiding scheme to improve the pixel-value differencing technique. In their scheme, a cover image is divided into several non-overlapping blocks 2×3 pixels in size. The center pixel in each block is a basic pixel that is used to compute differences between it and the other five pixels. These differences are used to conceal the secret message via the pixel-value differencing hiding method. In their scheme, the embedding capacity is limited by the size of the block. In this paper, we adjust the size of the block to increase the embedding capacity while maintaining the quality of the stego-image. Experimental results show that the embedding capacity of the proposed scheme is higher than that of Gulve and Joshis scheme, especially for a 2×2-sized block.

1 Introduction

With the rapid development of Internet technology, several people exchange or transfer their information on the Internet; therefore, digital data applications are widespread. However, transferred information may be tampered with or destroyed by hackers, so transferring data online is not entirely secure. To solve this problem, scholars have proposed an information hiding technique, wherein secret messages are embedded into multimedia, such as images, audios, texts, videos, or characters to produce stego-media. Stego-media is similar to the original media but is arranged such that embedded information can be securely transmitted without being noticed by the criminal.

Data hiding techniques can also be classified as involving the spatial, frequency, or compressed domains. Spatial domain schemes embed secret data into media using methods such as difference expansion, histogram shifting, or interpolation.

Difference expansion [9] expands the difference between two pixels in a pair to embed one secret bit. In histogram shifting, as proposed by Ni et al. [6], all

J.-S. Pan et al. (eds.), *Advances in Intelligent Information Hiding and Multimedia Signal Processing*, Smart Innovation, Systems and Technologies 63,
DOI 10.1007/978-3-319-50209-0_9

pixels in a cover image are calculated to produce a histogram. Then, a pair is selected that includes the peak and zero points in the histogram. The secret message is then embedded in the peak point and other pixels are shifted for recovery.

Frequency domain hiding schemes embed secret messages into the transformed coefficients of an image. Compressed domain schemes use a difference encoding strategy based on block features such as encoding a smooth area with the mean of the pixels in the block and using the rough area to produce the encoding block such as AMBTC [4].

In recent years, many scholars have proposed data hiding schemes based on spatial domain techniques, and the pixel value differencing scheme proposed by Wu and Tsai is the most famous. Many scholars have since proposed schemes based on pixel value differencing (PVD) to improve the capacity of Wu and Tsai's scheme. In 2015, Tyagi et al. [10] proposed a scheme that divides an image into several non-overlapping 2×2-sized blocks, and then summarizes the value of each pair of continuous pixels. If the summarized value S_i is smaller than (or equal to) 255, $S_i \leq 255$, then the scheme uses the pixel value differencing method to embed secret messages. If, on the other hand, the summarized value S_i is larger than 255, $S_i > 255$, the value of the two continuous pixels is modified before embedding. In addition, other researchers [5] [7] have proposed schemes in which the cover image is divided into different sizes to increase the hiding capacity or which combine distinct techniques to enhance the quality of the stego-image.

In 2015, Gulve and Joshi proposed the five pixel pair differencing and LSB substitution hiding scheme (FPPD), in which a cover image is divided into several non-overlapping 2×3-sized blocks. The center pixel in the block is designated as the basic pixel, and is used to compute its differences with the remaining five pixels. To increase the hiding payload, Gulve and Joshi combined this technique with the LSB replacement method to conceal more secret messages in the basic pixel.

In the FPPD scheme, the embedding capacity is limited by the size of the block. A 2×3-sized block may not be optimal for every image. Therefore, in this paper, we adjust the size of the block to several different sizes, including 1×3, 2×2, and 3×3, and then determine the effectiveness of the resulting embedding capacities and image qualities of the stego-images. Based on the image features, we can determine the appropriate block size for different images.

The rest of this paper is organized as follows. In section 2, we review the PVD technique, LSB replacement technique, and FFPD technique. In section 3, we describe the concept of the proposed scheme and use an example to demonstrate its performance. In section 4, we evaluate our experimental results, and we draw our conclusions in section 5.

2 Related Work

2.1 Pixel Value Differencing (PVD)

The PVD technique, proposed by Wu and Tsai in 2003 [11], uses a pixel pair that contains two continuous pixels px_i and px_{i+1} to compute difference value d_i, using the equation shown below:

$$d_i = |px_i - px_{i+1}|. \tag{1}$$

Based on the difference, we can identify the image features. If the difference value is large, this means that the pair is in a complex area. If the difference value is small, then we know that the pair is located in a smooth area.

The PVD scheme uses the difference value to determine the number of bits that can be embedded in the pixel. A quantization range table, shown in Fig. 1, is used to determine the number of hiding bits associated with each difference. In Wu and Tsai's scheme, the difference value is separated into six ranges, and each range has a corresponding upper bound u_i and lower bound l_i, and corresponding number of hiding bits.

$[u_i, l_i]$	[0,7]	[8,15]	[16,31]	[32,63]	[64,127]	[128,255]
Hiding bits	$log_2 8$ = 3	$log_2 8$ = 3	$log_2 16$ = 4	$log_2 32$ = 5	$log_2 64$ = 6	$log_2 128$ = 7
Range	R_1	R_2	R_3	R_4	R_5	R_6

Fig. 1. Quantization range table.

The following example illustrates the embedding processing of Wu and Tasi's PVD scheme. Suppose that the difference value is d = 9. The value d is located in R_2 of Fig. 1, for which the lower bound is 8, the upper bound is 15 and its corresponding hiding bits are calculated by:

$$k = \lfloor \log_2 (u - l + 1) \rfloor. \tag{2}$$

Hence, the hiding bit is k=$\lfloor \log_2 (15 - 8 + 1) \rfloor = 3$.

After extracting three bits from the secret sequence, the scheme converts the secret messages into a decimal value b_i. The new difference is computed by $d'_i = l_i + b_i$. The gap between the original difference d and the new difference d'_i is shared betweento the pixels px_i and px_{i+1} to generate the final stego-pixels px'_i and px'_{i+1}. The equation is shown below:

$$(px'_i, px'_i) = \begin{cases} (px_i + \lceil \frac{z}{2} \rceil, px_{i+1} - \lfloor \frac{z}{2} \rfloor) & if\, px_i \geq px_{i+1}\, and\, d'_i > d_i, \\ (px_i - \lceil \frac{z}{2} \rceil, px_{i+1} + \lfloor \frac{z}{2} \rfloor) & if\, px_i < px_{i+1}\, and\, d'_i > d_i, \\ (px_i - \lceil \frac{z}{2} \rceil, px_{i+1} + \lfloor \frac{z}{2} \rfloor) & if\, px_i \geq px_{i+1}\, and\, d'_i \leq d_i, \\ (px_i - \lceil \frac{z}{2} \rceil, px_{i+1} + \lfloor \frac{z}{2} \rfloor) & if\, px_i < px_{i+1}\, and\, d'_i \leq d_i. \end{cases} \tag{3}$$

In this way, the difference between the cover pixel and stego-pixel is small. In the equation, the value of z is the gap between d_i and d'_i.

In the extraction process, the receiver must first compute the difference d_i' between px_i' and px_{i+1}'. Next, the difference d_i' is mapped to the quantization range table, shown in Fig. 1, to obtain the lower bound l_i. Finally, the secret message is computed by $b = d_i' - l_i$ and then b is transformed into a binary string.

3 Proposed Method

Gulve and Joshi proposed the FPPD technique in 2015 [3], in which the block size is fixed at 2×3. The center pixel in the block is a basic pixel used to compute the differences between it and the other five pixels. The difference is the key factor in determining the length of the hidden secret bits. So, is the fixed 2×3 block size the optimal choice? In this paper, we adjust the size of a block to 1×3, 2×2, and 3×3 to test the performance of the hiding scheme for different block sizes. Finally, we generalize the size of the block based on the image features. In the next subsection, we set the block size to be 2×2 as an example.

3.1 Embedding Phase

In the proposed scheme, the cover image is a gray-scale image. First, we divide the image into several non-overlapping blocks of different sizes, including 1×3, 2× 2, and 3×3. The arrangements of the blocks of different sizes are shown in Fig. 2.

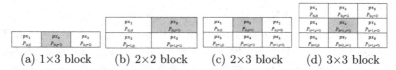

(a) 1×3 block   (b) 2×2 block   (c) 2×3 block   (d) 3×3 block

Fig. 2. Different block sizes and arrangements.

For example, the block in Fig. 2(b) is 2×2 in size and contains four pixels px_0, px_1, px_2, and px_3, where x and y are the pixel coordinates in the image. Fig. 3 shows an example of a 2×2-sized block.

96	160
63	166

Fig. 3. Example of a 2×2-sized block sized 2×2.

Let px_0 be a basic pixel. Using the LSB replacement technique, the scheme conceals three secret bits into the basic pixel px_0 to generate a prediction pixel $\widehat{px}_0$. The prediction pixel $\widehat{px}_0$ is then used to form three pixel pairs $(\widehat{px}_0, px_1)$, $(\widehat{px}_0, px_2)$, and $(\widehat{px}_0, px_3)$.

In Fig. 3, the basic pixel is $px_0 = 160 = (10100000)_2$. Using the LSB scheme to embedding 3 bits of secret messages to computing the prediction pixel $\widehat{px}_0$.

Then, the differences d_i between the prediction pixel $\widehat{px}_0$ and the other three pixels px_i are computed to determine the hiding bits.

Based on the range table in Fig. 1, the value d_i is used to determine the hiding bits k_i. In order to balance the hiding capacity of each pixel in a block, the scheme then computes the average values of k_i to obtain the appropriate hiding bits of each pixel. The average value is computed by:

$$avg = \frac{\sum k_i}{3}.$$ (4)

Next, the average value is used to narrow down the difference value to obtain the modified difference dl_i, using the equation:

$$dl_i = d_i \bmod 2^{avg}.$$ (5)

The offset difference OD_i is computed as $|d_i| - |dl_i|$ for each pair. The modified difference dl_i is then used to determined the number of hiding bits b_i that can be hiddened into each pair. The scheme then maps dl_i to determine the corresponding hiding bits and the lower bound l_i from Fig. 1 Finally, the candidate difference d_i' is calculated by:

$$d_i' = \begin{cases} OD_i + l_i + b_i \, if \, d_i \geq 0, \\ -(OD_i + l_i + b_i) \, if \, d_i < 0. \end{cases}$$ (6)

If d_i is a positive number, d_i' is calculated as $OD_i + l_i + b_i$. Otherwise, d_i' is calculated as $-(OD_i + l_i + b_i)$.

To identify a suitable reference pixel, the difference m_i between d_i' and d_i is then computed. The one with the smallest difference is chosen as the reference pixel. The scheme then computes the upper value pu_i and lower value pd_i of each pair, using the formula:

$$(pu_i, pd_i) = (\widehat{px}_0 - \left\lceil \frac{m_i}{2} \right\rceil, px_i + \left\lfloor \frac{m_i}{2} \right\rfloor).$$ (7)

where $\widehat{px}_0$ and px_i are the prediction pixel and the other three pixels. The upper value pu_i is value of the candidate pixels to which $\widehat{px}_0$ may be shifted. To achieve minimum distortion between the stego-pixel and the original pixel, the pixel with smallest distance between $\widehat{px}_0$ and pu_i is selected as the reference pixel. The distance between $\widehat{px}_0$ and pu_i is computed by $\widehat{px}_0$-$pu_i|$. The lower value is revised by:

$$pd_i' = pd_i + pu_{min} - pu_i.$$ (8)

At this point, the scheme has hidden secret messages into each pair. However, the reference pixel does not equal to the prediction pixel $\widehat{px}_0$, which means that three LSB bits of pu_1 is differ from those of $\widehat{px}_0$. Thus, the secret message cannot be extracted from pu_{min}. Hence, the lower values must be modified again to generate the stego-pixels. The stego pixel px_i' is computed by:

$$px_i' = pd_i' + (\widehat{px}_0 - pu_{min}).$$ (9)

The stego-pixel of the basic pixel is $px_0' = \widehat{px}_0$.

Finally, the embedding process is finished. The stego -block is shown in Fig. 4

97	167
58	167

Fig. 4. Stego-block of Fig. 3.

3.2 Extraction Process

In the extraction procedure, the stego-image is divided into several non-overlapping m×n-sized blocks. The values of m and n are assigned in the embedding procedure. For each block, the scheme computes the differences between the reference pixel px_0' and the other pixels. These differences are then used to extract the secret messages.

Consider the example shown in Fig. 4. The size of block is 2×2 and the reference pixel is $px_0' = 167$. Three secret bits are extracted from the three LSB bits of the reference pixel. In the example, the binary string of the reference pixel $px_0' = 167$ is $(10100111)_2$. The least three significant bits are 111, which are the concealed secret bits.

Then, the scheme computes the difference value d_i between the reference pixel px_0' and the neighboring pixels px_i'. These differences are then used to determine the hiding bits k_i from the range table in Fig. 1.

The hiding bits of each pair of Fig. 4 are $k_1=6$, $k_2=6$, $k_3=3$, respectively. The average value of the hiding bits is computed by $\left\lfloor \frac{(6+6+3)}{3} \right\rfloor = 3$. This value is then used to modify the difference dl_i' as $dl_i'=d_i' \bmod 2^{avg}$.

Then, the value of dl_i' is used to map the hiding bits and the corresponding lower bound l_i'. The secret message is extracted by using $b_i=|dl_i'| - |l_i'|$.

4 Experimental Results

Using six standard images, we compared the results of the proposed method with those of the FPPD method. The test images are shown in Fig. 5.

We used the peak signal-to-noise ratio (PSNR) to evaluate the performance of the stego-image, using the following equation:

$$PSNR = 10 \times \log_{10} \left[\frac{255^2}{\frac{1}{h \times w} \times \sum_{i=1}^{h} \sum_{j=1}^{w} (px_{i,j}' - px_{i,j})} \right]. \quad (10)$$

In the equation, h and w are the height and width of the image, respectively. $px_{i,j}$ and $px_{i,j}'$ are the original pixel and the stego-pixel, respectively. If the difference between the stego-pixel and the original pixel is small, this indicates that the PSNR value is high and the quality of the stego-image is good. In contrast, a low PSNR value means that the quality of stego-image is poor. In terms of embedding capacity, the formula used to evaluate the hiding payload of each pixel is:

$$bpp = \frac{C}{h \times w}. \quad (11)$$

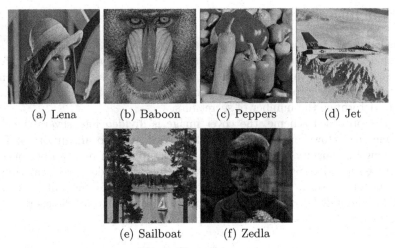

(a) Lena   (b) Baboon   (c) Peppers   (d) Jet

(e) Sailboat   (f) Zedla

Fig. 5. Six test images.

where C is the total number of hiding bits. A high bpp value indicates a good capacity. *Table 1* shows the bpp and PSNR values of the proposed scheme for different block sizes, 1×3, 2×2, and 3×3. The results show that the embedding capacity of the proposed scheme is more than 3 bpp when the block size is set to be 2×2 and the PSNR value of the stego-image is higher than 37.916 db. From the experimental results, we can see that the size of block set to be 2×2

Table 1. Performance evaluation in terms of capacity/distortion

Block size		Lena	Baboon	Jet	Sailboat	Peppers	Zedla
1×3	PSNR	38.569	37.819	38.596	37.935	38.004	37.997
	BPP	2.9911	3.0013	2.9920	2.9940	2.9909	2.9893
2×2	PSNR	38.050	37.584	38.014	37.914	38.004	37.930
	BPP	**3.0021**	**3.0281**	**3.0070**	**3.0072**	**3.0037**	**3.0025**
3×3	PSNR	38.099	37.858	38.153	38.028	38.082	38.004
	BPP	2.9774	2.9893	2.9782	2.9786	2.9778	2.9780
FFPD	PSNR	38.094	37.770	38.105	37.982	38.033	38.056
	BPP	2.9896	3.0049	2.9906	2.9923	2.9901	2.9896

can achieve better performance. Compared with the FFPD scheme, in which the block size is set to be 2×3, the hiding capacity of the proposed scheme with a block size of 2×2 is higher.

The experimental results also show that a block size of 2×2 can realize a higher hiding capacity whether the image is complex or smooth. Thus, the 2×2 block size is suitable for embedding any kind of image with large secret messages.

5 Conclusion

In this paper, we proposed an irreversible data hiding scheme based on the PVD technique. The proposed scheme improves the FPPD scheme by using a different block size to better control the hiding capacity and image quality. Based on the PVD technique, the proposed scheme also uses Wu and Tsai's quantization range table to determine the number of secret bits.

In the proposed scheme, the cover image is divided into several blocks of the same size. However, the features of a block are also an important factor influencing its capacity. For a complex block, more hiding bits can be concealed to increase the hiding payload. However, more hiding bits decrease image quality. Hence, in future work, we will try to identify the image features of a block and divide blocks of different sizes by analyzing the image characteristics to achieve the best capacity as well as maintain high image quality.

References

1. Chen, J.: A PVD-based Data Hiding Method with Histogram Preserving using Pixel Pair Matching. Image Communication. 29, 375–384 (2014)
2. Gulve, A.K., Joshi, M.S.: A High Capacity Secured Image Steganography Method with Five Pixel Pair Differencing and LSB Substitution. Graphics and Signal Processing. 7, 64–74 (2015)
3. Gulve, A.K., Joshi, M.S.: An Image Steganography Algorithm with Five Pixel Pair Differencing and Gray Code Conversion. Graphics and Signal Processing. 6, 488–493 (2014)
4. Lema, M. D., Mitchell, O. R.: Absolute Moment Block Truncation Coding and its Application to Color Image. IEEE Transactions on Communications. 32, 1148–1157 (1984)
5. Lee, Y.P., Lee, L.C., Chen, W.K., Chen, I.J., Chang, C.P., Chang,: High Payload Image Hiding with Quality Recovery using Tri-way Pixel-Value Differencing. Information Sciences. 191, 514–225 (2012)
6. NI, Z.C., Shi, Y.Q., Ansari, N.: Reversible Data Hiding. Circuits and Systems for Video Technology. 16, 354–362 (2003)
7. Swain, G., Lenka, S.K.: Pixel value differencing steganography using correlation of target pixel with neighboring pixels. In: 2015 IEEE Electrical, Computer and Communication Technologies, pp. 1–6. IEEE Press, Coimbatore (2015)
8. Swain, G.: Digital image steganography using nine-pixel differencing and modified LSB substitution. Science and Technology. 7, 1444–1450 (2014)
9. Tian, J.: Reversible Data Embedding Using a Difference Expansion. Circuits and Systems for Video Technology. 13, 890–896 (2003)
10. Tyagi, A., Changder, S., Roy, R.: High Capacity Image Steganography based on Pixel Value Differencing and Pixel Value Sum. In: 2015 Second IEEE Advances in Computing and Communication Engineering, pp. 488–493. IEEE Press, Dehradun (2015)
11. Wu, D.C., Tsai, W.H.: A Steganographic Method for Images by Pixel-Value Differencing. Pattern Recognition Letters. 24, 1613–1626 (2003)
12. Wu, H.C., Wu, N.I., Tsai, C.S., Hwang, M.S.: Image Steganographic Scheme Based on Pixel-Value Differencing and LSB Replacement Methods. Image and Signal Processing. 152, 611–615 (2005)

A Data Hiding Method based on Multi-predictor and Pixel Value Ordering

Yung-Chen Chou, Tzu-Chuen Lu, and Jhih-Huei Wu

168, Jifeng E. Rd., Wufeng District, Taichung, 41349 Taiwan
yungchen@gmail.com , tclu@cyut.edu.tw , s10214601@gm.cyut.edu.tw

Abstract. Pixel Value Ordering (PVO) is a commonly-used method for reversible data hiding. PVO divides a cover image into non-overlapping blocks and the minimum and maximum values in a block are used to embed the secret data. The visual quality and embedding capacity of the PVO method is closely related to the size of the blocks. Thus, the proposed data hiding method utilizes the modified prediction methods and integrates with Ou and Kims method to improve the performance of the visual quality and embedding capacity. Pixel prediction not only can be used in restoring missing pixels but can also be utilized to realize data embedding. The proposed method adopts the MED, NMI, INP, and CRS pixel prediction methods to embed the secret data, because we found that different pixel prediction methods can provide different information.

Key words: Multi-predictor, reversible data hiding, sorted pixel values, predicted value

1 Introduction

Tian presented a reversible data hiding method by using the difference expansion strategy [14]. Tian's method pairs the pixels in the cover image and expends the pixel difference twice for the pixel pair. After that, the stego pixels are generated by using the expended difference value plus the secret data. At the receiver side, the secret data is extracted by taking the last bit of the stego difference value of the pixel pair. Further, Alatter presented a difference expansion based reversible data hiding method [1], which takes four neighboring pixels each time to embed three secret bits.

Sachnev et al. proposed a Prediction Error Expansion (also called PEE) data embedding method [12] to achieve the reversibility of the data embedding method. The main idea of Sachnev's method is to use four closest neighboring pixels to determine the sequence for data embedding. The main purpose of Sachnev et al.'s method is to enhance the visual quality of the stego image. Li et al. combined the characteristic of an image region and the PEE concept to classify the image region into a Flat region or Rough region to propose a reversible data hiding method [7]. Li et al.'s method creates a different embedding capacity in

© Springer International Publishing AG 2017

J.-S. Pan et al. (eds.), *Advances in Intelligent Information Hiding and Multimedia*
Signal Processing, Smart Innovation, Systems and Technologies 63,
DOI 10.1007/978-3-319-50209-0_10

the Flat and Rough regions to achieve a good visual quality and embedding capacity. Furthermore, Gui et al. enhanced Li et al.'s method to create a reversible data embedding method [3]. Gui et al.'s method refines image region classification and assigns different embedding capacity to each image region type. Thus, Gui et al.'s method successfully improves the performance of the embedding capacity.

Histogram shifting is another commonly used reversible data embedding method. Ni et al. analyzed the pixel histogram of a cover image and embedded the secret data into the peak point pixel [11]. In Ni et al.'s method, the peak point and zero point information is retained for secret data extraction. After that, Wang et al. presented a reversible data embedding method [15] that uses the Markov Model to analyze pixel value distribution. Wang et al.'s method generates a histogram matrix and each cell in the matrix has different embedding capacity. Then, Chen et al. proposed a reversible data embedding method [2] by using the prediction error concept. Chen et al.'s method adopts the linear pixel prediction method to generate the maximum and minimum prediction error values that are used to generate the prediction error histograms. The different prediction error histograms use different embedding strategies to embed as much secret data as possible. Also, the stego image generated by Chen et al.'s method has good visual quality. Lu et al. proposed a reversible data embedding method [8] to improve Chen et al.'s method by combining Lukac prediction and Feng and Fan's prediction method to generate the maximum and minimum difference histograms. Also, the different histograms use different embedding strategies to embed the secret data.

On the other hand, Li et al. presented a data hiding method [6] by sorting pixels in a block and embedding the secret data into the pixel difference values. Li et al.'s main idea is that neighboring pixels have similar pixel value distribution, thus the difference between the minimum pixel and middle pixel and the difference between the maximum pixel and middle pixel are -1 and 1, respectively. Consequently, the data embedding capacity can be improved. Peng et al. proposed a pixel value ordering (PVO) based data embedding method [10] by modifying Li et al.'s pixel difference generation process. Peng et al.'s method subtracts a small value from a large value for pixel difference generation. Thus, the difference value will mostly fall on 0 or 1 to obtain a higher peak point (i.e. the embedding capacity can be improved). Wang et al. utilized dynamic pixel block partition strategy to segment the image and adopted an improved-pixel-value-ordering method to embed secret data into the cover image [16]. Ou and Kim proposed a pixel-based PVO (pixel value ordering) data embedding method [9] to embed more secret data into the cover image. Ou and Kim's method takes the maximum and minimum values from the adjoining pixels to embed the secret data.

As mentioned above, PVO, PPVO, and Wang's methods try to determine the maximum and minimum values from an image block. These methods achieve the goal of good visual quality for a stego image with acceptable embedding capacity. The PPVO method has a better data embedding capacity than others but the visual quality of the stego image can still be further improved. In this paper, a novel reversible data embedding method is presented. We are inspired by the Median Edge Detect, MED [17], Neighbor Mean Interpolation, NMI [4], Interpolation by Neighboring Pixels, INP [5], and High Capacity Reversible Steganography, CRS [13] methods to propose a data hiding method to further improve the embedding capacity and the visual quality of the stego image.

2 Related Works

2.1 Pixel-based Pixel Value Ordering (PPVO) [9]

To achieve high embedding capacity, Ou and Kim proposed the pixel-based Pixel-Value-Ordering (PPVO) method [9] to embed secret data into adjoining pixels. In PPVO, first, the current pixel $x_{i,j}$ is taken from an image block sized 4×4. Then, n neighboring pixels are taken from the block to form a reference set $C = \{c_1, c_2, \ldots, c_n\}$. The maximum and minimum reference pixels in C are chosen to embed the secret data using Eqs. 1, and 2.

$$x'_{i,j} = \begin{cases} x_{i,j} - s, \text{ if } x_{i,j} = C^{\min}, \\ x_{i,j} + s, \text{ if } x_{i,j} = C^{\max}, \\ x_{i,j} - 1, \text{ if } x_{i,j} < C^{\min}, \\ x_{i,j} + 1, \text{ if } x_{i,j} > C^{\max}, \\ x_{i,j}, \quad\quad \text{otherwise}, \end{cases} \tag{1}$$

$$x'_{i,j} = \begin{cases} x_{i,j} + s, \text{ if } x_{i,j} = C^{\min} = C^{\max} = 254, \\ x_{i,j} - s, \text{ if } x_{i,j} = C^{\max} = C^{\min}, \\ x_{i,j} - 1, \text{ if } x_{i,j} < C^{\min} = C^{\min}, \\ x_{i,j}, \quad\quad \text{otherwise}, \end{cases} \tag{2}$$

Where, $x'_{i,j}$ represents the stego pixel, s is the secret bit, and C^{max} and C^{min} are the maximum and minimum reference pixels in C, respectively.

2.2 Media Edge Detect (MED) [17]

Weinberger et al. presented a low computation cost median edge detection method [17] to effectively predict the missing pixel value located at the edge. The MED method uses a mask sized 2×2 to predict a suitable value for the missing pixel.

$$x_{i,j} = \begin{cases} \min(a, b), \text{ if } c \geq \max(a, b), \\ \max(a, b), \text{ if } c \leq \min(a, b), \\ a + b - c, \text{ otherwise}, \end{cases} \tag{3}$$

2.3   Data Hiding Method Using Image Interpolation (NMI) [4]

Jung and Yoo proposed a data embedding method [4] to embed secret data into
an enlarged image. Jung and Yoos method translates an image block sized 2×2
into an image block sized 3×3 and fills the missing pixels in the gaps using
interpolation (refer to Eq. 4).

$$P_{i,j} = \begin{cases} \lfloor \frac{x_{i,j-1}+x_{i,j+1}}{2} \rfloor, & \text{if } i = 2m, j = 2n + 1, \\ \lfloor \frac{x_{i-1,j}+x_{i+1,j}}{2} \rfloor, & \text{if } i = 2m + 1, j = 2n, \\ \lfloor \frac{x_{i-1,j-1}+P_{i-1,j}+P_{i,j-1}}{3} \rfloor, & \text{otherwise,} \end{cases} \tag{4}$$

Where m and n represent the height and width of the original image block.

2.4   Interpolating Neighboring Pixels (INP) [5]

Lee and Huang proposed a data embedding method [5] by using the difference
value between the original pixel and the predicted values. First, the cover image is
enlarged and the value for a missing pixel is calculated using the mean value and
weighted computation from two closest neighboring pixels. The pixel prediction
role is defined as shown in Eq. 5.

$$P_{i,j} = \begin{cases} \lfloor \frac{x_{i,j-1}+(x_{i,j-1}+x_{i,j+1})/2}{2} \rfloor, & \text{if } i = 2m, j = 2n + 1, \\ \lfloor \frac{x_{i-1,j}+(x_{i-1,j}+x_{i+1,j})/2}{2} \rfloor, & \text{if } i = 2m + 1, j = 2n, \\ \lfloor \frac{P_{i-1,j}+P_{i,j-1}}{2} \rfloor, & \text{otherwise,} \end{cases} \tag{5}$$

2.5   High Payload Reversible Data Embedding Method [13]

Tang et al. presented a data hiding method [13] that also adopts the pixel prediction strategy. Tang et al.'s method uses an image block sized 2×2 as one
process unit and determines the maximum value (denoted as x^{max}), minimum
value (denoted as x^{min}), and reference value AD from the block. Subsequently,
the predicted value can be obtained by applying Eq. 7.

$$AD = \frac{3 \times x^{min} + x^{max}}{4} \tag{6}$$

$$P_{i,j} = \begin{cases} \lfloor \frac{AD+(x_{i,j-1}+x_{i,j+1})/2}{2} \rfloor, & \text{if } i = 2m, j = 2n + 1, \\ \lfloor \frac{AD+(x_{i-1,j}+x_{i+1,j})/2}{2} \rfloor, & \text{if } i = 2m + 1, j = 2n, \\ \lfloor \frac{x_{i-1,j-1}+P_{i-1,j}+P_{i,j-1}}{3} \rfloor, & \text{otherwise,} \end{cases} \tag{7}$$

3   The Proposed Method

3.1   The Prediction Phase

Since the proposed method uses the predicted values as reference data for the
embedding procedure, the prediction phase combines the MED, NMI, INP, and

CRS methods to generate 16 predicted values. Pixels at the borders of a cover image are reserved, meaning that the pixels will not be used to embed the secret data. To easily describe the proposed method, the cover image is denoted as $X = \{x_{i,j} | i = 0, 1, \ldots, h - 1; j = 0, 1, \ldots, w - 1\}$ where h and w represent the height and width of the image X. $P_{MED} = \{p_i^{MED} | i = 1, 2, \ldots, 4\}$, $P_{NMI} = \{p_i^{NMI} | i = 1, 2, \ldots, 4\}$, $P_{INP} = \{p_i^{INP} | i = 1, 2, \ldots, 4\}$, and $P_{CRS} = \{p_i^{CRS} | i = 1, 2, \ldots, 4\}$ represent the predicted values generated by the MED, NMI, INP, and CRS methods, respectively. The prediction process uses a block sized 3×3 and the center pixel is the current pixel denoted as $x_{i,j}$. The prediction process is started from pixel $x_{1,1}$ and moves by one pixel at a time.

3.2 The Embedding Phase

After sixteen predicted values have been generated, all predicted values are collected to form the reference set $C = \{p_1^{MED}, p_2^{MED}, p_3^{MED}, p_4^{MED}, p_1^{NMI}, p_2^{NMI}, p_3^{NMI}, p_4^{NMI}, p_1^{INP}, p_2^{INP}, p_3^{INP}, p_4^{INP}, p_1^{CRS}, p_2^{CRS}, p_3^{CRS}, p_4^{CRS}\}$. Then, $C^{max} = \max(C)$ and $C^{min} = \min(C)$ values are chosen from the reference set C. The Eq. 1 is applied to embed the secret data. Fig. 1 shows a simple example for the proposed data embedding process. Pixels at the borders of the cover image X (shown in gray background color) are reserved, and the secret data is "10...". In the first block, the current pixel is $x_{1,1} = 52$, and the four corner pixels are $x_{0,0} = 55$, $x_{0,2} = 50$, $x_{2,0} = 51$, $x_{2,2} = 52$. After the prediction procession, the reference set $C = \{50, 55, 50, 55, 53, 54, 51, 50, 53, 54, 51, 50, 51, 52, 51, 51\}$ and $C^{max} = 55$, and $C^{min} = 50$. According to the embedding rule, $x_{i,j}$ is not used to embed the secret bit. Then the process block moves to the next pixel and $x_{1,2} = 50$, $x_{0,1} = 50$, $x_{0,3} = 50$, $x_{2,1} = 50$, $x_{2,3} = 50$. The reference set $C = \{50, 50, 50, 50, 50, 50, 50, 50, 50, 50, 50, 50, 50, 50, 50, 50\}$ is generated by applying the MED, NMI, INP, and CRS methods. According to the embedding rules, the current pixel $x_{1,2}$ will be changed to 49 to imply the secret bit '1'.

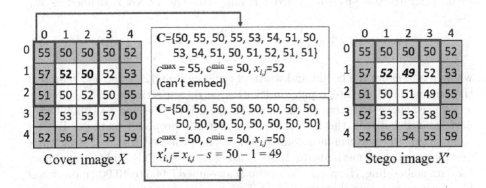

Fig. 1. A date embedding example

3.3   The Data Extraction Phase

The data extraction procedure is the reverse of the embedding phase. The data extraction process also takes a block sized 3×3 each time, starting from the last block in the stego image. For a stego block, the MED, NMI, INP, and CRS prediction methods are applied to generate the reference set C, c^{max} and c^{min}. Then, the secret bit can be extracted by using Eq. 8, and the original pixel can be restored by using Eq. 9.

$$s = \begin{cases} 0, \text{ if } x'_{i,j} = C^{max} \text{ or } x'_{i,j} = C^{min} \\ 1, \text{ if } x'_{i,j} = C^{max} + 1 \text{ or } x'_{i,j} = C^{min} - 1 \end{cases} \tag{8}$$

$$x_{i,j} = \begin{cases} x'_{i,j} + 1 \text{ if } x'_{i,j} < C^{min}, \\ x'_{i,j} - 1 \text{ if } x'_{i,j} > C^{max}, \\ x'_{i,j} \qquad \text{otherwise,} \end{cases} \tag{9}$$

3.4   Overflow and Underflow Handling

In cases where the pixel value equals to 255 or 0, then the stego pixel value might be changed to 256 or -1 after the embedding process. To avoid this situation, the proposed method will retain the pixel location information for the pixel whose value equals to 255 or 0. However, the pixel location information for pixels at the borders is redundant.

4   Experimental Results

Six test gray scale images were employed to be the test image (refer to Fig. 2). The visual quality and the embedding capacity are two most import factors for evaluating the performance of a data embedding method. To evaluate the visual quality of a stego image, which is generated by the proposed method, the Peak-Signal-to-Noise-Ration (PSNR) is adopted. The PSNR is defined as Eq. 10.

$$psnr = 10 \times \log_{10} \left[\frac{255^2}{\frac{1}{h \times w} \sum_{i=1}^{h} \sum_{j=1}^{w} (x'_{i,j} - x_{i,j})^2} \right] \text{ (dB)}, \tag{10}$$

where h and w are the height and width of the image, and $x_{i,j}$ and $x'_{i,j}$ represent the original pixel and stego pixel in location (i, j), respectively.

We set the parameters to the same as the PPVO method to compare the performance between the PPVO method and our proposed method. Table 1 summarizes the performance comparison of the proposed method using three and four prediction methods. In Table 1, using the NMI, INP, and CRS methods for data embedding, the proposed method can embed 5,000 to 10,000 more secret bit into a cover image than the PPVO (i.e., $n = 12$) method. In cases where four prediction methods are used in the proposed method, our visual quality is worse than the PPVO method at around 1 dB but we can embed more secret bits into a cover image than the PPVO method.

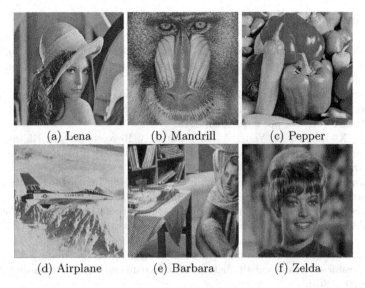

(a) Lena   (b) Mandrill   (c) Pepper

(d) Airplane   (e) Barbara   (f) Zelda

Fig. 2. The test images

Table 1. Performance comparison results (using three or four prediction methods)

Method		Lena	Mandrill	Pepper	Airplane	Barbara	Zelda
Propose Method	Capacity	28,742	10,042	28,159	43,975	23,240	28,029
(MED+NMI+INP)	PSNR	53.54	52.83	53.25	53.68	53.15	53.67
Propose Method	Capacity	29,655	10,332	29,090	45,135	24,264	29,626
(MED+NMI+CRS)	PSNR	53.40	52.67	53.09	53.53	53.01	53.47
Propose Method	Capacity	28,857	10,071	28,259	43,959	23,446	28,526
(MED+INP+CRS)	PSNR	53.18	52.77	53.19	53.61	53.11	53.60
Propose Method	Capacity	34,509	12,320	33,788	51,705	27,016	33,963
(NMI+INP+CRS)	PSNR	52.45	51.6	52.28	52.55	51.87	52.72
Propose Method	Capacity	28,738	10,002	28,181	43,911	23,308	28,095
(MED+NMI+INP+CRS)	PSNR	53.55	52.83	53.26	53.68	53.16	53.68
PPVO	Capacity	24,195	7,564	20,224	37,317	19,503	22,561
($n=12$)	PSNR	53.98	54.89	54.27	54.27	54.23	54.09
PPVO	Capacity	22,480	6,963	18,818	34,868	18,362	21,311
($n=15$)	PSNR	54.30	55.34	54.56	54.57	54.61	54.31

5   Conclusion

The proposed method utilizes multiple prediction methods to generate many predicted pixel values and incorporates Ou and Kim's PPVO data hiding technique to embed the secret data into a cover image. The experiment results show that the proposed method has better performance than other works in terms of embedding capacity and visual quality. In addition, the MED prediction method has good visual quality but the embedding capacity is limited. The proposed method uses the CRS method to obtain high embedding capacity with acceptable visual quality.

References

1. A.M. Alattar, Reversible Watermark Using the Difference Expansion of a Generalized Integer Transform, IEEE Transactions on Image Processing, Vol. 13, pp. 1147-1156, 2004.
2. X. Chen, X. Sun, H. Sun, Z. Zhou, and J. Zhang, Reversible Watermarking Method Based on Asymmetric-Histogram Shifting of Prediction Errors, Journal of Systems and Software, Vol. 86, pp. 2620-2626, 2013.
3. X. Gui, X. Li, and B. Yang, A High Capacity Reversible Data Hiding Scheme Based on Generalized Prediction-Error Expansion and Adaptive Embedding, Signal Processing, Vol. 98, pp. 370-380, 2014.
4. K.H. Jung and K.Y. Yoo, Data Hiding Method Using Image Interpolation, Computer Standards and Interfaces, Vol. 31, pp. 465-470, 2009.
5. C.F. Lee and Y.L. Huang, An Efficient Image Interpolation Increasing Payload in Reversible Data Hiding, Expert Systems with Applications, Vol. 39, pp. 6712-6719, 2012.
6. X. Li, J. Li, B. Li, and B. Yang, High-fidelity Reversible Data Hiding Scheme Based on Pixel-value-ordering and Prediction-error Expansion, Signal Processing, Vol. 93, pp. 198-205, 2013.
7. X. Li, B. Yang, and T. Zeng, Efficient Reversible Watermarking Based on Adaptive Prediction-Error Expansion and Pixel Selection, IEEE Transactions on Image Processing, Vol. 20, pp. 3524-3533, 2011.
8. T.C. Lu, C.Y. Tseng, and J.H. Wu, Asymmetric-histogram Based Reversible Information Hiding Scheme Using Edge Sensitivity Detection, Journal of Systems and Software, in Press.
9. X. Ou and H.J. Kim, Pixel-based Pixel Value Ordering Predictor for High-fidelity Reversible Data Hiding, Signal Processing, Vol. 111, pp. 249-260, 2015.
10. F. Peng, X. Li, and B. Yang, Improved PVO-based Reversible Data Hiding, Digital Signal Processing, Vol. 25, pp. 255-265, 2014.
11. Z. Ni, Y.Q. Shi, N. Ansari, and W. Su, Reversible Data Hiding, IEEE Transactions on Circuits and Systems for Video Technology, Vol. 16, No. 3, pp. 354-362, 2006.
12. V. Sachnev, H.J. Kim, J. Nam, S. Suresh, and Y.Q. Shi, Reversible Watermarking Algorithm Using Sorting and Prediction, IEEE Transactions on Circuits and Systems for Video Technology, Vol. 19, pp. 989-999, 2009.
13. M. Tang, J. Hu, and W. Song, A High Capacity Image Steganography Using Multilayer Embedding, International Journal for Light and Electron Optics, Vol. 125, pp. 3972-3976, 2014.
14. J. Tian, Reversible Data Hiding Using a Difference Expansion, IEEE Transactions on Circuits and Systems for Video Technology, Vol. 13, No. 8, pp. 890-896, 2003.
15. C.T. Wang and H.F. Yu, A Markov-based Reversible Data Hiding Method Based on Histogram Shifting, Journal of Visual Communication and Image Representation, Vol. 23, pp. 798-811, 2012.
16. X. Wang, J. Ding, and Q. Pei, A Novel Reversible Image Data Hiding Scheme Based on Pixel Value Ordering and Dynamic Pixel Block Partition, Information Sciences, Vol. 310, pp. 16-35, 2015.
17. M. J. Weinberger, G. Seroussi, and G. Sapiro, The LOCO-I Lossless Image Compression Algorithm: Principles and Standardization into JPEG-LS, IEEE Transactions on Image Processing, Vol. 9, No. 8, pp. 1309-1324, 2000.

Advanced Center-Folding based Reversible Hiding Scheme with Pixel Value Ordering

Tzu-Chuen Lu *, Chi-Quan Lin, Jia-Wei Liu and Yu-Ci Chen

Department of Information Management, Chaoyang University of Technology,
Taichung 41349, Taiwan

tclu@cyut.edu.tw
s9914624@cyut.edu.tw
s10414609@cyut.edu.tw

Abstract. Dual image techniques have been broadly applied to reversible information hiding in recent years. Center folding strategy proposed by Lu et al. in 2015 compresses the secret symbol and then utilizes averaging to embed it into two stego-images. This effectively sustains the image quality and achieves a high embedding capacity with relatively little distortion. This study uses the center folding strategy to fold the secret message and embed them separately into two stego-images. Follow by embedding the remaining secret message the two stego-images again using the pixel value ordering method to increase the embedding capacity in the image.

Keywords: Center-Folding, Reversible Hiding, Pixel Value Ordering

1 Introduction

Information hiding technique uses text, images, or other types of multimedia information as a carrier to embed secret information within it in order to achieve information encryption. Information hiding techniques can be divided into reversible and irreversible, based on whether restoration of original image is achievable or not. After extracting the hidden information from reversible information hiding, the original carrier can be restored. This paper uses reversible information hiding as a basis to embed secret information into images.

The difference-expansion technique uses two pixels as a group and employed pixel value calculation to expand the differences in pixel values several times, before performing secret information embedding. In 2003, Tian proposed the difference-expansion technique. This method calculates the difference between two pixels and expands the difference value by two folds before embedding one bit of secret information [1]. In 2004, researcher Alatter further improved Tian's method by using a vector orientated approach to replace the original difference in pixel value matching method and calculate the difference in values between four neighboring pixels. The difference is then expanded by two folds and 3 bits

© Springer International Publishing AG 2017
J.-S. Pan et al. (eds.), *Advances in Intelligent Information Hiding and Multimedia Signal Processing*, Smart Innovation, Systems and Technologies 63,
DOI 10.1007/978-3-319-50209-0_11

of secret information are embedded [2]. In 2011, Li et al. proposed an adaptive embedding method. This method divides the image into an even region and a complex region depending on the complexity of the pixel value. Information embedding is then carried out according to the errors between pixel values and the regional complexity [3]]. In 2014, Gui et al. improved method proposed by Li et al. By increasing types of pixel value complexity as partitions, amount of information can be embedded in the prediction error increase thus improve overall embedding capacity [4].

Pixel value ordering (PVO) divides images into several identically sized regions. All the pixels within a region are ordered and maximum or minimum value will be used for information embedding. In 2013, Li et al. proposed the first pixel value ordering method. This method uses pixel values within a region and performs ordering. The largest and second largest or the smallest and second smallest values obtained are used as a group and predicted error value obtained from subtraction between pixel values will be used for information embedding [5]. In 2015, Qu and Kim improved upon the PVO method and proposed the pixel based pixel value ordering method (PPVO). This method uses pixel values as a foundation instead of regions. The neighboring pixel values are used as reference, and the maximum and minimum values are used for information embedding [6].

Besides the three methods mentioned above, dual image techniques are also commonly used. These methods replicate two copies of stego-image which have the same size as the original image and information embedding is then performed such that the amount of embedded information can be increase effectively. In 2007, Chang et al. proposed combining dual image techniques with modulus function hiding method. After establishing a 256×256 function matrix, two secret messages are grouped and transformed into a quinary based secret symbol. Two secret symbols are then grouped as one to perform information embedding. Finally, stego-pixel values of the two stego-images can be obtained from the left and right diagonals from the function matrix [7]. In 2015, Lu et al. proposed a dual image hiding scheme based on least significant bit matching. This method uses least significant bit matching on the original image to generate pixel values for two stego-images. Two pixels are then grouped as one and averaging is used to determine whether or not the stego-pixel can be restored to the original pixel value.[8]. In the same year, Lu et al. also proposed dual image technique which exploited center folding strategy. This effectively reduces secret information distortion and therefore allows more information to be embedded and better image quality [9].

2   Related Works

Dual image techniques have been broadly applied to reversible information hiding in recent years. Center folding strategy proposed by Lu et al. in 2015 compresses the secret symbol and then utilizes averaging to embed it into two stego-images. This effectively sustains the image quality and achieves a high embedding capacity with relatively little distortion.

2.1 Pixel Value Ordering (PVO)

PVO is a reversible information hiding technique proposed by Li et al. in 2013 [5]. This method separates the image into several regions of the same size, and performs ordering on the pixel values within each region. The largest and the second largest pixel value is then used (or the smallest and the second smallest pixel value) to find the error between the two PEmax or PEmin as reference for embedding.When the error value PEmax equals 1 or when PEmin equals -1, secret information embedding can be performed. When $PE_{max}=1$, adding the largest value with the secret information b will yield the stego-pixel value. When $PE_{min}=-1$,subtracting the secret information b from the smallest value will yield the stego-pixel value.

2.2 Center folding strategy combined with dual image technique

In 2015, Lu et al., proposed the center folding dual image information hiding method [9].This method first compresses the secret symbol by using center folding before embedding it into two stego-images. First, every k bits in the original secret message are grouped as a group and transformed into a decimal based secret symbol and the range of the secret symbol is transformed from the original $R = \{0,1,2,3,4,5,\ldots,2^{k-1}\}$ into the new range of $\bar{R} = \{-2^{k-1}, -2^{k-1}+1, \ldots, -1, 0, 1, \ldots, 2^{k-1}-2, 2^{k-1}-1\}$ by folding. $\bar{d}$ is the secret symbol after folding, 2^{k-1} is median within range. With the use of equation, original secret symbol d is transformed into the folded secret symbol $\bar{d}$. If $d < 2^{k-1}$, this means the new secret symbol is smaller than 0 and the folded secret symbol will have a negative sign. When $d = 2^{k-1}$, secret symbol $\bar{d}$ will be represented as 0. When $d > 2^{k-1}$, this means the new secret symbol is greater than 0 and the folded secret symbol $\bar{d}$ will have a positive sign.

The folded secret symbol $\bar{d}$ will then be embedded into two stego-images using averaging. This effectively improves image quality. $\bar{d}_1$ and $\bar{d}_2$ represent $\bar{d}$ when it is divided into two equal parts and they are embedded into original pixel value $x_{i,j}$, to form two stego-images $x'_{i,j}$ and $x''_{i,j}$. Repetition of the steps above will complete the embedding.

3 The Proposed Scheme

This study uses the center folding strategy and PVO to fold the secret message and embed them separately into two stego-images. Follow by embedding the remaining secret message the two stego-images again using the pixel value ordering method to increase the embedding capacity in the image.

3.1 Embedding Processing

Before embedding the hidden information, pre-processing using center folding process is required. The secret message takes every k bits as a group, and transforms them into a secret symbol d. In order to avoid image distortion, center folding method is used to compress d, and change the range of secret symbol from $R = \{0,1,2,...,2^{k-1}\}$ to $\bar{R} = \{-2^{k-1}, -2^{k-1}+1, ..., 0, ..., -2^{k-1}-2, -2^{k-1}-1\}$. The formula is as follows:

$$\bar{d} = d - 2^{k-1}, \tag{1}$$

where $\bar{d}$ is the secret symbol after folding and the median value is 2^{k-1} .Secret symbol d is transformed into the center folded secret symbol $\bar{d}$ using equation (1). When $d < 2^{k-1}$, the folded secret symbol $\bar{d}$ has a negative sign. Conversely when $d > 2^{k-1}$, the secret symbol $\bar{d}$ has a positive sign. In order to improve image quality, this paper used dual image method to embed the folded secret symbol $\bar{d}$ into two stego-images separately. The image embedding formulas are as follows:

$$\begin{cases} \bar{d}_1 = \left[\frac{\bar{d}}{2}\right], \\ \bar{d}_2 = \left\lceil\frac{\bar{d}}{2}\right\rceil, \end{cases} \tag{2}$$

$$\begin{cases} x'_{i,j} = x_{i,j} - \bar{d}_1, \\ y_{i,j} = x_{i,j} + \bar{d}_2. \end{cases} \tag{3}$$

Where $\bar{d}_1$ and $\bar{d}_2$ are obtained by dividing $\bar{d}$ into half and embedded into pixel value $x_{i,j}$ to generate two stego-pixel values $x'_{i,j}$ and $x''_{i,j}$. While generating the stego-pixels, overflow problems may arise due to the size of the secret symbols. After completing the first stage of embedding, the second stage PVO embedding is performed. The two stego-images are divided into several non-overlapping 1?3 blocks. Let $X' = \left\{x'_{i,j-1}, x'_{i,j}, x'_{i,j+1}\right\}$, the median pixel value is then subtracted by the left and right side pixel values. The formula is as follows:

$$\begin{cases} e_{i,j-1} = x'_{i,j-1} - x'_{i,j}, \\ e_{i,j+1} = x'_{i,j+1} - x'_{i,j}. \end{cases} \tag{4}$$

The obtained difference value is then shifted and the secret message is embedded into this difference. If the difference value equals 0 or 1, this means information embedding can take place. Otherwise one more action needs to be taken. The displacement formula is as follows:

$$\bar{e}_{i,j-1} = \begin{cases} e_{i,j-1} - 1, if\ e_{i,j-1} \le -1 \\ e_{i,j-1} + 1, if\ e_{i,j-1} \ge 2 \end{cases} \tag{5}$$

$$\bar{e}_{i,j+1} = \begin{cases} e_{i,j+1} - 1, if\ e_{i,j+1} \le -1 \\ e_{i,j+1} + 1, if\ e_{i,j+1} \ge 2 \end{cases} \tag{6}$$

The secret message is embedded into the difference values of $\bar{e}_{i,j-1}$ and $\bar{e}_{i,j+1}$. When the difference value is 0 and the secret message is also 0, no actions are taken. When the secret message is 1, 1 is taken away from the pixel value. Conversely, when the error value is 1, and the secret message is 0, no action is taken. On the other hand, when the secret message is 1, 1 is added to the error value. The formulas are as follows:

$$e'_{i,j-1} = \begin{cases} \bar{e}_{i,j-1} - 1, & \bar{e}_{i,j-1} = 0\ and\ b = 1, \\ \bar{e}_{i,j-1} + 1, & \bar{e}_{i,j-1} = 1\ and\ b = 1, \\ \bar{e}_{i,j-1}, & otherwise. \end{cases} \tag{7}$$

$$e'_{i,j+1} = \begin{cases} \bar{e}_{i,j+1} - 1, & \bar{e}_{i,j+1} = 0\ and\ b = 1, \\ \bar{e}_{i,j+1} + 1, & \bar{e}_{i,j+1} = 1\ and\ b = 1, \\ \bar{e}_{i,j+1}, & otherwise. \end{cases} \tag{8}$$

Finally the difference value after displacement and embedding is added to pixel value $x'_{i,j}$ to obtain the stego pixel values for the second phase. The formula is as follows:

$$\begin{cases} Y'_{i,j-1} = e'_{i,j-1} + x'_{i,j}, \\ Y'_{i,j+1} = e_{i,j+1} + x'_{i,j}. \end{cases} \tag{9}$$

Obtained stego-image region is $Y' = \left\{ Y'_{i,j-1}, x'_{i,j}, Y'_{i,j+1} \right\}$. Summing up the above, the embedding process is as follows:

(1) Let the original image to be $X = \left\{ x_{0,0}, \ x_{0,1}, \ \cdots x_{i,j} \cdots, \ x_{m,n} \right\}$.

(2) Determine whether or not the pixel value $x_{i,j}$ is between the range of 2^{k-1} and $256 - 2^{k-1}$). If so, then move on to step (3). If not, assume stego-pixels are $x'_{i,j} = x_{i,j}$ and $x''_{i,j} = x_{i,j}$, and move on to the next pixel.

(3) Extract 4 secret messages and convert them to decimal based secret symbol d. Use formula (1) to obtain the folded secret symbol $\bar{d}$.

(4) Substituting the folded secret symbol $\bar{d}$ into equation (2), and break down $\bar{d}$ into $\bar{d}_1$ and $\bar{d}_2$, equation (3) can be used to obtain the stego-pixels $x'_{i,j}$ and $x''_{i,j}$.

(5) In a similar fashion, repeat step (1) to step (4) until all the secret symbols have been embedded into the image.

(6) Separate the stego-image X' into several non-overlapping 1?3 regions $B = \left\{ x'_{i,j-1}, x'_{i,j}, x'_{i,j+1} \right\}$.

(7) Us equation (4) to obtain the difference value e between the left and right pixels and the median value.

(8) Use equation (5) and (6) to displace the pixel difference value e to obtain e'.

(9) Substitute secret message b into equation (7) and (8). Embed it into the error value $\bar{e}$ to obtain e'.

(10) Use equation (9) to obtain the pixel values $y' = \left\{ y'_{i,j-1}, y'_{i,j}, y'_{i,j+1} \right\}$ of the stego-image.

(11) Repeat steps (6) to (10) on the stego-image x'' until all the information is embedded into the image and the stego pixels y'' are obtained.

3.2 Extraction Processing

During the stage of hidden information extraction, the stego-images y' and y'' are seperated into several non-overlapping 1×3 regions, letting $B' = \left\{ y'_{i,j-1}, y'_{i,j}, y'_{i,j+1} \right\}$ as 1?3 stego regions. Calculating the difference value between the median pixel value and the left and right pixel values we get equation (10) as follows:

$$\begin{cases} e'_{i,j-1} = Y'_{i,j-1} - Y'_{i,j}, \\ e_{i,j+1} = Y'_{i,j+1} - Y'_{i,j}. \end{cases} \tag{10}$$

If the difference value equals -1 or 2, this means the embedded secret message is 1. If the difference value is 0 or 1, then this means the embedded message is 0, other values represent no embedded messages. After message extraction, displacement is performed to obtain the original difference value e. The formulas are as follows:

$$e_{i,j-1} = \begin{cases} e'_{i,j-1} - 1, & if \ e'_{i,j-1} \geq 2, \\ e'_{i,j-1} + 1, & if \ e'_{i,j-1} \leq -1, \\ e'_{i,j-1}, & e'_{i,j-1} = 1 \ or \ e_{i,j-1} = 0. \end{cases} \tag{11}$$

$$e_{i,j-1} = \begin{cases} e'_{i,j+1} - 1, & if \ e'_{i,j+1} \geq 2 \,, \\ e'_{i,j+1} + 1 \,, & if \ e'_{i,j+1} \leq -1 \,, \\ e'_{i,j+1}, & if \ e'_{i,j+1} = 1 \ or \ e'_{i,j+1} = 0. \end{cases} \quad (12)$$

$$b = \begin{cases} 1 \,, & if \ e' = 2 \ or \ e' = 1, \\ 0, & if \ e' = 0 \ or \ e' = -1. \end{cases} \quad (13)$$

After using equations (11), (12), and (13) to extract the secret message and recover displacement, the difference value is substituted into equation (14) to obtain the stego-image after the first stage of embedding. The formula is as follows:

$$\begin{cases} x'_{i,j-1} = e_{i,j-1} + y'_{i,j}, \\ x'_{i,j} = y'_{i,j}, \\ x'_{i,j+1} = e_{i,j+1} + y'_{i,j}. \end{cases} \quad (14)$$

After undergoing the above process, stego-image y' will be restored to x' while stego-imago y'' will also be restored to the original x'' using the same manner to obtain the second restored image x' and x''. Next will be the second stage of restoration. For one, whether or not the pixel value $x'_{i,j}$ and $x''_{i,j}$ are equal is determined. If $x'_{i,j}=x''_{i,j}$ and the value isn't within the range of 2^{k-1} and $256 - 2^{k-1}$, this means there are no hidden messages and the original pixel is $x_{i,j} = x'_{i,j}$. Conversely, if one of the values is within the range, this means that there is a hidden message and the stego-image $x'_{i,j}$ and $x''_{i,j}$ will be subtracted from each other to obtain the secret symbol $\bar{d}$. With the addition of the median value 2^{k-1}, the folded secret symbol is restored to the decimal based secret symbol d. The extraction and restoration formulas are as follows:

$$\bar{d} = x'_{i,j} - x''_{i,j} \quad (15)$$

$$d = \bar{d} + 2^{k-1} \quad (16)$$

Substituting into equation (15) to obtain secret symbol $\bar{d}$ and substituting folded secret symbol $\bar{d}$ into equation (16) to obtain secret symbol d. The secret symbol d is converted to a binary based symbol such that the secret message can be immediately obtained. For image restoration, after averaging the stego pixels $x'_{i,j}$ and $x''_{i,j}$, the original image pixel value $x_{i,j}$ can be restored. The restoration formula is as follows:

$$x_{i,j} = \frac{x'_{i,j} + x''_{i,j}}{2}. \quad (17)$$

3.3   Overflow Problem

During the embedding process of the center folding strategy, if the original image is not within the range of 2^{k-1} to $256-2^{k-1}$, it may cause overflow or underflow problems. In order to solve these problems, we do not perform any actions on these pixels. During secret message restoration or image restoration, suppose the two stego-images have the same pixel values and they both exceeded the range of 2^{k-1} to $256 - 2^{k-1}$, this means that the stego-image pixels are the same as the original pixels. In PVO embedding, when the left and right pixel value is 0 or 255, although messages can be embedded, overflow problems will occur. Therefore, a location map is needed to record the positions that will overflow.

During extraction, the location map is used to determine whether or not overflow occurs after embedding.

4   Experiment results

This study used 512×512 sized grayscale images for the experiment and are shown in Fig. 1. The paper utilizes peak signal to noise ratio (PSNR) to evaluate

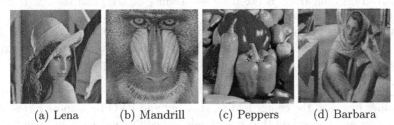

(a) Lena   (b) Mandrill   (c) Peppers   (d) Barbara

Fig. 1. Standard grayscale images

the difference between original image and stego-image. The PSNR formula is as follows:

$$\mathrm{PSNR} = 10log_{10}\left[\frac{255^2}{\frac{1}{h \times w} \times \sum_{i=1}^{h}\sum_{j=1}^{w}\left(x'_{i,j} - x_{i,j}\right)^2}\right]. \tag{18}$$

The smaller the difference between two images, the better the image quality and the higher the PSNR value. n order to evaluate the embedding ability of an information hiding technique, Bit Per Pixel, (BPP) is used to evaluate the amount of hidden information in the stego-image. The BPP formula is as follows:

$$\mathrm{bpp} = \frac{C}{2 \times h \times w}. \tag{19}$$

Where C is the sum of the embedding capacity of both stego-images. The larger the bpp value, the more information can be embedded. Conversely, smaller bpp value implies worse embedding ability of the image.From Table 1, we find that embedding capacity of the proposed method is extremely high and qualities of the two stego-images are able to be maintained at above 40 db. The embedded capacity is also above 3 bpp. In order to compare with other methods, we

Table 1. Comparison of total embedding capacity and image quality of different images

	Lena	Mandrill	Barbara	Peppers
PSNR(1)	40.2815	40.7191	40.5098	40.5128
PSNR(2)	40.2731	40.7221	40.5245	40.5142
bpp	3.0858	3.02612	3.053	3.0104
Payload	1,617,897	1,586,451	1,601,030	1,577,161

used the method proposed by Lu et al., to perform comparison. Fig.2 shows the comparison results of the average embedding capacity and the average image quality (PSNR) of the proposed method and the method proposed by Lu et al. When the embedding capacity is the same (bpp=2), the image quality from the proposed method is superior to the method proposed by Lu et al.Although the proposed method produces images of lower quality than the one proposed by Lu et al., at a low embedding capacity, at high embedding capacities, the proposed

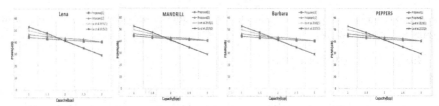

Fig. 2. Average comparison between the embedding capacity and the PSNR of proposed method and method proposed by Lu et al.

method produces images that have quality of around 40 db, with only a little bit of distortion. Hence, this proves the PVO method is to able to shift parts of the distorted pixels back to close to their original image pixel values.

5 Conclusions

This paper has used the center folding strategy and PVO to propose a reversible information hiding method that has a high embedding capacity and good image quality under the framework of dual image techniques. In order to reduce distortion, center folding strategy was used to fold the secret symbol and evenly shared between two images. PVO was then used according to the size of the difference values to perform embedding and increased the embedding capacity. Some of the pixels can even be shifted back to position close to its original pixel. For grayscale images, the proposed method produces better image quality and has a higher embedding capacity than method proposed by Lu et al.

References

1. Tian, J.:Reversible Data Hiding Using a Difference Expansion. IEEE Transactions on Circuits and Systems for Video Technology, 13, 890-896(2003).
2. Alattar, A.M.:Reversible Watermark Using the Difference Expansion of a Generalized Integer Transform. IEEE Transactions on Image Processing,13, 1147-1156 (2004).
3. Li, X., Yang B., and Zeng, T.:Efficient Reversible Watermarking Based on Adaptive Prediction-Error Expansion and Pixel Selection. IEEE Transactions on Image Processing,20,3524-3533(2011).
4. Gui, X., Li X., and Yang B.:A High Capacity Reversible Data Hiding Scheme Based on Generalized Prediction-Error Expansion and Adaptive Embedding. Signal Processing,98,370-380(2014).
5. Li, X., Li, J., Li, B., and Yang, B.:High-fidelity Reversible Data Hiding Scheme Based on Pixel-value-ordering and Prediction-error Expansion. Signal Processing, 93,198-205(2013).
6. Qu, X. and Kim, H.J.:Pixel-based Pixel Value Ordering Predictor for High-fidelity Reversible Data Hiding. Signal Processing, 111,=249-260(2015).
7. Chang, C.C., Kieu, T.D., and Chou, Y.C.:Reversible Data Hiding Scheme Using Two Steganographic Images. Proceedings of IEEE Region 10 Intermational Conference (TENCON),1-4(2007).
8. Lu, T.C., Tseng, C.Y., and Wu, J.H.:Dual Imaging-based Reversible Hiding Technique Using LSB Matching. Signal Processing, 108, 77-89(2015).
9. Lu, T.C., Wu, J.H., and Huang, C.C.:Dual-Image-Based Reversible Data Hiding Method Using Center Folding Strategy. Signal Processing, 115, 195-213(2015).

A Large Payload Webpage Data Embedding Method Using CSS Attributes Modification

Jing-Xun Lai, Yung-Chen Chou*, Chiung-Chen Tseng, and Hsin-Chi Liao

Department of Computer Sciences and Information Engineering,
Asia University, Taichung 41354, Taiwan
(Email: {kevin50406418, yungchen, ilxgaau, hsinciliao}@gmail.com)

Abstract. The Internet provides an ideal environment for users to communicate with one another. Digitized data can be transmitted over the Internet almost instantly. More and more confidential data has been delivered and the amount will keep on increasing. How to achieve the security of confidential data delivery is a very important issue in this digital era. Data encryption may help to ensure the cipher data is difficult to decode by an unknown user but it may not guarantee the delivery of the confidential data. Data hiding technique is a solution for confidential data delivery. In this paper, an HTML file is used as the cover medium for delivering confidential data. By utilizing the flexibility of the HTML webpage creation, the proposed method uses different CSS attribute settings to encode the secret message. Because there are various ways to set the CSS margin and padding attributes, the proposed method not only helps to achieve the goal of secret data delivery but also provides a large embedding payload. The simulation results demonstrate that a CSS tag with margin and padding attributes can embed up to 112 secret bits.
Keywords: HTML, Data Hiding, CSS Attribute.

1 Introduction

As Internet usage becomes ever more widespread and more information is available online, a user can use a search engine to look for information or manage digital data on the data cloud environment anywhere. Unfortunately, the Internet is insecure. Any user can easily obtain or intercept a datagram on the Internet. In addition, private data delivery and copyright protection for digital information are important issues.

For private data (also known as secret data) [6] protection can be achieved by using a cipher system. However, the secret data delivery might fail if a user terminates the data transmission unexpectedly. The cipher data is made to look like random noise to not attract the attention of any user who might attempt to intercept the data or terminate the data transmission with malicious intent. As an alternative solution to the problem mentioned previously, steganography can also achieve the goal of secret data delivery. The main idea of steganography is to embed the secret data into a cover medium before sending the data to the receiver. For security purposes, the secret message should be encrypted using

J.-S. Pan et al. (eds.), *Advances in Intelligent Information Hiding and Multimedia
Signal Processing*, Smart Innovation, Systems and Technologies 63,
DOI 10.1007/978-3-319-50209-0_12

a cipher system. In general, multimedia data is a good medium to act as the cover medium for embedding the secret message [8,9,10,11,12]. A multimedia file contains a lot of redundant space which can be removed or modified. Because of this property, the cipher data can be embedded into the redundant space of a multimedia file. Many researchers utilize this important property to design rules to embed secret data.[7,13,14,15]

Huang, et al.,[3] presented a data hiding method to conceal secret data into HTML by setting different attribute arrangements of a tag. As mentioned previously, a web browser can easily parse the settings of tag attributes and display the content properly for users. Thus, different attribute arrangements will not distort the content of a webpage. Yang and Yang proposed a data hiding method by using different quotation marks for the attribute settings to embed the secret message. Since HTML attribute setting allows the use of single quotation, double quotation, or no quotation, every attribute setting can be used to carry one secret data bit. Furthermore, Lee and Tsai [2] found that the space between two words can be coded by using the special blank character. Lee and Tsai collected some characters to define the data embedding rule and used these characters to embed the secret message. To enhance the embedding capacity of Lee and Tsai's method, Chou, et al., adopted the Cartesian production to pair the special space character groups. However, Lee and Tsai's method and Chou et al.'s method [5] are more focused on the blank space between words which is limited to Roman language webpages.

In this paper, a webpage based data hiding method is presented by modifying CSS attribute settings. We found that the margin and padding attribute settings are flexible. That is, there are more than one way to code a webpage. Thus, we take advantage of this property to create the rules for embedding the secret message. The experimental results demonstrate that the proposed method provides better embedding payload than others.

2   Related works

2.1   Huang, et al.,'s method

Huang, et al., presented a data hiding method by setting different tag attributes to embed the secret message [3]. An HTML tag may contain several attributes for creating a colorful webpage. For example, the <font> tag contains three attributes:*face*, *size*, and *color*. The *face* attribute is used to set the font family. The *size* and *color* attributes are used to set the size of context and context's color, respectively. For example, <font size=1 face= "Times New Raman" color=blue> Asia University</font> sets the context "Asia University" in blue color with the size number 1, and the font family is "Times New Roman". Huang's method pre-arranges the attribute sequence to embed the secret message.

2.2   Yang and Yang's method

Yang and Yang's method [1] uses HTML tags attribute setting style to embed the secret data. HTML tag attribute settings can be set in different ways. For example, the webpage font size can be set by using "<font size = 4>", "<font size = '4'>", or "<font size = "4">. The tag attribute setting allows the use of single quotes, double quotes, or no quote. Yang and Yang adopted this characteristic to embed secret data into HTML tag attribute settings. For example, in an HTML file, "<font size='4'color = "black">" is used to embed secret data "01". When a user browses to the stego webpage, the user cannot easily detect the distortion on the stego webpage with the naked eye.

2.3   Lee and Tsai's method

Lee and Tsai's method [2] uses different blank space codings to embed the secret message in an HTML file. In the HTML file, a blank space between words can be coded by typing a space, " ", or " ", etc. That is, a user cannot tell any difference when browsing to the stego webpage. For example, a sentence "A HTML File Data Hiding by Using Modification CSS Border Attribute" is used to carry the secret message "011 101 100 110 001 101 111 001 000 010". The HTML source code is "A&#32HTML File Data Hiding by Using Modification CSSBorder Attribute."

2.4   Chou et al.'s method

Chou, et al., presented a webpage-based data hiding method by using special space code data hiding to embed secret message in the space between words [4].Chou used many special blank character codes and adopted the Cartesian production to create a combination of the codes. Secret data can be embedded by selecting a code combination (i.e., a code combination sequence number that equals to the secret message in decimal) to code the space of words.

3   The proposed method

With the advanced development of webpage design technique, a colorful webpage can be created by using HTML5 with CSS styles. Our proposed method uses the "padding" and "margin" CSS style setting to embed the secret message. First, we use the "em", "%", "px" and "pt" units in CSS styles. "em" inherits the size of the parent element and then defines itself by superposition. A unit of "%" is equivalent to a unit of "em"; that is 100% equals to 1em. Unit "px" corresponds to the pixel, which is the smallest unit that can be displayed on the screen. The "pt" unit represents the a point for printing, which has a fixed size and cannot be resized.

3.1   Data embedding phase

Based on the CSS style rules, "margin" is used to set the four outside borders (i.e., margin-top, margin-right, margin-bottom, margin-left) of an object. Because of the flexibility of webpage design, margin setting can done in various ways. For example, <span style= "margin: 1em 1.5em;> is the same as <span style= "margin: 1em 1.5em 1em 1.5em;> and <span style= "margin-top: 1em; margin-bottom: 1em; margin-right: 1.5em; margin-left: 1.5em;>. On the other hand, padding is used to set the four inside borders (i.e., padding-top, padding-right, padding-bottom, and padding-left) of an object. Again, the padding of a webpage can be done in one of several ways. For example, <span style= "padding: 1em 1.5em;> is the same as <span style= "padding: 1em 1.5em 1em 1.5em;> and <span style= "padding-top: 1em; padding-bottom: 1em; padding-right: 1.5em; padding-left: 1.5em;>. Tables 5 and 6 summarize the different ways to set the margin and padding for webpage design, respectively.

Table 1. CSS margin setting

Margin attribute setting	Description
<span style= "margin: 1em;">	Top margin is 1em
<span style= "margin: 1em 1em 1em 1em;">	Right margin is 1em
<span style= "margin-top: 1em; margin-top:1em; margin-bottom: 1em; margin-left: 1em; mrgin-right: 1em;">	Bottom margin is 1em Left margin is 1em
<span style= "margin: 1em 1.5em;">	Top margin is 1em
<span style= "margin: 1em 1.5em 1em 1.5em;">	Right margin is 1.5em
<span style= "margin-top: 1em; margin-bottom: 1em; margin-right: 1.5em; margin-left: 1.5em;">	Bottom margin is 1em Left margin is 1.5em
<span style= "margin: 1em 1.5em 2em;">	Top margin is 1em
<span style= "margin: 1em 1.5em 2em 1.5em;">	Right margin is 1.5em
<span style= "margin-top: 1em; margin-bottom: 2em; margin-right: 1.5em; margin-left: 1.5em;">	Bottom margin is 2em Left margin is 1.5em

Table 2. CSS padding setting

Padding attribute setting	Description
<span style= "padding: 1em;">	Top padding is 1em
<span style= "padding: 1em 1em 1em 1em;">	Right padding is 1em
<span style= "padding-top: 1em; padding-top: 1em; padding-bottom: 1em; padding-left: 1em; padding-right: 1em;">	Bottom padding is 1em Left padding is 1em
<span style= "padding: 1em 1.5em;">	Top padding is 1em
<span style= "padding: 1em 1.5em 1em 1.5em;">	Right padding is 1.5em
<span style= "padding-top: 1em; padding-bottom: 1em; padding-right: 1.5em; padding-left: 1.5em;">	Bottom padding is 1em Left padding is 1.5em
<span style= "padding: 1em 1.5em 2em;">	Top padding is 1em
<span style= "padding: 1em 1.5em 2em 1.5em;">	Right padding is 1.5em
<span style= "padding-top: 1em; padding-bottom: 2em; padding-right: 1.5em; padding-left: 1.5em;">	Bottom padding is 2em Left padding is 1.5em

The proposed method utilizes the flexibility of webpage design to create a mapping table to embed the secret message. The four different size units and different margin and padding setting rules are combined to create the mapping rules for embedding the secret message. For example, to embed the secret message "1011", the CSS tag "<span style="margin: 16px; padding: 100%;">" is used to code the webpage.

Table 3. An example for the proposed embedding rules

Data embedding rule	Bit
<span style="margin: 1em; padding: 0em;">	0000
<span style="margin: 1em; padding: 0pt;">	0001
<span style="margin: 1em; padding: 0px;">	0010
<span style="margin: 1em; padding: 0%;">	0011
<span style="margin: 12pt; padding: 0em;">	0100
<span style="margin: 12pt; padding: 0pt;">	0101
<span style="margin: 12pt; padding: 0px;">	0110
<span style="margin: 12pt; padding: 0%;">	0111
<span style="margin: 16px; padding: 0em;">	1000
<span style="margin: 16px; padding: 0pt;">	1001
<span style="margin: 16px; padding: 0px;">	1010
<span style="margin: 16px; padding: 0%;">	1011
<span style="margin: 100%; padding: 0em;">	1100
<span style="margin: 100%; padding: 0pt;">	1101
<span style="margin: 100%; padding: 0px;">	1110
<span style="margin: 100% padding: 0%;">	1111

The key steps of the proposed method are summarized as follows:

Input: An HTML file H, secret message S, and the predefined embedding rules ER with n rules
Output: A stego HTML file H'
Step 1: Encrypt secret message S using any well-known cipher system.
Step 2: Take $\lfloor log_2 n \rfloor$ bits from S and convert them to a decimal value x.
Step 3: Use the x-th embedding rule to code the webpage component.
Step 4: Repeat **Steps 2-3** until all the secret message bits have been embedded.
Step 5: Output stego html file H'.

3.2   Data extraction phase

At the receiver side, the secret message can be extracted by parsing the source code of the stego webpage. Because the sender and the receiver use the same data embedding rules, the receiver only needs to parse the tags in the stego webpage source code to find the tags that are used to embed the secret message. Then, the secret message can be extracted by referring to the date embedding rules.

At this time, the extracted data is a cipher text message, the real secret message can be obtained by applying the decryption method with a pre-shared private key. The key steps of the proposed secret message extraction are summarized as follows:

Procedure: Data extraction

Input: A stego HTML File H'
Output: Secret message S
Step 1: Access the stego webpage and to take the source code of the stego webpage.
Step 2: Check the tags in the source code of the stego webpage.
Step 3: If a tag is one of the embeddable tags, extract the secret message from the tag by referring to the embedding rules.
Step 4: Repeat **Steps 2 and 3** until all the tags have been parsed.
Step 5: Decrypt the secret message by applying the decryption algorithm with the private key
Step 6: Output the actual secret message.

4   Experimental results and analysis

To evaluate the performance of the proposed method, we implement the proposed method, Huang, et al.,'s method, Yang and Yang's method, Lee and Tsai's method, and Chou, et al.,'s method by using Octave 3.6.4 on Windows 7 operating system. The webpage based data embedding method can be classified into 1) embedding secret data into context; 2) embedded secret data into the HTML tags. Lee and Tsai's method and Chou's method embed the secret data into the context. Huang's method, Yang and Yang's method, and our proposed method embed the secret message into the HTML tags.

Here, we use the text "United States Declaration of Independence" as the context of the webpage. Unfortunately, there is no objective evaluation mechanism to evaluate the performance of the visual quality of the stego webpage. Thus, we use some popular web browsers (e.g., IE, Firefox, Chrome, and Safari) to access the stego webpages and evaluate the visual quality through the naked eye. Figure 8 shows the visual quality of the stego webpage that is generated by the proposed data embedding method. As we can see, there has no visible distortion on the stego webpage.

Fig. ?? illustrates the visual quality of the stego webpages that are generated by Huang's method, Yang and Yang's method, Lee and Tsai's method, Chou's method and the proposed method, respectively. All the stego webpages do not show any visible distortion that can be detected with the naked eye. Fig. 1 shows the performance of the embedding payloads of the simulated data hiding methods. The embedding payload of the proposed method is the best, as compared with the other data embedding methods. In addition, the embedding payload of the proposed method is closely related to the number of tags that can be used to set the margin and padding attributes. Thus, the proposed method is more suitable for the modern webpage design style.

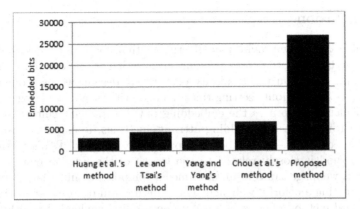

Fig. 1. Capacity comparison using "United States Declaration of Independence" stego HTML file

We analyze the combination of size units and setting styles, and the case of margin or padding attribute settings, there are only 4^9 possible combinations, which can be used to embed $\lfloor log_2(4^9) \rfloor = 18$ bits. Furthermore, if margin and padding attributes are used at the same time, there are 4^{56} possible combinations. This means that the two tag attributes can be used to embed $\lfloor log_2(4^{56}) \rfloor = 112$ bits. Fig. 2 shows the possible combinations for using margin and padding attributes at the same time.

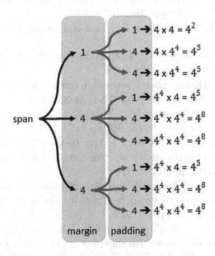

Fig. 2. Analysis of total margin and padding attributes setting combinations

5   Conclusion

Data hiding technique provides an effective method for secret message delivery. We use a webpage as the cover medium because it has low computational cost requirement. Based on the flexibility of webpage design, we uses different attribute units and attribute setting rules to create the data embedding rules. The proposed method improves the embedding payload for data embedding. For a CSS tag with margin and padding attribute settings, a single tag can embed up to 112 secret bits. The embedding payload of Lee and Tsai's method and Chou's method is closely related to the length of the English context on the webpage. A webpage can contain multimedia components and a large number of tags for creating a colorful webpage. The experimental results demonstrate that the proposed method offers a better data embedding payload than others.

References

1. Yang, Y. J. and Yang, Y. M: 'An Efficient Webpage Information Hiding Method Based on Tag Attributes', *FSKD*, 2010, **3**, pp. 1181-1184
2. I. S. Lee and W.H. Tsai: 'Secret communication through webpages using special space codes in html files', *IJASE*, 2008, **6**, pp. 141-149
3. Huang, H. J., Zhong, S. H., and Sun, X. M.: 'An Algorithm of Webpage Information Hiding Based on Attributes Permutation', *IIHMSP*, 2008, pp. 257-260
4. Chou, Y. C. and Huang, C. Y. and Liao, H. C.: 'A reversible Data Hiding Scheme Using Cartesian Product for HTML File', *ICGEC*, 2012, pp. 153-156
5. Chou, Y. C. and Liao, H. C.: 'A Webpage Data Hiding Method by Using Tag and CSS Attribute Setting', *IIHMSP*, 2014
6. Fabien A. P. Petitcolas, Markus G. Kuhn and Ross J. Anderson: 'Information Hiding-A Survey', *IEEE*, 1999, pp.1062-1078
7. Dey, S., Al-Qaheri, H., and Sanyal, S.: 'Embedding Secret Data in HTML Web Page', *IPCC*, 2009, pp.474-481
8. Zhong, S., Cheng, X., and Chen, T.: 'Data Hiding in a Kind of PDF Texts for Secret Communication', *IPCC*, 2009, pp.474-481
9. Wu, X. and Sun, W.: 'High-capacity reversible data hiding in encrypted images by prediction error', *signal processing*, 2014, **104**, pp.387-400
10. Liu, Y., Li, Z., Ma, X. and Liu, J.: 'A robust data hiding algorithm for H.264/AVC video streams', *signal processing*, 2013, **86**, pp.2174-2183
11. Lee, C. W. and Tsai, W. H.: 'A data hiding method based on information sharing via PNG images for applications of color image authentication and metadata embedding', *JSS*, 2013, **93**, pp.2010-2025
12. Ilchev, S. and Ilcheva, Z.: 'A secure high capacity full-gray-scale-level multi-image information hiding and secret image authentication scheme via Tchebichef moments', *signal processing-Image*, 2013, pp.531-552
13. S.M. Elshoura, and D.B. Megherbi: 'Modular Data Hiding as an Alternative of Classic Data Hiding for Web-based Applications', *ITC*, 2012, pp.9-15
14. Sui, X. G. and Luo, H.: 'A New Steganography Method Based on Hypertext', *APRASC*, 2004, pp.181-184
15. Zhang, X., Zhao, G. and Niu , P.: 'A Novel Approach of Secret Hiding in Webpage by Bit Grouping Technology', *JSW*, 2012, pp.2614-2621

The Study of Steganographic Algorithms Based on Pixel Value Difference

Chin-Feng Lee[*,1], Jau-Ji Shen[2], Kuan-Ting Lin[3]

[*,1]Department of Information Management, Chaoyang University of Technology,
Taichung 41349, Taiwan, R.O.C.
lcf@cyut.edu.tw
[2]Department of Management Information Systems, National Chung Hsing University
Taichung 40227, Taiwan, R.O.C.
jjshen@nchu.edu.tw
[3]Department of Information Management and Finance, National Chiao Tung University
Hsinchu 300, Taiwan, R.O.C.
inazuma184@hotmail.com

Abstract. Wu and Tsai proposed PVD (pixel-value differencing) data hiding method in 2003. PVD partitions a cover image into non-overlapping blocks of two consecutive pixels and calculated the difference value from the two pixels in each block. This method provides an easy way to produce a more imperceptible result. Afterwards, some scholars have proposed PVD-related improvements in image quality and embedding capacity. This paper classifies and discusses the recent PVD-based methods. We will compare each PVD related method in terms of the embedding capacity and PSNR.

Key words: Steganography, data hiding, pixel value difference, PVD

1 Introduction

Information security is becoming more and more important. Applications such as covert communication, copyright protection, etc. stimulate the research of information hiding techniques; therefore, in order to ensure that data are not vulnerable to be attack, scholars developed data hiding that embedded secret data in the multimedia, such as digital image, video or audio, etc. It can be monitored by unauthenticated viewers who will only notice the transmission of an image without discovering the existence of the hidden message. After the receiver obtains the resulting stego-image, the extraction algorithm uses the same key to retrieve the message.

Irreversible Data Hiding Techniques has the advantage of high embedding capacity. Least Significant Bit substitution (LSB substitution) [1], Exploiting Modification Direction (EMD) [2], and Pixel-Value Differencing (PVD) [3] are well-known and common methods of irreversible data hiding.

© Springer International Publishing AG 2017

J.-S. Pan et al. (eds.), *Advances in Intelligent Information Hiding and Multimedia Signal Processing*, Smart Innovation, Systems and Technologies 63,
DOI 10.1007/978-3-319-50209-0_13

Least Significant Bit replaces the last unimportant bits to embed secret data. But LSB substitution can't resist the attack of statistical analysis [4] [5] [6]. In 2006, Mielikainen proposed the method of LSB Matching Revisited [7]. LSB Matching Revisited reduces the amount of pixel modification which provides better imperceptibility for the stego-image than that of LSB substitution; thereby increasing the difficulty of detection. Zhang and Wang et al. proposed a new data hiding scheme called exploiting modification direction (EMD). This scheme has largest theoretical embedding capacity about 1.16 bpp (bits per pixel). In 2003, Wu and Tsai proposed Pixel Value Differencing, also called PVD. PVD partitions the cover image into non-overlapping blocks. Each block contains a pair of adjacent pixels. Wu and Tsai designed a range table an instructor of data embedding size. PVD allows a greater modification when a larger difference between adjacent pixels values thus providing high embedding capacity. Besides, PVD has the ability of resisting statistical analysis attack. Thus there are many researches to improve PVD scheme. That is why various approaches based on PVD method have been proposed to achieve better embedding capacity and image quality.

In 2005, Wu et al. based on the original PVD method, proposed a data hiding scheme that combines PVD and LSB substitution. In this research, we called the method of Wu et al. [8] as L-PVD for short. This scheme sets up a threshold denoted as *Div* to determine which region belongs to smooth or edge area. LSB substitution method is applied onto the smooth regions but PVD method to the complex area for enhancing the amount of embedding capacity and preserving stego-image quality. In 2007 Chang et al. [9] in order to increase the embedding capacity and to improve falling-off boundary problem, Tri-way Pixel-value Differencing method (Tri-PVD for short) was proposed. In 2008, an improved PVD method were proposed by Yang et al. [10] that can improve the embedding capacity; so in this study we called it as AdvPVD. At first, AdvPVD calculates the difference between a pair of adjacent pixels, and then uses the original PVD method, and the difference between adjacent pixels is divided into low-level, middle-level and high-level. According to the pixel located at different intervals, using LSB method to hide 3, 4, 5 bits of secret message.

In order to improve the cover image distortion of PVD method, 2008 Wang et al. [11] based on the original PVD technology architecture proposed a data hiding technology with high-quality effects, in this study called HqPVD. This method is based on the calculation modulus function to obtain the sum of the pixels of the remainder and use the remainder value to record the secret message. In 2015, Shen et al. [12] proposed a novel data hiding method that combined PVD and EMD, in this study called E-PVD. Since the human eyes tolerate more changes in edge areas than in smooth areas, the method exploits PVD to estimate the base of digits to be embedded into pixel pairs and using the optimal problem to solve falling-off boundary problem.

2 Related Works

2.1 Pixel Value Differencing (PVD)

The details of data hiding steps are described as follows.

1. Calculate the difference d_i for each block of two consecutive pixels p_i and p_{i+1} such that $d_i = |p_{i+1} - p_i|$.

2. Find the range R_i where the difference d_i falls in.

3. Compute the range width k_i which means the number of secret bits to be hidden depending on the range table $RT = \{R_j = [l_j, u_j]\} = \{ [0, 7], [8, 15], [16, 31],$ $[32, 63], [64, 127], [128, 255] \}$, where l_j and u_j are the lower and upper bound values of R_j).

4. Convert k_i-bit secret data into a decimal value s_i and using Eq. (1) to calculate a new difference d_i^*.

$$d_i^* = \begin{cases} l_i + s & if\ d \geq 0 \\ -(l_i + s) & if\ d < 0 \end{cases} \tag{1}$$

5. Use Eq. (2) to adjust p_i and p_{i+1} and obtain a new pair of stego-pixels (p'_i, p'_{i+1}).

$$(p'_i, p'_{i+1}) = \begin{cases} \left(p_i - \left\lfloor \frac{d^*-d}{2} \right\rfloor, p_{i+1} + \left\lceil \frac{d^*-d}{2} \right\rceil\right), if\ d\ is\ odd \\ \left(p_i - \left\lceil \frac{d^*-d}{2} \right\rceil, p_{i+1} + \left\lfloor \frac{d^*-d}{2} \right\rfloor\right), if\ d\ is\ even \end{cases} \tag{2}$$

2.2 Pixel Value Differencing with LSB Replacement (L-PVD)

The details of data hiding steps are described as follows.

1 Predefine a threshold Div which is to partition the range table into two levels, i.e., the lower-level and higher-level.if $d_i \leq Div$ then it belongs to lower-level, otherwise d_i belongs to higher-level.

2. Calculate the difference $d_i = |p_{i+1} - p_i|$ for each non-overlapping block of two consecutive pixels p_i and p_{i+1}.

3. Find the range R_i where the difference d_i belongs to.

4. If R_i belongs to the higher-level, then the PVD method is used to embed a secret message; otherwise LSB substitution is applied.

5. Adjust the new difference by Eq. (3) as follows.

$$(p'_i, p'_{i+1}) = \begin{cases} (p'_i - 8, p'_{i+1} + 8), & if\ p'_i \geq p'_{i+1} \\ (p'_i + 8, p'_{i+1} - 8), & if\ p'_i < p'_{i+1} \end{cases} \tag{3}$$

Assume that a pixel pair (33, 18) and a binary secret message 111000_2. First, L-PVD method assumes a Div. For example, let $Div =15$. Then calculate the difference d_i for (33, 18) and get $d_i = |33 - 18| = 15$. Because $d_i \leq 15$, the pixel pair (33, 18)$\in R_2$ which belongs to a lower-level. Therefore LSB substitution is applied, L-PVD transforms (33, 18) into binary value (100001, 010010) and hides secret message $(111000)_2$. After hiding, we get a new pixel pair (39, 16), but the difference d'_i

for (39, 16) belongs to a higher-level. As a result, we have to re-adjust (39, 16) by using Eq. (3) and finally we get a new pixel pair (31, 24).

2.3 Tri-way Pixel-Value Differencing (TriPVD)

The details of data hiding steps are described as follows. Assume that a 2×2 pixel block (100,126,115,107) and a binary secret message $S = (1101010100)_2$. TriPVD method makes the block into three pixel pairs: P_0=(100, 126), P_1=(100, 115), P_2 (100, 107) and calculates three differences d_t ($t = 0, 1, 2$) where d_0=26, d_1=15, d_2=7. Then apply PVD method to compute the number of secret bits k_i to be hidden according to d_t. In this case, k_0=4, k_1=3, and k_2=3. Read k_t bits ($t = 0, 1, 2$) from the binary secret data which are $(1101)_2$=13, $(010)_2$=2, $(100)_2$=4, respectively and transform the bit sequence into decimal values. Then calculate the new pixel differences d^*_t which are d^*_0=16 +13 =29, d^*_1=8 +2=10, d^*_2=0 +4=4. At last, adjust the pixel pairs using Eq. (4) where $m_t = |d^*_t - d_t|$.

$$(p'_i, p'_{i+1}) = \begin{cases} (p_i - \left\lceil \frac{m}{2} \right\rceil, \ p_{i+1} + \left\lceil \frac{m}{2} \right\rceil), & \text{if } p_{i+1} \geq p_i \ and \, d^*_t \geq d_t \\ (p_i + \left\lceil \frac{m}{2} \right\rceil, \ p_{i+1} - \left\lceil \frac{m}{2} \right\rceil), & \text{if } p_{i+1} \geq p_i \ and \ d^*_t < d_t \\ (p_i + \left\lceil \frac{m}{2} \right\rceil, \ p_{i+1} - \left\lfloor \frac{m}{2} \right\rfloor), & \text{if } p_{i+1} < p_i \, and \ d^*_t \geq d_t \\ (p_i - \left\lfloor \frac{m}{2} \right\rfloor, \ p_{i+1} + \left\lceil \frac{m}{2} \right\rceil), & \text{if } p_{i+1} < p_i \, and \ d^*_t < d_t \end{cases} \tag{4}$$

After   re-adjusting   we   have   three   pixel pairs $P_0 = (98, 127), P_1 = (103, 113)$ and $P_2 = (102, 106)$. Then MSE (Mean-Square-Error) is to determine the reference point $P_2(102, 106)$. Finally we can get a stego-pixel block $(102, 131, 112, 106)$.

2.4 Adaptive PVD in Edge Areas (AdvPVD)

The details of data hiding steps are described as follows. Assume that a pixel pair (64, 47) and a binary secret message $S = (10100000)_2$. First, AdvPVD calculates the difference d_i for (64, 47) and get $d_i = |65 - 57|$=8.we division range table to three-level, lower-level where $d_i \leq 15$, middle-level $16 \leq d_i \leq 31$, higher-level $32 \leq d_i$, the pixel pair (64, 47) belongs to the middle-level which indicates it can carry 4-bit data. Therefore we use LSB substitution method to embed a 4-bits sub-message $(1010)_2$ and change (64, 47) into the pixel pair (74, 32). For a distortion reduction consideration, the pixel pair is adjusted as $(74 - 2^4, 32 + 2^4) = (58, 48)$. Then AdvPVD calculates the new difference $d'_i = |58 - 48| = 10$ belonging to the lower-level. However, d'_i does not belong to the middle-level so we have to adjust the pair (58, 48). The strategy of adjusting pixel value should satisfy two conditions: the difference of stego-pixel pair still falls in the middle-level and the rightmost 3 LSBs of each pixel can not be changed otherwise the secret data can not be recovered. Therefore, either $(58 + 2^4, 48)$ or $(58, 48 + 2^4)$ satisfies the two requirements. Because $(58 + 2^4, 48)$ belongs to the middle-level, as a result (74, 48) is a better choice.

2.5 PVD Using Modulus Function (HqPVD)

HqPVD modifies the remainder of two consecutive pixels p_i and p_{i+1} for considering better stego-image quality. The details of data hiding steps are described as follows.

Assume that a pixel pair (p_i, p_{i+1}) = (28, 32) and a binary secret message S = $(110)_2$. First, HqPVD calculates the difference d_i for (28, 32) and get $d_i = |32-28| = 4$. According to the given range table, the pixel pair $(28, 32) \in R_1$ which indicates it can carry 3-bit data. Then HqPVD calculates the remainder values $p_{rem(i)}$, $p_{rem(i+1)}$ and $F_{rem(i)}$ of (28, 32) by the following equations, $p_{rem(i)} = p_i \bmod 2^{k_i}$, $p_{rem(i+1)} = p_{i+1} \bmod 2^{k_i}$, $F_{rem(i)} = (p_i + p_{i+1}) \bmod 2^{k_i}$, therefore we have $p_{rem(i)} = 4$, $p_{rem(i+1)} = 0$, $F_{rem(i)} = 4$. To hide the secret message 110_2 into (28, 32) and achieve minimal pixel distortion, we modify (28, 32) by the 6th case of Eq. (5) and get the stego-pixel pair (29, 33).

Case 1: $F_{rem(i)} > k'_i$, $m \le 2^{k_i-1}$ and $p_i \ge p_{i+1}$
$$\left(p'_i, p'_{i+1}\right) = (p_i - \lceil m/2 \rceil, p_{i+1} - \lfloor m/2 \rfloor)$$
Case 2: $F_{rem(i)} > k'_i$, $m \le 2^{k_i-1}$ and $p_i < p_{i+1}$
$$\left(p'_i, p'_{i+1}\right) = (p_i - \lfloor m/2 \rfloor, p_{i+1} - \lceil m/2 \rceil)$$
Case 3: $F_{rem(i)} > k'_i$, $m > 2^{k_i-1}$ and $p_i \ge p_{i+1}$
$$\left(p'_i, p'_{i+1}\right) = (p_i + \lfloor m/2 \rfloor, p_{i+1} + \lceil m/2 \rceil)$$
Case 4: $F_{rem(i)} > k'_i$, $m > 2^{k_i-1}$ and $p_i < p_{i+1}$
$$\left(p'_i, p'_{i+1}\right) = (p_i + \lceil m/2 \rceil, p_{i+1} + \lfloor m/2 \rfloor \tag{5}$$
Case 5: $F_{rem(i)} \le k'_i$, $m \le 2^{k_i-1}$ and $p_i \ge p_{i+1}$
$$\left(p'_i, p'_{i+1}\right) = (p_i + \lfloor m/2 \rfloor, p_{i+1} + \lceil m/2 \rceil)$$
Case 6: $F_{rem(i)} \le k'_i$, $m \le 2^{k_i-1}$ and $p_i < p_{i+1}$
$$\left(p'_i, p'_{i+1}\right) = (p_i + \lceil m/2 \rceil, p_{i+1} + \lfloor m/2 \rfloor)$$
Case 7: $F_{rem(i)} \le k'_i$, $m > 2^{k_i-1}$ and $p_i \ge p_{i+1}$
$$\left(p'_i, p'_{i+1}\right) = (p_i - \lceil m/2 \rceil, p_{i+1} - \lfloor m/2 \rfloor)$$
Case 8: $F_{rem(i)} \le k'_i$, $m > 2^{k_i-1}$ and $p_i < p_{i+1}$
$$\left(p'_i, p'_{i+1}\right) = (p_i - \lfloor m/2 \rfloor, p_{i+1} - \lceil m/2 \rceil)$$

2.6 PVD Exploiting Modification Directions (E-PVD)

Assume that a pixel pair (p_i, p_{i+1}) = (3, 243) and a binary secret message $S = (11110)_2$. First, E-PVD calculates the difference $d_i = |243 - 3| = 240$. According to range table, the pixel pair $(3, 243) \in R_5$, with the width w_j=128. Then E-PVD computes the number of bits to be embedded $k_i = \lfloor log_2 x_i^2 \rfloor$ where $x_i = \lfloor log_2 w_j \rfloor = \lfloor log_2 128 \rfloor = 7$, $k_i = \lfloor log_2 x_i^2 \rfloor \lfloor log_2 7^2 \rfloor = 5$. E-PVD reads 5 bits from S and transforms it into a decimal number $m_i = 30$ in a 7^2 − ary system. The result of $F(3, 243)$ is 4 which is also a digit in a 7^2 − ary system. According to the Case 2 of Eq. (6), because $m_i = (30)_{49} > F(p_i, p_{i+1}) = (4)_{49}$, we get $p'_i = 3 - [30 - 4] \bmod 7 = -2$ and $p'_{i+1} = 243 + [(30 - 4)/7] + (30 - 4) \bmod 7 = 251$. Since the pixel $p'_i - 2 < 0$, we adjust (-2, 251) by using the optimization problem. The optimization problem is to obtain the unknown (x, y) by Minimizing $(p_i - x)^2 + (p_{i+1} - y)^2$; subject to: $F(p_i, p_{i+1}) = m_i$, where $d_i = |p_{i+1} - p_i|$, $\overline{d_i} = |x - y|$. After adjusting the stego-pixel pair is (5, 252).

Case 1: If $m_i = F(p_i, p_{i+1})$ then $(p'_i, p'_{i+1}) = (p_i, p_{i+1})$

Case 2: If $m_i > F(p_i, p_{i+1})$

$$p'_i = p_i - \left[m_i - F_{(p_i, p_{i+1})} \right] \bmod x_i$$

$$p'_{i+1} = p_{i+1} - \left[\frac{m_i - F_{(p_i, p_{i+1})}}{x_i} \right] + \left[m_i - F_{(p_i, p_{i+1})} \right] \bmod x_i$$

Case 3: If $m_i < F(p_i, p_{i+1})$ (6)

$$p'_i = p_i + \left[F_{(p_i, p_{i+1})} - m_i \right] \bmod x_i$$

$$p'_{i+1} = p_{i+1} - \left[\frac{F_{(p_i, p_{i+1})} - m_i}{x_i} \right] - \left[F_{(p_i, p_{i+1})} - m_i \right] \bmod x_i$$

3   Experiment Results and Discussion

This section focuses on conducting a comprehensive discussion on the PVD data hiding method in recent years. Discussion key is to compare Embedding capacity and PSNR of PVD methods. Fig. 1 is our classification diagram of the PVD.

First, according to Fig. 2, we found that HqPVD has the best image quality, it stego-image in Lena, Baboon, Pepper has more than 40dB PSNR. E-PVD has the second highest PSNR. These two methods solve falling-off boundary problem. That is the reason why HqPVD and E-PVD have the better image quality than the other four PVD methods. HqPVD uses re-adjust the remainder of pixel pair, E-PVD use the optimization problem to solve falling-off boundary problem. However, these two methods use modulus to adjust pixel pair, it will cause decrease of embedding capacity but HqPVD and E-PVD still have more capacity than PVD. According to Fig 2 we can found that L-PVD has the lowest PSNR, the reason is L-PVD using the LSB matching in smooth area.

According to Fig. 3 and Fig. 4, we can found that AdvPVD has the highest embedding capacity and payload. Lena, Baboon, Pepper all have high payload more than 3bpp. L-PVD has the second highest payload; its payload in Lena, Baboon, Pepper are also has more than 2.7bpp. The above two use LSB data hiding method in smooth area, because original PVD can't hide lots of information in smooth area. According to Fig. 3, AdvPVD and L-PVD can hide more 1bpp than original PVD.

Tri-PVD divides the image into 2×2 pixel blocks, and hides secret message S with three different directional edges. Payload of Tri-PVD in Lena, Baboon, Pepper are respectively 2.31bpp, 2.51bpp, 2.31bpp, which is 1bpp more than the original PVD method. Fig. 4 shows that PVD, HqPVD, E-PVD have the similar embedding capacity and payload.

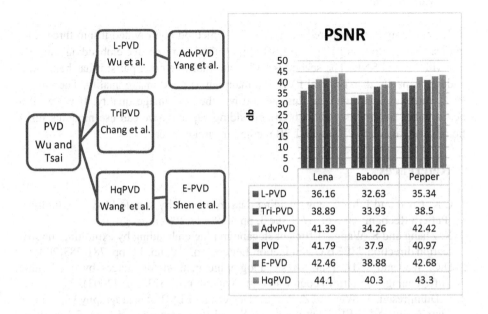

Fig. 1 PVD-based methods   Fig. 2 PSNR of each PVD methods

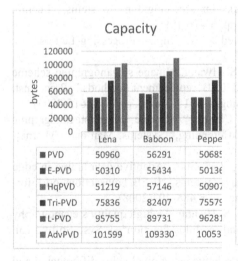

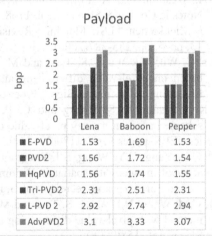

Fig. 3 Capacity of each PVD methods   Fig. 4 Payload of each PVD methods

4   Conclusions

In this study, we compared the recent PVD method and divided it into three categories. The first one is PVD with LSB, this method can improve embedding capacity but with lower PSNR. The second is PVD with multiple pixel pairs; it has better embedding capacity than original PVD and maintains a good image quality. The third is PVD with modulus function, this method has the best image quality. If we want a better image quality as well as high embedding capacity, we can use modulus function in edge area and PVD with LSB method in smooth area.

References

1. C.K. Chan and L.M. Cheng, "Hiding data in images by simple LSB substitution," Pattern Recognition, vol. 37, Issue 3, pp. 469-474(2004).
2. X. Zhang and S. Wang, "Efficient steganographic embedding by exploiting modification direction," IEEE Commination Letters, vol. 10, no. 11, pp. 781-783(2006).
3. D.C. Wu, and W.H. Tsai, "A steganographic method for images by pixel-value differencing," Pattern Recognition Lett., vol. 24, pp. 1613-1626(2003).
4. S. Dumitrescu, X. Wu, and Z. Wang, "Detection of LSB Steganography via Sample Pair Analysis," IEEE Transactions on Signal Processing, vol. 51, no. 7, pp. 1995-2007(2003).
5. J. Fridrich, M. Goljan, and D. Rui, "Detecting LSB steganography in color, and gray-scale images," IEEE Multimedia, Vol. 8, No 4, pp. 22-28(2001).
6. A. Westfeld and A. Pfitamann, "Attacks on Steganographic Systems," Lecture Notes in Computer Science, vol. 1768, pp. 61-76(1999).
7. J. Mielikainen, "LSB Matching Revisited," IEEE Signal Processing Letters, vol. 13, no. 5, pp. 285-287(2006).
8. H.C. Wu, N.I. Wu, C.S. Tsai, and M. S. Hwang, "Image steganographic scheme based on pixel-value differencing and LSB replacement methods," Proc. Inst. Elect. Eng., Vis. Images Signal Processing, vol. 152, no. 5, pp. 611–615(2005).
9. K.C. Chang, P. S. Huang and C. P. Chang, "Adaptive image steganographic scheme based on tri-way pixel-value differencing," Systems, Man and Cybernetics, IEEE, pages 1165–1170(2007).
10. C.H. Yang, C.Y. Weng, S.J. Wang and H.M. Sun, "Adaptive data hiding in edge areas of images with spatial LSB domain systems". IEEE Transactions on Information Forensics and Security, vol. 3, no. 3, p. 488-497(2008).
11. C.M. Wang, N.I. Wu, C.S. Tsai and M. S. Hwang, "A high quality steganography method with pixel-value differencing and modulus function", J. Syst. Softw., vol. 81, pp. 150-158(2008).
12. S.Y. Shen, L.H. Huang, "A data hiding scheme using pixel value differencing and improving exploiting modification directions" Computers & Security., vol.48, pp. 131 (2015).

Improvement of the Embedding Capacity for Audio Watermarking Method Using Non-negative Matrix Factorization

Harumi Murata[1] and Akio Ogihara[2]

[1] Department of Information Engineering, School of Engineering,
Chukyo University, Toyota, Japan
[2] Department of Informatics, Faculty of Engineering,
Kindai University, Higashi-Hiroshima, Japan

Abstract. We propose a watermarking method using nonnegative matrix factorization (NMF) for audio signals. NMF is applied to the host signal, and the amplitude spectrogram of the host signal is factorized into the basis matrix and the activation matrix, which are nonnegative matrices. The notes are estimated from the activation matrix, and the estimated notes are regarded as root notes. In the existing method, one-bit watermark is embedded into the dominant note corresponding to the root note. However, it is difficult to say that the embedding capacity is enough. Hence, in the proposed method, watermarks are embedded into the mediant and leading notes in addition to the dominant note. Up to three-bit watermarks can be embedded into one note while maintaining a detection accuracy.

Keywords: Audio watermarking, Nonnegative matrix factorization, Diatonic chord

1 Introduction

Digital watermarking is a technique to embed other digital data into digital content such as music, images, and videos. For audio signals, the sound quality of the stego signal should not deteriorate. With current methods [1], [2], high sound quality means that the difference between the host and stego signals is small. In these methods, watermarks are embedded by operating on the components of the host signal. Therefore, the noise resulting from embedding watermarks tends to be perceived as an annoying sound.

However, host signals are not always shown to users in actual systems that apply information hiding technology; in fact, it is believed that host signals are not shown in many cases. Therefore, there is no problem even if another sound, not in the host signal, is perceived when the sound of the stego signal is musical. In this case, we can regard the sound quality of the stego signal as high in this paper.

Accordingly, we decided to focus on chords of music theory and embed watermarks in consonance with the host signal. A consonance is defined as chords

© Springer International Publishing AG 2017
J.-S. Pan et al. (eds.), *Advances in Intelligent Information Hiding and Multimedia Signal Processing*, Smart Innovation, Systems and Technologies 63,
DOI 10.1007/978-3-319-50209-0_14

in which the frequency ratio between notes consisting of a chord is represented as simple whole numbers or their approximate values. We use a diatonic chord as consonance. A diatonic chord is the most basic chord, and it is composed of root, mediant, and dominant notes. In a tetrad, the leading note is included as one of the notes making up a diatonic chord in addition to these notes.

Moreover, we use nonnegative matrix factorization (NMF) [3] for embedding watermarks. NMF is an algorithm that factorizes a nonnegative matrix that is an amplitude spectrogram of a target signal into two nonnegative matrices that correspond to the spectral patterns of the target signal and the activation of each spectrum. A group of spectral patterns of the target signal and the intensity variation of each spectral pattern are represented with the basis matrix and the activation matrix, respectively. In this study, the watermarks are embedded by modifying the activation matrix coefficients obtained by NMF.

Furthermore, the root notes are estimated from the activation matrix coefficients, and the key of the host signal is identified. In an embedding method using NMF [4], the watermarks are embedded by modifying the activation coefficients of dominant notes corresponding to the estimated root notes. However, it is difficult to say that embedding capacity of watermarks is enough because a one-bit watermark is embedded into one note. Therefore, the watermarks are embedded into the mediant and leading notes in addition to the dominant note. Up to three-bit watermarks can be embedded into one note; hence, an improvement in embedding capacity is expected.

2   Nonnegative matrix factorization (NMF)

NMF is one of the techniques used for separation of an audio mixture that consists of multiple instrumental sources. The following equation represents the factorization of a simple NMF.

$$Y \simeq AB, \tag{1}$$

where Y is an observed nonnegative matrix, which represents the time-frequency amplitude spectral components obtained via a short-time Fourier transform (STFT), and A and B are nonnegative matrices. In addition, the matrix A is called the basis matrix; it represents spectral patterns of the observed spectrogram Y. B is called the activation matrix; it involves activation information for A.

The multiplicative update algorithms of the standard NMF based on the Euclidean distance are shown in Eqs.(2) and (3).

$$a_{m,k} = \frac{[YB^{\mathrm{T}}]_{m,k}}{[ABB^{\mathrm{T}}]_{m,k}} a_{m,k}, \tag{2}$$

$$b_{k,n} = \frac{[A^{\mathrm{T}}Y]_{k,n}}{[A^{\mathrm{T}}AB]_{k,n}} b_{k,n}, \tag{3}$$

where $a_{m,k}$ and $b_{k,n}$ are the coefficients of matrices A and B.

3   Proposed audio watermarking method using NMF

In this section, we explain the conventional audio watermarking method using NMF [4] and show how to improve the embedding capacity of watermarks.

3.1   Conventional embedding and extracting watermarks

The host signal is divided into frames of L samples and is transformed to the STFT domain with a 50% overlap between successive frames. The amplitude spectrogram Y is factorized into the basis matrix A and the activation matrix B. Here, the objective function is obtained using the square of the Euclidian distance between each column of Y and its approximation AB for simplicity.

The basis matrix A would preferably be fixed to prevent large changes in the factorization results of the embedding and extracting processes. Therefore, NMF is applied to the spectrogram of specific instrumental sound in the host signal as prior learning, and the factorized spectral basis matrix is used as teaching information. The teaching information is regarded as the basis matrix A, and NMF is applied to the amplitude spectrogram Y. An activation matrix B is obtained as in Fig.1, and it is used for embedding watermarks.

Next, the onset time and offset time are estimated from the obtained activation matrix B. The duration between the estimated onset and offset times is defined as one note. The estimated notes are regarded as root notes, and we modify the activation matrix coefficients corresponding to notes which become consonant with the root note. In this method, the watermarks are embedded by operating on the activation matrix coefficients of the dominant note. A dominant note is the fifth note from the root note, and the root and dominant notes have a relationship of consonance. Hence, even if notes that are not included in the host signal are perceived, we consider that there is no problem in case that these notes are arranged based on music theory.

Before embedding watermarks, the quantization level of the activation coefficients is made coarse in order to improve tolerance against attacks, as shown in Fig.2. Even if an attack changes the maximum coefficient value of Fig.2 (b) from

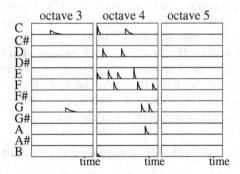

Fig. 1. Example of the activation matrix.

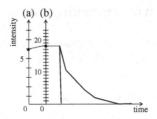

Fig. 2. Quantization level of the activation coefficients: (a) after modification, (b) before modification.

18 to 16, the value of the coefficient of Fig.2 (a) remains the same. Next, the maximum activation coefficient of the dominant note is calculated, and the embedded watermark is the evenness or oddness of this value. That is, watermark bit '0' is embedded if the maximum activation coefficient of the dominant note is even, and watermark bit '1' is embedded if the maximum activation coefficient of the dominant note is odd. If watermark bit '0' would like to be embedded and the maximum activation coefficient of the dominant note is odd, the above embedding rule is not satisfied. Hence, the activation coefficients are amplified by adding its root note's activation coefficients which change the magnification. Similarly, if watermark bit '1' would like to be embedded and the maximum activation coefficient of the dominant note is even, the activation coefficients of the dominant note are amplified until the embedding rule is satisfied. The waveform of the watermark signal can be maintained by using root note's activation coefficients even if the dominant note is not played in the host signal.

After embedding watermarks in all notes, the amplitude spectrogram Y' is obtained using Eq.(4).

$$Y' = AB', \tag{4}$$

where B' is the activation matrix after embedding the watermarks. The amplitude spectrogram Y' is inverse short-time Fourier transformed, and the stego signal is obtained. In addition, the stego signal uses the phase information of the host signal.

In the extracting process, the stego signal is divided into frames of L samples and is transformed to the STFT domain with a 50% overlap between successive frames. The amplitude spectrogram Y' is factorized into the basis matrix A and the activation matrix B'. Here, the basis matrix A uses the same matrix as the embedding process. The root notes are estimated from the coefficients of activation matrix B', and the dominant note is specified. If the maximum activation coefficient of the dominant note is even, the watermark bit '0' is extracted. Otherwise, the watermark bit '1' is extracted.

3.2 Improvement of embedding capacity of watermarks

In the embedding method of section 3.1, a one-bit watermark is embedded into one note, and it is difficult to say whether the embedding capacity of the watermarks is enough. Thus, multiple bits are embedded into one note in order to improve embedding capacity.

In the proposed method, the watermarks are embedded into the mediant and leading notes in addition to the dominant note. A mediant is the third note from the root note and a leading note is the seventh note from the root note. Moreover, a diatonic chord consists of the root, mediant, dominant, and leading notes. Because a diatonic chord is a consonance, it is considered that there is no problem even if these notes are perceived.

The embedding process of the watermarks is the same as in the case of the dominant note described in section 3.1. An example of embedding the watermark bits '001' is shown in Fig.3. The conditions of the mediant and leading notes are satisfied, and no operation is performed. On the other hand, the condition of the dominant note is not satisfied with the embedding rule, and the activation coefficients of dominant note are amplified using its root note's activation coefficients for the same reason as described in section 3.1.

However, as shown in Fig.4, it is possible for the embedding positions of multiple root notes to be the same. For example, suppose that the watermark bits '001010' are embedded into the estimated notes C and G. In the case of Fig.4, the leading note of the root note C and the mediant note of the root note

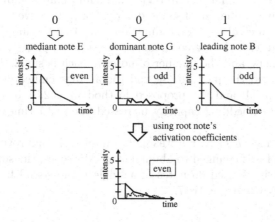

Fig. 3. Modification of the activation matrix coefficients in the case of embedding the watermark bits '001'.

watermark bits : 0 0 1 0 1 0 ...

estimated tones   candidate for the watermark signal

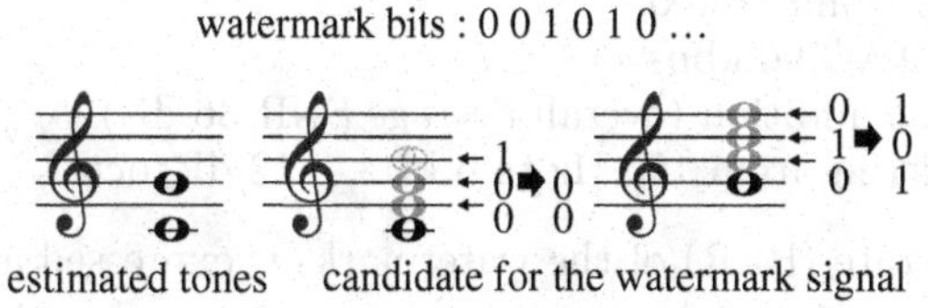

Fig. 4. Example in which watermarks cannot be embedded.

G are B. Moreover, the watermark bit '1' has to be embedded into the leading note of C, and the watermark bit '0' has to be embedded into the mediant note of G. However, multiple bits cannot be embedded into same note, and we decide whether the watermark should be embedded into the note or not in order of consonance. Namely, the watermarks are embedded in the order of dominant note, mediant note, and leading note. Hence, in Fig.4, the watermark bit '1' is not embedded into the leading note of C, and the watermark bits '00101' are embedded into C and G. Therefore, the number of embedding bits will differ depending on notes, and up to three-bit watermarks can be embedded into one note.

In the extracting process, root notes are estimated from the activation coefficients of the stego signal. The mediant, dominant, and leading notes are specified from the estimated root notes. The extracted notes are decided in order of consonance when the extraction positions of multiple root notes are the same. The watermarks are extracted from the determined notes in accordance with the extracting rules of section 3.1.

4 Experiments

We conducted an evaluation of the proposed method's tolerance against various attacks based on the evaluation criteria for audio information hiding technologies [5] and a subjective evaluation of sound quality. For testing, we used ten pieces of music selected from [6] of 60 seconds duration at a 44.1-kHz sampling rate, 16-bit quantization, and with the stereo channel. The selected pieces are simple piano etudes for beginners, and the number of notes in each is small. The embedding watermark bits used the payload [5], and the average embedding capacities of the watermarks in [4] and the proposed method were 2.51 bps and 6.74 bps, respectively. The embedding capacity increased about 2.7 times in comparison with [4].

The frame length L of STFT was 8192 samples. To improve tolerance, the quantization level was changed to about 11-bit. Moreover, the supervised signal of the basis matrix A used 36 notes of a single piano sound from octave 3 to octave 5 made by Cubase Artist7.

4.1 Tolerance against attacks

We examined the tolerance of the proposed method against the following attacks.

- MP3 128 kbps joint stereo
- MPEG4 HE-AAC 96 kbps
- Gaussian noise addition (overall average SNR 36 dB)
- Bandpass filtering from 100 Hz to 6 kHz, −12 dB/oct

The bit error rate (BER) of the watermarks is expressed as

$$\mathrm{BER} = \frac{\text{number of error bits}}{\text{number of embedding bits}} \cdot 100 \ [\%]. \tag{5}$$

Table 1. Music titles and BER [%] results: (A) MP3, (B) MPEG4 HE-AAC, (C) additive Gaussian noise, and (D) bandpass filtering.

number	title	BER [%]			
		(A)	(B)	(C)	(D)
1	Pachelbel's Canon	8.55	8.79	1.20	0
2	Toccata and Fugue in D minor	6.97	9.83	1.16	9.69
3	Four Seasons, Concerto No.1 in E	9.39	7.71	0	3.91
4	Prelude and Fugue No.1	9.18	4.72	1.81	1.39
5	Aria (Goldberg Variations)	8.29	4.58	4.25	9.87
6	Minuet in G major	9.36	3.70	0.25	4.05
7	Jesu, Joy of Man's Desiring	5.31	7.01	0	9.16
8	Air on the G String	9.62	8.88	1.39	5.87
9	Lascia ch'io pianga	6.70	1.64	0.58	4.29
10	Hallelujah	7.38	5.52	0.21	3.66
	average	8.07	6.24	1.08	5.19

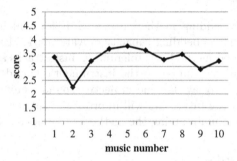

Fig. 5. MOS results from ten test subjects.

The BER must be less than 10 % according to the criteria [5]. Table 1 lists the BER results. The embedding strength of each note was adjusted recursively such that the BERs against all attacks became less than 10%. The results listed in the table confirm that the BERs against all attacks were satisfied with the criterion. However, it is difficult to say that the tolerance against attacks is high.

4.2 Sound quality

In our approach, we want to evaluate the sound quality of the stego signals from the musical viewpoint. Hence, we used the mean opinion score (MOS) in a subjective evaluation of sound quality involving ten test subjects. Each test subject listened to the stego signal of the proposed method. The subject then scored the sound on a scale from 1 to 5, a score of 1 being lowest quality and a score of 5 being highest quality.

The average scores of all test subjects are shown in Fig.5. The criterion of subjective sound quality is that the score of each piece of stego signals should be higher than 2.5, the same as that of the objective sound quality [5], in this paper. From this result, the average scores of nine stego signals were satisfied with the criterion.

The played range in No.2 is from octave 2 to octave 6, and it is different from the range of basis matrix. The watermarks were embedded from octave 3 to octave 5, even if the played note was below octave 3 or above octave 5. Moreover, the host signal and watermark signal became dissonant when the estimated notes were different from the played notes. This seems to be the cause of the sound quality deterioration. Hence, the range of the bases should be extended, and the accuracy of the pitch estimation should be improved.

However, the timbre of the watermark signal was similar to the instrumental sound, because the basis matrix A was represented by the spectral patterns of the piano sound in the prior learning. Therefore, it is considered that the sound quality of the stego signal is not necessarily low.

5   Conclusions

We proposed an embedding method using NMF for real instrumental sound and for improving the embedding capacity of watermarks. Experimental results indicated the proposed method was tolerant against attacks and that the timbre of watermark signal was similar to the instrumental sound of the host signal. However, it is difficult to say that the subjective sound quality of the stego signal was good. Hence, in the future, we will improve the subjective sound quality of the stego signal while maintaining the method's tolerance against attacks.

Acknowledgments

This work was supported by JSPS KAKENHI Grant Numbers JP26870681, JP26330214.

References

1. Lie, W.N., Chang, L.C.: Robust and High-Quality Time-Domain Audio Watermarking Based on Low-Frequency Amplitude Modification. IEEE Trans. on Multimedia. vol.8, no.1, pp.46–59 (2006)
2. Murata, H., Ogihara, A., Uesaka, M.: Sound Quality Evaluation for Audio Watermarking Based on Phase Shift Keying Using BCH Code. IEICE Trans. Inf. & Syst.. vol.E98-D, no.1, pp.89–94 (2015)
3. Lee, D.D., Seung, H.S.: Algorithms for Non-negative Matrix Factorization. Neural Inf. Process. Syst.. vol.13, pp.556–562 (2001)
4. Murata, H., Ogihara, A.: Digital Watermark for Real Musical Instrument Sounds Using Non-negative Matrix Factorization. In: International Technical Conference on Circuits/Systems, Computers and Communications, pp.677–680 (2016)
5. IHC Evaluation Criteria and Competition, http://www.ieice.org/iss/emm/ihc/IHC_criteriaVer4.pdf, Accessed December 21, 2015.
6. Classic Anthology 100 Played Easily for Adults. RittorMusic (2009)

Digital audio watermarking robust against Locality Sensitive Hashing

Kotaro Sonoda and Kentaro Morisaki

Graduate school of engineering, Nagasaki University, Bunkyo-machi 1-14, Nagasaki, 852-8521 Nagasaki, Japan
sonoda-iihmsp16@cis.nagasaki-u.ac.jp

Abstract. Audio fingerprinting is an audio feature vector and a technique to identify the music source by comparing with original fingerprints in a fingerprint database. The fingerprints are often hashed to shorten codes. The content similarities between fingerprints are maintained in the hashed code. Such a hash function family is called Locality Sensitive Hashing (LSH). Because audio fingerprint using LSH is known to be resistant against CPO (Content Preserving Operations, perceptively acceptable manipulations) such as compression, noise adding, mean filtering, it is possible to identify the original source even if the source was slightly modified.

On the other hand, mixed arrangement (mashup) of several music sources is allowed as legitimate artistic expression. In the conventional fingerprint based retrieval system, the mixed arrangements could identify the origins segmentally but the arranger's authorization is ignored.

In this report, we propose an audio watermarking method robust against LSH coding. That is, the arranger information is watermarked in the audio signal and it is detectable from not only stego audio signal but also stego audio fingerprint of LSH.

Keywords: Audio Fingerprinting, Locality Sensitive Hashing, Audio watermarking

1 Introduction

1.1 Identification of the music source

In regard to using a similar music search system, to identify the music source, an audio fingerprinting technique is often used. The audio fingerprint is an inherent feature of the audio signal like a human fingerprint or biometric feature. As the audio feature vector, for example, one can utilize the short-time Fourier transform (STFT) coefficients, the chroma codes, and zero-crossing intervals and so on.

The similarity is evaluated in the distance norm between the vectors in the feature domain space. However, in the case that the vector dimension is high, the calculation of the norm also becomes high. Therefore, the feature vector is

© Springer International Publishing AG 2017   115
J.-S. Pan et al. (eds.), *Advances in Intelligent Information Hiding and Multimedia Signal Processing*, Smart Innovation, Systems and Technologies 63,
DOI 10.1007/978-3-319-50209-0_15

shortened by hashing technique which fulfills both keeping the dimension small and yet providing a high level of accuracy in similarity measurement.

Fridrich et al. proposed a robust hash function for still image fingerprinting [2]. In their study, the still image hash is constructed by projecting its each descrete cosine transform (DCT) block on the zero-mean random patterns. Their hash function is robust against CPO (Content Preserving Operations, visually acceptable manipulations) such as compression, noise adding, and mean filtering.

Radhakrishnan et al. proposed audio signature hashing by applying Fridrich's hashing method to the audio spectrogram [4].

1.2 Detection of arrangement

By using the above mentioned audio fingerprinting techniques, the music's original sources are identified in the hashing domain. However, we should consider arrangements of music source. How should the arranged music be differenciated from the original music sources? Particularly, how should we treat mashup music (e.g. one made by concatenating several music sources)?

There are three options that may be considered.

- One possible option is to ignore the original sources and fingerprint the mashup as an independent music sources.
- The second option is to ignore the mashup and preserve the original fingerprints of the original sources separately.

In these two options, both the arranger and composer cannot coexist. We would propose another third option.

- Detecting every original fingerprint of the original sources separately as well as detect the watermark of the arrangement from the fingerprints.

In the third options, one can detect both the originals and the arrangement as shown in Fig.1.

In this concept, the arranger S3 embeds his watermark into the arranged source before broadcasting it. In the authorization checking system, the hashed audio fingerprints are extracted from the uploaded arranged source and the segmental fingerprints identify the original sources S1 and S2 as in conventional fingerprint retrieval. Moreover as in conventional watermarking method, the arranger S3's watermark is detected by analyzing the uploaded music signal before the hashing. Providing that the watermarking scheme is robust against the hashing, the hashed audio fingerprints preserve the arranger's watermark and the arranger S3's watermark is detected from the hashed fingerprints.

To utilize this third option, we propose a digital audio watermarking scheme that is robust against Locality Sensitive Hashing in this paper. In section 2, detail of the Locality Sensitive Hashing of the audio fingerprinting is introduced. And we propose a watermarking scheme that is robust against LSH in section 3. The inaudibility and the robustness of the proposed method is evaluated in section 4. The paper is brought to a conclusion in section 5.

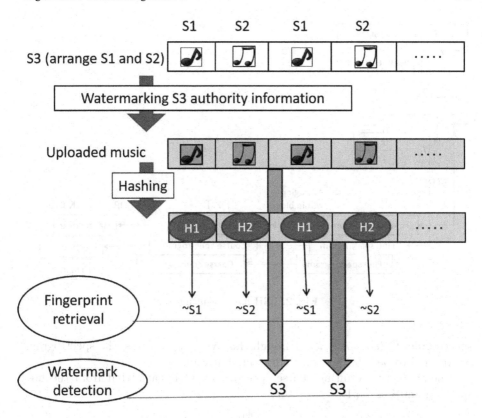

Fig. 1. Proposed system: Identification of each original and watermark detection of the arrangement

2 Audio fingerprinting based on STFT and Local sensitive hashing (LSH)

In this section, we introduce the audio fingerprinting scheme based on STFT and LSH which is proposed by Regunathan et al. [4]. Fig.2 shows the block diagram of Regunathan's LSH audio fingerprinting.

2.1 Audio feature vector based on STFT spectrogram

The audio signal is segmented into chunks (time length of a chunk is T_{ch}) with overlapping T_o. Similarity measurement is carried out with chunk as an unit. In every chunk, segmented signal $(X_i, i = 1, 2, \ldots, T_{ch})$ is transformed to short-time Fourier coefficients and concatenated. The concatenated coefficient matrix (fine spectrogram **S**) is segmented to $F \times T$ blocks and their blocks are averaged in both the time and frequency domains, and as a result we can get a coarse

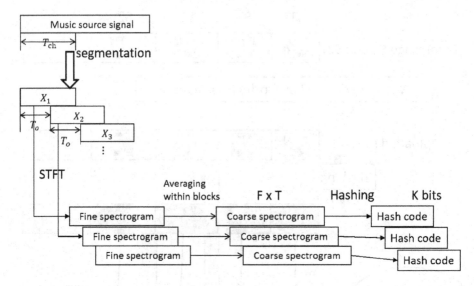

Fig. 2. LSH fingerprinting

spectrogram $\mathbf{Q}$ (size: $F \times T$). Through this averaging process, the spectrogram is expected to be robust against slight modifications.

Therefore, the resulting coarse spectrogram $\mathbf{Q}$ is calculated from the fine spectrogram $\mathbf{S}$ as Eq.1:

$$Q_{k,l} = \frac{1}{W_f * W_t} \sum_{i=(k-1)W_f}^{k \cdot W_f} \sum_{j=(l-1)W_t}^{l \cdot W_t} S_{i,j},$$

$$k = 1, 2, ...F; \quad l = 1, 2, ...T, \tag{1}$$

where W_t and W_f are the matrix length and width of an average block size in the fine spectrogram, respectively.

This coarse spectrogram $\mathbf{Q}$ is produced for every chunk.

2.2 Locality Sensitive Hashing

The coarse spectrogram $\mathbf{Q}$ is an audio feature vector and it can be used to evaluate the audio similarity by calculating the vector distance norm from other audio content. However, the precise similarity requires a $\mathbf{Q}$ with a higher dimension of F and T. Therefore, the feature vector is transformed to be short in keeping with their geometric features using the Locality Sensitive Hashing technique. Radhakrishnan et al. proposed a hashing scheme for music signal [4].

In their method, the coarse spectrogram $\mathbf{Q} \in \mathbb{R}^{F \times T}$ is hashed to a K-bits code $\mathbf{H} \in \mathbb{B}^K$. Initially K random matrices $\mathbf{P}^{(k)} \in \mathbb{B}^{F \times T}, k = 1, 2, ...K$ are prepaired. They are equally distributed in $[-1, 1]$. Using these random matrices, K-bits hashed code $\mathbf{H}$ is calculated as following Eq.3,

$$h_k = \sum_{i=1}^{F} \sum_{j=1}^{T} Q_{i,j} * P_{i,j}^{(k)} \tag{2}$$

$$H_k = \begin{cases} 1 & \text{if} \quad h_k > \text{Median}(\mathbf{h}), \\ 0 & \text{otherwise} \end{cases} \tag{3}$$

3   Audio watermarking method robust against LSH

As our goal, we wished to detect the original music sources and simultaneously detect their arrangement authority by analyzing hashed audio features. Moreover, we are required to detect them by analyzing the raw signal before hashing.

In conventional watermarking methods, it is possible to embed and detect robustly against some signal processing attacks, such as MP3 coding, noise adding and resampling. However the hashing process had never been considered.

The hashing process is a mapping process whereby the original signal is transformed by a random matrices set of variables, and their mapped components are fully quantized. Because of that, segment-based watermarks are randomized and can't be identified autonomously.

The hashing processes every segment by using the same hashing matrices set. While the hashed codes are randomized in a segment, the similarity relations between each couple of segments are expected to be preserved in the similarity measurement of hash codes after LSH. Therefore, the watermarking embedded in the relationship between time series segments is expected to be preserved. As a such type of watermarking method, the simplest one is the echo hiding method [3].

In this study, we apply conventional echo hiding to the system.

3.1   Embedding process

In the conventional echo hiding method, the host signal is added slight gained echo on about 10 ms delay in order to maintain inaudibility, and the echo kernels are exchanged as according to the embedding rate.

Robust hash echo hiding we propose is the same process as the conventional method but it adds slight echoes on about over segment unit (e.g., two segments length delayed echo is added for watermark bit '0' and three segments length for bit '1') and the embedding bit rate is about 1 bps. That is the delay times of echo are large compared to the conventional echo hiding. Such a large delay time of echoes distort the audio signal. However we study the robust detection method rather than inaudibility in this report.

3.2   Detection process

Before LSH coding, the detection process is also the same as that of conventional echo hiding. And its robustness is inherently the same as the conventional method.

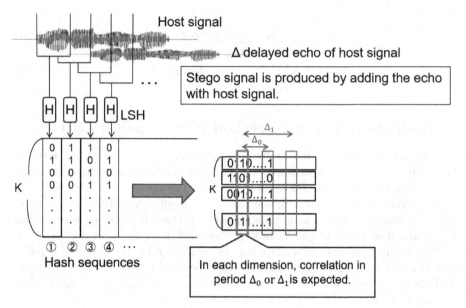

Fig. 3. Detection process

After LSH coding, each segment is mapped to orthogonal K-dimensional codes. When Δ segments delayed echo was added, a similar bit is expected to appear every Δ segments in the hashed codes. In this study, we tried to detect the watermark by finding peaks in the averaged auto-correlation of each K's dimension as shown in Fig. 3

4 Experiment

To evaluate the robustness against LSH coding, we tried to detect watermarks from LSH coded sequences of the music sources. The testing music source is embedded random payload by echo hiding and converted to the LSH code sequences.

The LSH configurations are the same as Radhakrishan's settings [4]:

– Segment length (Chunk Size T_{ch}): 133 ms (6400 samples at 48 kHz sampling)

– Segmentation stepping length (T_0): 10 ms (512 samples at 48 kHz sampling)

– Spectrogram resolution : Fine (128×25) $\Rightarrow$ Coarse (F:20 $\times$ T:10)

– Hash code size (K) : 18

Testing music source is selected from SQAM (Sound Quality Assessment Material) set and 8 tracks are used [1]. Table 1 shows the evaluated-source list. In this report, any source mixing arrangement is not carried out.

Table 1. Evaluated source lists from SQAM

Track	SQAM
27	Castanets
32	Triangles
35	Glockenspiel
40	Harpsichord
65	Orchestra
66	Wind ensemble
69	ABBA
70	Eddie Rabbit

Table 2. Configurations for LSH robust echo hiding

Gain α	0.3
Delay time for watermark "0" Δ_0	25 segments
Delay time for watermark "1" Δ_1	40 segments
Frame length for 1 bit embedding L	62 segments
Embedding bit rate	1.15 bps

The configurations for echo hiding are listed in Table 2. Note the delay times for watermarks are set as very long as the length of 40 segments (523 ms). Such a long delay echo may deteriorate inaudibility, but it is overlooked in this experiment.

As a result of experiment, the averaged detection bit error rate (BER) is 46.2%. All the eight testing stego signals show almost similar detection bit error rates. This is clearly unacceptably high and only slightly below the chance level.

However some frames were successfully embedded and detected. Such an example of correctly detected correlation peak is shown in Fig.4. Therefore, further improvement of the embedding and detecting method is clearly essential.

5   Conclusion

Audio fingerprinting generates a hash code for every segment based on an audio feature vector. However, the mixed arrangement is ignored in the audio hash. In this report, we proposed to detect an additional arrangement authority by embedding a watermark. To utilize this protocol, we tried to apply the echo hiding watermarking method that is robust against Locality Sensitive Hashing.

However the proposed watermarking method outlined in this report is not yet full functional. Further improvement of the embedding and detecting method is clearly essential.

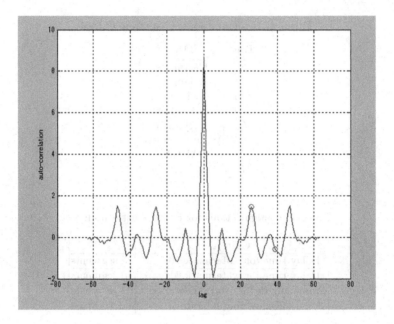

Fig. 4. Averaged auto correlation graph of successfully detected frame. Higher peak value at $\Delta_0 = 25$ or $\Delta_1 = 40$ (two red circles) indicates bit "0" or "1". In this example's case, Peak is found at Δ_0 (watermark bit "0").

References

1. European Broadcasting Union (EBU): SQAM (2008), http://tech.ebu.ch/publications/sqamcd
2. Fridrich, J., Goljan, M.: Robust hash functions for digital watermarking. In: In Proceeding of ITCC (May 2000)
3. Gruhl, D., Lu, A., Bender, W.: Echo hiding. In: In Proceeding of International Workshop on Information Hiding. pp. 295–315 (1996)
4. Radhakrishnan, R., Bauer, C., Cheng, C., Terry, K.: Audio signature extraction based on projections of spectrograms. In: In proceeding of IEEE International Conference on Multimedia and Expo (ICME). pp. 2110–2113 (2007)

On a Synchronization Method for Audio Watermarks Robust Against Aerial Transmission Using Embedded Random Synchronization Pattern

Kazuhiro Kondo and Joji Yamada**

Graduate School of Science and Engineering, Yamagata University,
4-3-16 Jonan, Yonezawa, Yamagata 9928510, Japan

Abstract. We evaluated the effect of synchronization on audio water-marks. Synchronization signal was generated by using an M-sequence for the phase, and shaping its magnitude relative to the host signal spectrum. Synchronization at the detector was achieved by calculating a cross-correlation function with the same M-sequence, and detecting the peak in the correlation. The synchronization signal was added to a spread-spectrum watermarked audio signal. This signal was then transmitted over analog channels, by simply connecting the D/A output to the A/D input using a cable, and also playing out the analog signal through a loudspeaker and recording this signal using a microphone. The addition of the synchronization signal enabled the detection of the watermark through analog channels at acceptable accuracy. The synchronization rate was also shown to strongly influence the watermark bit error rate (BER).

Keywords: Audio watermark, frame synchronization, M-sequence, cross-correlation, aerial transmission

1 Introduction

Watermarking methods for audio signals have been studied for some time now. Initially, these audio watermarks were intended to be used to control the illegal distribution of copyrighted material. However, some recent proposals are intended to be used to hide supplementary data into audio. For example, Ito et al. [5] and Aoki [1] attempt to hide data that will facilitate concealment of lost audio segments, or extend the bandwidth of the signal by extrapolating the high-frequency components. Matsuoka [7] attempts to hide the URL of the web site of the artist of the host audio signal, which can be used to give additional promotional information about the host audio signal.

Some of the above data hiding methods are used in applications that do not go through analog channels. For example, digital music signals may be distributed only on digital channels (e.g., the Internet), and may never be transmitted through analog channels. However, there are applications that require

---

** Currently with Alpine Giken Inc.

© Springer International Publishing AG 2017   123
J.-S. Pan et al. (eds.), *Advances in Intelligent Information Hiding and Multimedia
Signal Processing*, Smart Innovation, Systems and Technologies 63,
DOI 10.1007/978-3-319-50209-0_16

the transmission of watermarked audio through analog channels. For instance, the work listed above by Matsuoka [7] was intended to be used on smart phones. In his work, music signals with embedded URLs, most likely played out from loudspeakers, are picked up by the microphones on the smart phones, and the promotional material of the song on the artist's web page is displayed. Nishimura also attempted to embed the lyrics of a Karaoke song using data hiding [8]. The lyrics can then be decoded and displayed at the exact timing of the song when the lyrics should be sung.

The detection of hidden data from a watermarked audio signal that go through analog channels may be a challenge for a number of reasons. Various types of noise may be added during the analog transmission. Non-linear distortion can also be added to the signal due to the analog circuitry associated with the A/D and D/A conversion. Timing skews also may be added due to unstable clock signals. Since most watermarks embed data in frames, maintaining frame position at the detector is essential, and may lead to a large number of errors in the detected watermark if not recovered accurately. However the framing within the host signal is lost due to analog conversion, and the frame position must be recovered at the detector.

So far, there has been very few efforts reported on the investigation of the effect of analog transmission on embedded watermarks. For example, Hiratsuka et al. investigated the addition of synchronization codes to the host signal for frame alignment of a patchwork watermark [4]. They evaluated the synchronization accuracy with analog loop back (ALB), but not aerial transmission (AT). Karnjana et al. have recently evaluated the effectiveness of synchronization codes embedded in the LSBs of the host signal, and also addition of random synchronization patterns [6]. However, they have not evaluated their methods in analog channels.

In this paper, we investigate a simple frame position recovery method which adds random synchronization pattern in the phase of the host signal. We study the effectiveness of this method by trying to recover the frame position from received signals after analog transmission, including AT. We also study the effect of frame position recovery on the BER of the recovered watermark data.

This paper is organized as follows. The next section will describe the frame synchronization method investigated in this paper. This will be followed by evaluation of robustness to analog channels, and its results. Finally, conclusions and issues will be given.

2 Frame Synchronization for Audio Watermarks

In this paper, we investigate a frame synchronization method for audio watermarks based on additive M-sequence synchronization pattern. As shown in Fig. 1, a synchronization signal is added to the host signal. The phase of the additive signal is the M-sequence pattern, while the envelope of the signal is shaped according to the host signal, improving the imperceptibility of the added signal.

The recovery processing of the frame position is depicted in Fig. 2. The received signal with unknown frame position is split into interim frame positions with a fixed frame length (which is assumed to be known), and cross-correlation between these frame signals and the same M-sequence added in Fig. 1 is calculated. This cross-correlation is averaged over multiple frames, smoothed to remove variations due to noise signals, and the peak position is picked from this signal to detect the frame starting position.

The cross-correlation, $g_{xy}(\tau)$ between the received signal with added synchronization pattern, $x'(t)$, and the synchronization pattern (M-sequence), $y(t)$, can be defined as follows:

$$g_{xy}(\tau) = \int x'(t)y(t+\tau)\,dt \tag{1}$$

where τ is the lag time.

$G_{xy}(\omega)$, which is the Fourier transform of the above, is the cross-power spectrum between x' and y. The generalized cross-correlation function (GCCF) can then be defined as follows.

$$R_{xy} = \int \Psi(\omega)G_{xy}(\omega)e^{i\omega t}\,d\omega \tag{2}$$

where $\Psi(\omega)$ is the generalized weighting function.

The weighting function needs to be optimized so that the GCCF will show a distinctive peak at the corresponding start position of the frame in the received signal $x'(t)$. Through some preliminary experiments, we decided to use the Cross-power Spectrum Phase analysis (CSP) method, which uses the generalized weighting function given below:

$$\Psi(\omega) = \frac{1}{|G_{xy}(\omega)|} \tag{3}$$

The CSP method normalizes the cross-correlation function by the magnitude of the cross-correlation itself, thereby being influenced mostly by the relative phase of the two signals, and not the magnitude of the signals.

The resultant GCCF signal still contains a large number of false peaks due to the channel noise, and also the host signal itself. Thus, we smoothed the GCCF in order to atte   isons between

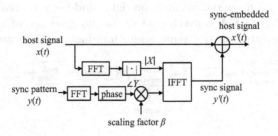

Fig. 1. Synchronization Pattern Embedding Method

some popular smoothing methods, we decided to employ the Weighted Moving Average (WMA) which uses linearly decreasing weights for the samples to be averaged in both the forward and backward direction from the center sample.

3   Robustness of Proposed Frame Synchronization with Analog Conversion

We conducted evaluation experiments to find out the robustness of the frame synchronization method to analog transmission. We investigated the robustness to ALB (without AT) and to AT. The frame synchronization recovery rate as well as bit error rate of the watermark embedded simultaneously with the synchronization pattern was evaluated.

3.1   Experimental Conditions

We selected 70 clips from the SQAM database [3]. All samples were sampled at 44.1 kHz, 16 bits/sample, monaural. We selected 10 s from each clip.

M-sequences with a length of 2047 samples were used for the synchronization pattern. The synchronization signal was prepared from these M-sequences as described in Section 2. and added to the host signal with a frame length of 1 s. The scale factor was determined from preliminary experiments for a balance between imperceptibility and robustness, and was fixed at 0.08. The average Objective Difference Grade (ODG), estimated with the ITU-R BS.1387 Recommendation (PEAQ) [9], over selected 20 samples in the SQAM database was -1.92 with this scale factor.

The CSP method was used to calculate the GCCF, smoothed with WMA, and the frame position was detected from the peaks in the resultant GCCF, as described in Section 2.

We simultaneously embedded watermark data to the host signal using the classic spread-spectrum method [2]. This watermark was arbitrary chosen to test the effect of frame synchronization on a conventional watermark method. The watermark data was spread with an M-sequence with a length of 4095 samples and a chip rate of 10, and added to the host signal along with the synchronization signal. The scale factor of the spread spectrum signal was also fixed at 0.08, chosen empirically.

Experimental configuration for the ALB experiment is shown in Fig. 3. The test signal, with both the synchronization signal and the watermark, is converted to analog using the D/A converter on the audio interface of a PC, fed back to the A/D converter on the same audio interface with an audio cable, and

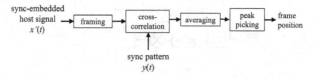

Fig. 2. Frame Synchronization Recovery Method

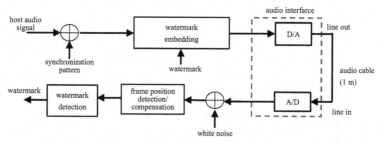

Fig. 3. Experimental Setup for ALB

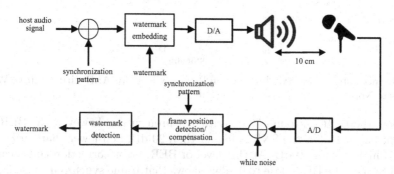

Fig. 4. Experimental Setup for AT

synchronization detection as well as the watermark detection was attempted on this signal. White noise was added to the signal after A/D conversion to simulate additive noise in the analog channel. Noise was added after the A/D conversion to have better control over the level of the noise added. The level of the added noise was adjusted to obtain signals with SNRs ranging from 0 to 50 dB.

We also attempted to evaluate the effect of AT on the test signal. The configuration of this experiment is shown in Fig. 4. We used a female speech sample with read English sentence, selected from the SQAM database for the test signal. The sampling rate was 44.1 kHz, 16 bits/sample, monaural, and the length was 10 s. The test signal is converted to an analog signal with the D/A converter, played out from a loudspeaker, and recorded using a microphone placed 10 cm from the loudspeaker. The recorded analog signal is then converted to digital using the A/D converter, white noise is added, and then synchronization and watermark detection was attempted.

3.2 Results and Discussions

Analog Loop Back (ALB) Fig. 5 shows the frame synchronization rate with CSP and WMA. As can be seen, a frame synchronization rate close to 100% can be achieved to about SNR of 20 dB. Below this SNR, the synchronization rate gradually decreases, to about 65% at SNR 0 dB.

Fig. 6 shows the BER of the detected watermark, both with and without the synchronization pattern addition. As can be seen, without frame synchro-

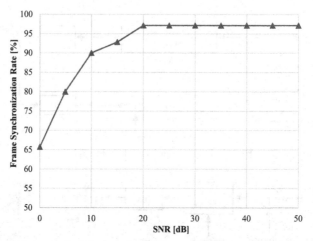

Fig. 5. Successful Frame Synchronization Recovery Rate for ALB with Additive White Gaussian Noise

nization, the BER is around 50%. With synchronization, however, the BER can be kept below 15% if the SNR is higher than 20 dB, and lower than 10% if the SNR is higher than 30 dB. At this level of BER, error correction codes can be added to keep the BER close to 0. This shows that frame synchronization is very effective in maintaining the BER of the embedded watermark within acceptable levels. It can also be observed that the increase in the frame synchronization rate corresponds well with the decrease in the BER, indicating that frame synchronization plays a major role in the detection accuracy of the watermark. Although these results were for spread spectrum, which is known to be affected by frame alignment accuracy, other watermark methods are also known to be affected by frame alignment, and so the BER performance should be equally improved by the use of synchronization methods.

Aerial Tranmission (AT) Fig. 7 compares the BER for AT and ALB, both with frame synchronization. As can be seen, AT shows about 10% higher BER at all SNRs than with ALB alone. However, it is surprising that the increase in the BER can be kept this low since AT potentially includes both an electro-acoustic and acoustic-electrical transducers which may include significant non-linear characteristics.

Table 1 tabulates the BER with and without frame synchronization for AT and ALB. No white noise was added. Again, the BER is about 10% higher for AT with synchronization compared to ALB. However, it is clear that the addition of synchronization improves the BER in both ALB and AT significantly.

4   Conclusion

We evaluated the effect of synchronization on watermark BER. Synchronization signal was added to the host signal. The phase of the synchronization signal

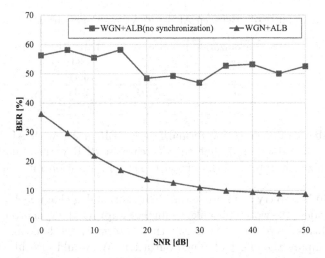

Fig. 6. BER for ALB with Additive White Gaussian Noise

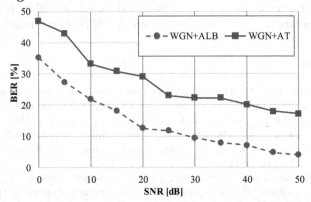

Fig. 7. BER for AT and ALB with Additive White Gaussian Noise

was a known M-sequence pattern, while the magnitude was shaped according to the host signal spectrum. This signal was added to the host signal after scaling. Synchronization at the detector was achieved by calculating the GCCF with the M-sequence of the synchronization signal, and detecting the peak in the smoothed GCCF. In order to investigate the effect of synchronization on water-marked signals, synchronization signal was added to a watermarked host signal. The classic spread spectrum watermark was used in this work. We evaluated the synchronization rate with and without the addition of synchronization. We also evaluated the BER of the watermark, with and without synchronization. The watermarked signal was transmitted over analog channels. We tested ALB, which simply connects the D/A output with the A/D input using a cable, and AT, in which we played out the analog signal through a loudspeaker, and recorded this signal using a microphone placed 10 cm away.

It was found that without synchronization, the detection of the watermark is not possible, with BER over 50% in all cases. With synchronization, however,

Table 1. BER Comparison Between ALB and AT with and without Synchronization

synchronization	transmission	BER [%]
yes	ALB	2.34
	AT	16.41
no	ALB	52.34
	AT	50.78

it was possible to detect the watermark at a BER of 2% with ALB, and 16% with AT. It was also found that the decrease in BER corresponds well with the synchronization rate, i.e., the higher the synchronization rate, the lower the BER.

We would like to try to improve the BER over analog channels by using error correction codes. We would also like to further improve the synchronization rate with other forms of cross-correlations. Other watermark methods also need to be tested to compare the effect of synchronization. We would also like to test the synchronization with actual additive noise in the analog channels. Especially, the addition of reverberation needs to be tested since it is well known that these forms of degradation have a large impact on the BER of watermarks.

References

1. Aoki, N.: Multimedia Information Hiding Technologies and Methodologies for Controlling Data, chap. Enhancement of Speech Quality in Telephone Communication by Steganography, pp. 164–181. IGI Global, Hershey, PA, USA (Oct 2012)
2. Boney, L., Tewfik, A.H., Hamdy, K.N.: Digital watermarks for audio signals. In: Proc. International Conf. on Computing and Systems. pp. 473–480. IEEE, Hiroshima, Japan (June 1996)
3. European Broadcasting Union: Sound quality assessment material (SQAM). CD
4. Hiratsuka, K., Kondo, K., Nakagawa, K.: On the accuracy of estimated synchronization positions for audio digital watermarks using the modified patchwork algorithm on analog channels. In: Proc. International Conf. on Intelligent Inf. Hiding and Multimedia Sig. Process. pp. 628–631. IEEE, Harbin, China (Aug 2008)
5. Ito, A., Suzuki, Y.: Multimedia Information Hiding Technologies and Methodologies for Controlling Data, chap. Advanced Information Hiding for G.711 Telephone Speech, pp. 129–163. IGI Global, Hershey, PA, USA (Oct 2012)
6. Karnjana, J., Nhien, P., Wang, S., Ngo, N.M., Unoki, M.: Comparative study on robustness of synchronization information embedded into an audio watermarked frame. In: IEICE Technical Report. No. EMM2015-83, IEICE, Yakushima, Kagoshima, Japan (March 2013)
7. Matsuoka, H.: Multimedia Information Hiding Technologies and Methodologies for Controlling Data, chap. Acoustic OFDM Technology and System, pp. 90–103. IGI Global, Hershey, PA, USA (Oct 2012)
8. Nishimura, A.: Audio data hiding that is robust with respect to aerial transmission and speech codecs. International journal of innovative computing, information & control 6(3(B)), 1389–1400 (March 2010)
9. Thiede, T., Treurniet, W.C., Bitto, R., Schmidmer, C., Sporer, T., Beerends, J.G., Colomes, C., Keyhl, M., Stoll, G., Brandenburg, K., Feiten, B.: PEAQ - the ITU standard for objective measurement of perceived audio quality. J. Audio Eng. Soc. pp. 3–29 (Jan-Feb 2000)

Tampering Detection in Speech Signals by Semi-Fragile Watermarking Based on Singular-Spectrum Analysis

Jessada Karnjana[1,2], Masashi Unoki[1],
Pakinee Aimmanee[2], and Chai Wutiwiwatchai[3]

[1] School of Information Science, Japan Advanced Institute of Science and Technology
1-1 Asahidai, Nomi, Ishikawa 923-1292 Japan
{jessada,unoki}@jaist.ac.jp
[2] Sirindhorn International Institute of Technology, Thammasat University
131 Moo 5, Tiwanont Rd., Bangkadi, Muang, Pathumthani, 12000 Thailand
pakinee@siit.tu.ac.th
[3] National Electronics and Computer Technology Center
112 Thailand Science Park, Phahonyothin Rd., Khlong Luang,
Pathumthani 12120 Thailand
chai.wutiwiwatchai@nectec.or.th

Abstract. To solve the problem of unauthorized modification in speech signals, this paper proposes a novel speech-tampering-detection scheme by using the semi-fragile watermarking based on the singular-spectrum analysis (SSA). The SSA is used to analyze the speech signals of which the singular spectra are extracted. The watermark (e.g., signature information) is embedded into those signals by modifying some parts of the singular spectra according to the watermark bit. By comparing the extracted watermark with the original one, the tampered segments of the speech signals are identified and located. The evaluation results show that the proposed scheme is fragile to several malicious attacks but robust against other signal-processing operations. It also satisfies the inaudibility criteria. The proposed scheme not only can locate the tampered locations, but it also can make a prediction about the tampering types and the tampering strength.

Keywords: singular-spectrum analysis, singular values, speech-tampering detection, semi-fragile watermarking, inaudible watermarking

1 Introduction

The problem of authentication in speech signals or digital work has become serious in our modern society. Advanced technologies, such as voice conversion or manipulation [1, 2], enable any people to tamper easily with the speech signals in the way the tampering is difficult to detect. The problem could be more serious if the speech signals contain important information, for instance, police evidence in the investigation signal. One possible and promising solution to solve this

J.-S. Pan et al. (eds.), *Advances in Intelligent Information Hiding and Multimedia Signal Processing*, Smart Innovation, Systems and Technologies 63,
DOI 10.1007/978-3-319-50209-0_17

problem is to embed signature information into the signals [3]. That is, speech or audio watermarking techniques with some specific requirements can be applied to solve the problem.

The speech tampering can be classified based on purposes into tampering with speech content (what a speaker said), with the speaker individuality (who the speaker is), and with non-linguistic information (how the speaker said) [4]. Thus, any signal-processing operations that change the speech content, the speaker individuality, or the non-linguistic information should be considered as the malicious attacks. To correctly detect the tampering, the watermark-detection scheme should be fragile to the malicious attacks but robust against other signal-processing operations. This property is called the *semi-fragility*. In addition, it is desirable to locate and to discriminate the unintentional modifications [5]. Besides the semi-fragility, the watermarking schemes that are used to detect the tampering must be inaudible. In short, the *inaudibility* and the semi-fragility are two mandatory requirements for the tampering-detection watermarking schemes.

In reported studies, Wu *et al.* proposed two speech fragile-watermarking schemes based on the exponential-scale odd/even modulation and based on the linear addition [6]. Both schemes used the distributions of the tamper assessment function to identify the content replacement. Even though the results were reasonable, their work considered only the tampering with the content. Yan *et al.* proposed the semi-fragile speech-watermarking scheme based on the quantization of the linear prediction parameters [7]. It was inaudible and robust against the constant amplitude scaling. Despite the fact that the parameters used in the scheme were selected by trail and error, it could correctly detect unwatermarked signals. Similar to the work of Wu *et al.*, their work focused only on the tampering with the content.

Unoki *et al.* proposed the scheme based on the characteristic of cochlear delay [4]. It could detect not only the tampering with the content but also that with the speaker individuality and the non-linguistic information. However, it was quite fragile to the G.726 speech companding. Also, the proposed scheme could not differentiate the different types of tampering nor the degrees of tampering. Wang *et al.* proposed the tampering-detection method based on the formant enhancement [8, 9]. Their results showed that the scheme was robust against the G.711 and G.726. Their work seemed to satisfy both inaudibility and semi-fragility. However, it was too robust to some malicious attacks, such as a single-echo addition, and too fragile to some non-malicious attacks, such as the shifting down in pitch by only 4%. Also, their work could not predict the types and degrees of tampering. Although there is no clear differentiation between the malicious attacks and the unintentional modifications, these published schemes showed that satisfying the semi-fragility had never been an easy task. In addition, to our knowledge, the challenge to determine the types and strength of tampering has never been reported anywhere before either.

Recently, we proposed the audio-watermarking scheme based on the singular-spectrum analysis (SSA) [10, 11], which deploys the singular value decomposition

(SVD) to extract the algebraic features called the singular values from the audio signals. The watermark is embedded into the signals by changing some singular values with respect to the watermark bit. Although our previously proposed scheme is robust, it is flexible enough to modify to serve different purposes.

This work aims to show that the SSA-based framework can be utilized to detect the tampering in the speech signals by adapting the watermark-embedding and extraction rules. With appropriate rules, it is possible to make the robust scheme semi-fragile. We propose a novel speech-tampering-detection method which can identify the tampered speech segments, the attack types, and also their strength.

2 Proposed Scheme

The proposed scheme is partly based on the robust and inaudible SSA-based audio-watermarking scheme, which is proposed by Karnjana et al. [10, 11], but is adapted to embed the watermark into speech signals. We used the basic SSA [12] to analyze the signals and to embed the watermark by modifying the singular spectra of the analyzed signals. We discovered that the SSA-based watermarking scheme could be made robust, fragile, or semi-fragile depending upon the modified parts of the singular spectrum. The following subsections describe how the method is done to achieve the semi-fragility.

2.1 Embedding Process

The embedding process consists of six following steps. First, the host signal is segmented into non-overlapping frames, where the number of frames is equal to the number of the watermark bits since one bit is embedded into one frame. Second, we construct the trajectory matrix X to represent each frame. Third, the SVD is performed on X to obtain the singular spectrum $\{\sqrt{\lambda_1}, \sqrt{\lambda_2}, ..., \sqrt{\lambda_q}\}$, where $\sqrt{\lambda_i}$ for $i = 1, ..., q$ are called the *singular values*, and they are sorted in descending order. An example of singular spectrum is shown in Fig. 1(a).

Fourth, the watermark bit is embedded by slightly modifying some singular values. We experimentally discovered that when the singular values of which its value is less than $0.0125 \times \max(\{\sqrt{\lambda_i}\})$ are modified, the sound quality of the modified signal is hardly perceptually-distinguishable from that of the unmodified one. We made use of this finding to hide the watermark inaudibly.

Let $\sqrt{\lambda_p}$ denote the largest singular value that is less than $0.0125 \times \max(\{\sqrt{\lambda_i}\})$. The modification rule is as follows:

$$\sqrt{\lambda_j} = \begin{cases} \left(\frac{\sqrt{\lambda_q} - \sqrt{\lambda_p}}{q - p} \right) \times (j - p) + \sqrt{\lambda_p}, & \text{if the watermark bit is 1,} \\ \\ 0, & \text{if the watermark bit is 0,} \end{cases} \qquad (1)$$

for $j = p, ..., q$. An example of the level of $0.0125 \times \max(\{\sqrt{\lambda_i}\})$ is illustrated by the red line in Fig. 1. The singular values under this red line are modified with

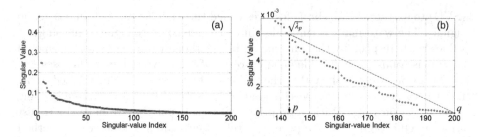

Fig. 1. Singular spectrum: (a) all singular spectrum obtained from one frame and (b) the zoomed-in singular values from indice 140 to 200. The red line indicates the level of $0.0125 \times \max(\{\sqrt{\lambda_i}\})$. The blue dashed line connects $\sqrt{\lambda_p}$ and $\sqrt{\lambda_q}$.

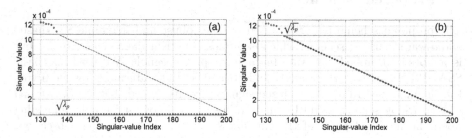

Fig. 2. Modified singular values: (a) all singular values under the red line are set to 0 when the watermark bit is 0 and (b) they are set to the corresponding values on the blue dashed line when the watermark bit is 1.

respect to the watermark bit. The blue dashed line in this figure connects two singular values $\sqrt{\lambda_p}$ and $\sqrt{\lambda_q}$. This line defines a new value of each singular value when the watermark bit is 1. Figure 2 shows two examples of the proposed singular-value modification rule.

Fifth, the modified matrix is reconstructed by SVD reversion, and then it is converted to the watermarked frame. Finally, the watermarked signal is produced by stacking these resulting frames.

2.2 Tampering-Detection Process

The tampering-detection process consists of five steps. The first three steps are the same as the first three steps of the embedding process. After these steps, we obtain the singular spectra which look very similar to those of the host signals, except that if we zoom in to investigate the singular values that have the value less than $\sqrt{\lambda_p}$, we will find that they look differently. For example, two singular spectra with different embedded watermark bits are shown in Fig. 3. We found that the extracted singular values are farther away from the blue dashed line in the case of embedding 0 than they are in the case of embedding 1. We utilize this observation in the remaining two steps to detect the tampering in the speech signals.

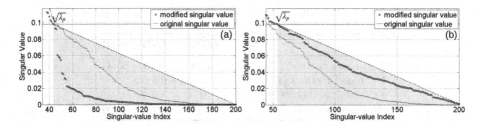

Fig. 3. Extracted singular values (black circles) in the tampering-detection process: (a) when the watermark bit is 0 and (b) when the watermark bit is 1. The red curves illustrate the original singular values as shown in Fig. 1(b).

We first calculate the ratio

$$r = \frac{1}{Q} \sum_{j=p}^{q} \sqrt{\lambda_j}, \qquad (2)$$

where Q is the area between the blue dashed line and the index axis, as shown by the shadowed triangles in Fig. 3. As r reaches 1, the extracted singular values get close to the blue dashed line. In this work, we use the ratio r to predict the watermark bit. That is, if $r < \frac{2}{3}$, the extracted bit is 0; otherwise, the extracted bit is 1. Finally, the extracted watermark is compared with the original watermark in order to determine the tampering and to identify the tampered location in the speech signals.

3   Evaluation

In this work, we evaluated the proposed scheme in three aspects: the sound quality of the watermarked signals, the semi-fragility, and the tampering identification. Twelve speech signals from the ATR databased (B set) [13], which are Japanese sentences uttered by six men and six women, were used. All have a sampling rate of 16 kHz, 16-bit quantization, and one channel. The watermark was embedded into the speech signals starting from the initial frame. The frame size was 400 samples or 25 ms. Thus, the embedding capacity was 40 bps. Three hundreds and twenty bits of the watermark were embedded into each signal in total. The duration for each signal was therefore 8 seconds.

The proposed scheme was compared with the least-significant-bit replacement (LSB) method [14]. We also compared our results with those of the cochlear-delay-based (CD) method [4] and the formant-enhancement-based (FE) method [8, 9] when they are available.

3.1   Sound-Quality Evaluation

Three distance measures were used to evaluate the sound quality: the perceptual evaluation of speech quality (PESQ), the log-spectral distance (LSD), and the

Table 1. Sound-quality evaluations: the proposed scheme vs. the other methods.

	ODG	LSD	SDR
LSB-based method [14]	4.49	0.19	65.35
CD-based method [4]	∼3.1-4.3	∼0.6-0.8	-
FE-based method [8, 9]	∼3.9	∼0.4	-
Proposed method	3.64	0.69	30.96

signal-to-distortion ratio (SDR). The PESQ measures the degradation of the watermarked signal compared with the host signal. It covers a scale, called the *objective difference grade* (ODG), from −0.5 (very annoying) to 4.5 (imperceptible) [15]. The LSD is a distance measure between the power spectra of the host and the watermarked signals [16]. The lower LSD is better in sound quality. The SDR is the power ratio of the amplitudes between the signal and the distortion, where the distortion is defined as the difference in amplitude between the original and the watermarked signals [10, 11]. The higher SDR is better.

Our evaluation criteria for the good sound-quality are as follows. The ODG should be greater than 3 (slightly annoying), the LSD should be less than 1 dB, and the SDR should be greater than 25 dB.

The results are shown in Table 1. All methods satisfied the sound-quality evaluation. The LSB-based method was the best in inaudibility, whereas the others were comparable.

3.2   Semi-Fragility Evaluation

The robustness/fragility was measured by the watermark extraction precision, which is represented by the bit-error rate (BER). The BER is defined as the number of error bits divided by the total number of embedded bits.

Our criterion for the robustness is that the BER should be less than 10%. The speech signals are said to be tampered when their BER is greater than 20%. If the BER is between 10% and 20%, the signals are presumably unintentionally modified or tampered with a low degree.

Ten signal-processing operations were performed on the watermarked signals: Gaussian-noise addition (AWGN) with average signal-to-noise ratio (SNR) of 15 and 40 dB, G.711 speech companding, G.726 companding, band-pass filtering (BPF) with 100-6000 Hz and −12 dB/Octave, MP3 compression with 128 kbps, MP4 compression with 96 kbps, pitch shifting (PSH) by ±4%, ±10%, and ±20%, single-echo addition with −6 dB and the delay time of 20 and 100 ms, replacing 1/3 and 1/2 of the watermarked signals with an unwatermarked segment, and ±4% speed changing (SCH). Note that some of these operations can be considered as the malicious attacks, whereas the others may not be. The discussion in detail will be provided in Section 4.

The BER results are shown in Table 2. The LSB-based method was fragile to almost all operations even though to the ones that are suppose not to be the malicious attacks. Therefore, it was too fragile. The rest including our proposed

Table 2. BER(%): the proposed scheme vs. the other methods.

	LSB-based method	CD-based method	FE-based method	Proposed method
No attack	0.00	~0-1	0.00	0.49
G.711	0.00	~4	0.00	0.49
G.726	51.77	~10-25	0.00	27.66
MP3	50.49	-	-	3.69
MP4	49.53	-	-	32.79
BPF	50.83	-	-	50.23
AWGN (15, 40 dB)	50.70, 49.53	-	~54	49.69, 24.53
PSH (-4%, -10%, -20%)	35.64, 35.33, 4.08	-	~31, -, -	10.58, 22.03, 47.83
PSH ($+4\%$, $+10\%$, $+20\%$)	34.42, 34.36, 38.03	-	-	12.44, 15.33, 20.47
Echo (20, 100 ms)	50.18, 51.34	-, ~50	-, ~5	15.76, 20.33
Replace (1/3, 1/2)	16.51, 24.97	-	~57, -	17.08, 25.78
SCH (-4%, $+4\%$)	49.47, 48.72	-	~20, -	47.00, 47.19

method were semi-fragile. Thus, they are good candidates for the tampering detection. We will discuss in Section 4 that our proposed scheme is robust against non-malicious operations but fragile to the malicious attacks, and in some senses it has advantages over the CD-based and FE-based methods. However, there were two drawbacks in our proposed scheme, i.e., it was quite fragile to the G.726 and the MP4 compression. The FE-based method was the only one that can survive the G.726 companding.

3.3 Tampering-Identification Evaluation

To intuitively realize the use of the SSA-based watermarking for tampering detection, the experiment was conducted as follows. First, a 96×21 binary image shown in Fig. 4(a) was chosen to be the watermark. Then, it was embed into a speech signal of 400×96×21 samples (50.4 seconds). Second, the watermark signal was divided into three equal segments, where the middle segment was subjected to the signal-processing operations described in Section 3.2. Last, the extracted watermark image was reconstructed and compared with the original one.

The results are shown in Fig. 4. It can be seen that the reconstructed images can be used not only to locate the tampered locations precisely but also to predict the types of attacks. On one hand, for example, adding the noise or high-frequency components causes the tampered area white, such as those in Figs. 4(d) and 4(p). On the other hand, removing the high-frequency components or changing to lower-pitched voice causes it black, such as those in Figs. 4(e) and 4(o). Figures 4(i) and 4(j) also show a unique appearance when the middle segments were replaced by the unwatermarked ones.

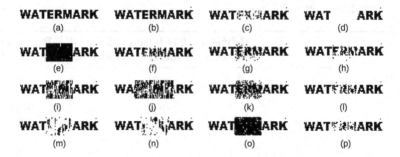

Fig. 4. Results of the tampering detection. Original image (a) and the reconstructed images after performing the following signal-processing operations: (b) G.711, (c) G.726, (d) AWGN (15 dB), (e) BPF, (f) Echo (100 ms), (g) PSH −4% , (h) PSH +4%, (i) Replace (1/3), (j) Replace (1/2), (k) PSH −10%, (l) PSH +10%, (m) SCH −4%, (n) SCH 4%, (o) PSH −20%, and (p) PSH +20%.

4   Discussion

In this section, we first discuss the proper evaluation of tampering-detection schemes and then the prediction of the tampering types and strength.

First, comparing the effectiveness among the tempering-detection schemes is not straightforward as comparing that among the robust watermarking ones. Because in the latter case we can reasonably conclude that the scheme with the lower BER is better in robustness. The conversion is not true for the tampering detection since the scheme with the greater BER is not necessary to be the better scheme. We first have to categorize the signal-processing operations into two groups: the malicious attacks and the non-malicious operations. The malicious attacks are the ones that (i) change the speech content, such as the replacement with unwatermarked segments, (ii) change the speaker individuality or the non-linguistic information of the speech signal, such as the band-pass filtering, the speed changing, and the pitch shifting, and (iii) change the speaker environment, such as the echo and noise additions. The other operations, such as the companding and compression, should not be considered as the malicious attacks. As shown in Table 3.2, besides the G.726 and MP4, our results satisfied this criterion, whereas the FE-based method failed to detect the echo addition. This problem is quite serious because, with such the echo, we can easily perceive the voice as if the speaker speaks in the different ambient environment. In addition, the FE-based method might produce a false alarm with the −4% PSH because the BER of 31% was too high when we consider the fact that the shifting down in pitch by only 4% does not perceivably change the speaker individuality. Note that the embedding capacity of the FE-based method is 10 times lower than that of our proposed scheme. That is, the number 31% can be increased to more than 40% if the FE-based method embeds the watermark without embedding 10 repetitions for each watermark bit.

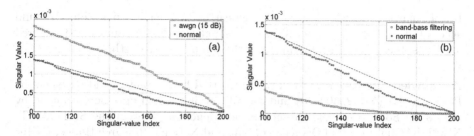

Fig. 5. Comparison of the singular values: (a) when adding the Gaussian-noise with the SNR of 15 dB and (b) when performing the band-pass filtering.

Second, compared with the CD-based and FE-based methods, the reconstructed images from our proposed scheme can be used to predict the tampering types and the tampering strength. Adding the noise or high-frequency components causes the singular values within the embedding region to increase their values, as shown in Fig. 5(a), hence the ratio r is shifted toward 1. Therefore, the tampered areas get white. Turning a normal voice to the higher-pitched one, in effect, is also similar to adding the higher-frequency components. Our results show that the higher the pitch, the more white areas can be found in the reconstructed images, as demonstrated by those in Figs. 4(h), 4(l), and 4(p). Thus, our proposed scheme can predict not only the types but also the strength of the tampering. On the contrary, changing the normal voice to the lower-pitched one or cutting its high-frequency components causes the singular values within the embedding region to decrease their values, as shown in Fig.5(b), hence the ratio r is shifted far away from 1. Therefore, the tampered areas get dark, as shown by those in Figs. 4(g), 4(k), and 4(o). We can see the strength of tampering in these figures as well. Even though the reconstructed images from the CD-based and FE-based methods can be used to locate the tampered locations correctly, but they can hardly be used to specify both types and strength of the tampering. For example, from the results of the FE-based method, the reconstructed images after the +4% SPC and the −4% PSH look very similar despite the fact that the previous one severely affects the speaker individuality, whereas the latter is hardly detectable by listening. Therefore, our proposed scheme is better than both CD-based and FE-based in this aspect.

The final remark on this work is that the tampering-detection process assumes to know the frame positions in advance. This assumption might not hold in some situations. Thus, besides improving the robustness against some companding and compression, solving this problem will be our future work.

5   Conclusion

This paper proposed the tampering-detection scheme based on the SSA-based watermarking technique. The watermark was embedded by modifying some less-significant singular values. We showed that the proposed scheme was fragile to the malicious attacks but robust against non-malicious operations. It could successfully detect the tampering as well as make predictions about the tampered locations, the types of tampering, and the tampering strength.

Acknowledgments. This work was supported by a Grant-in-Aid for Scientific Research (B) (No. 23300070) and an A3 foresight program made available by the Japan Society for the Promotion of Science, a grant in the SIIT-JAIST-NECTEC Dual Doctoral Degree Program, and National Research Budget (NRu) from Thailand.

References

1. Toda, T., Black, A. W., Tokuda, K.:Voice conversion based on maximum-likelihood estimation of spectral parameter trajectory. IEEE Transactions on Audio, Speech, and Language Processing, 15(8), 2222-2235. (2007)
2. STRAIGHT, http://www.wakayama-u.ac.jp/~kawahara/STRAIGHTadv/index_e.html [Accessed 11 June 2016]
3. Cox, I., Miller, M., Bloom, J., Fridrich, J., Kalker, T.: Digital watermarking and steganography. Morgan Kaufmann. (2007)
4. Unoki, M., Miyauchi, R.: Detection of tampering in speech signals with inaudible watermarking technique. In IIHMSP, pp. 118–121. (2012)
5. Cvejic, N. (ed.): Digital Audio Watermarking Techniques and Technologies: Applications and Benchmarks: Applications and Benchmarks. IGI Global. (2007)
6. Wu, C. P., Kuo, C. C.: Fragile speech watermarking for content integrity verification. In ISCAS, pp. 436–439. (2002)
7. Yan, B., Lu, Z. M., Sun, S. H., Pan, J. S. (2005, September). Speech authentication by semi-fragile watermarking. In KES, pp. 497–504. (2005)
8. Wang, S., Unoki, M., Kim, N. S.: Formant enhancement based speech watermarking for tampering detection. In INTERSPEECH, pp. 1366–1370. (2014)
9. Wang, S., Miyauchi, R., Unoki, M., Kim, N. S.: Tampering detection scheme for speech signals using formant enhancement based watermarking. J. Inform. Hiding Multimed. Sign. Process, 6, 1264-1283. (2015)
10. Karnjana, J., Unoki, M., Aimmanee, P., Wutiwiwatchai, C.: An Audio Watermarking Scheme Based on Singular-Spectrum Analysis. LNCS, 9023, 145–159. (2014)
11. Karnjana, J., Aimmanee, P., Unoki, M., Wutiwiwatchai, C.: An audio watermarking scheme based on automatic parameterized singular-spectrum analysis using differential evolution. In APSIPA, pp. 543–551. (2015)
12. Golyandina, N., Nekrutkin, V., Zhigljavsky, A.: Analysis of Time Series Structure: SSA and related techniques. Chapman and Hall/CRC, United States (2001)
13. Takeda, K., Sagisaka, Y., Katagiri, S., Abe, M., Kuwabara, H.: Speech database users manual. ATR Interpreting Telephony Research Laboratories, TR-I-0028. (1988)
14. Bassia, P., Pitas, I.: Robust audio watermarking in the time domain. In EUSIPCO, pp. 25–28. (1998)
15. ITU Recommendation P.862: Perceptual evaluation of speech quality, http://www.itu.int/rec/T-REC-P.862/en [Accessed 11 June 2016]
16. Log-spectral distance (LSD), https://en.wikipedia.org/wiki/Log-spectral_distance [Accessed 11 June 2016]

Copyright Protection Method based on the Main Feature of Digital Images

Xinyi Wang[1], Zhenghong Yang[1], Shaozhang Niu[2]

[1]School of Science, China Agricultural University, Beijing, China
[2] Beijing Key Lab of Intelligent Telecommunication Software and Multimedia, Beijing University of Posts and Telecommunications, Beijing, China
[1] xawangxy@163.com, [1]yangjohn@cau.edu.cn,[2] szniu@bupt.edu.cn

Abstract. With the rapid development of information technology, images can be easily copied, modified, and re-released. This paper proposes a copyright protection method based on the main feature of digital images by using SIFT algorithm. It aims to prevent the image which main feature information is stored in database from being abused. The innovation point is to store feature information instead of images with much redundant information, which can accelerate matching. It can be judged whether a picture infringes copyright by extracting its main feature information, then comparing with those in database. The experiment results show that this method can resist tampering attack such as JPEG compression, noise, and geometric distortion.

Keywords: digital image copyright protection; SIFT algorithm; main feature information; image tampering; search matching.

1   Introduction

The Internet has been widely used in every corner of the world. More and more people can use this extensive platform to publish, transmit and display their digital image. However, the image can be easily unauthorized copied, modified and redistributed. The copyright owner is facing the unprecedented threat of piracy. Their intellectual property rights cannot be effectively protected. Sometimes they even suffer huge economic losses. Therefore, the copyright protection of digital image has become the significant research in the field of information security.

This paper extracts the discriminative feature information of an image to represent it, and stores their digital information into database. In this way the information database involves the feature information of pictures that need to be protected [1]. Therefore, accurate extraction and description of the main image feature is the key point in the process of protecting copyright in this paper.

The core issue of image feature description is invariance (robustness) and the ability to distinguish.   Among many feature descriptors, SIFT feature descriptor is most widely used in detecting tampered image. SIFT algorithm is a kind of feature extraction algorithm proposed by Lowe in 1999[2], refined and summarized in 2004 [3]. It has advantage in robustness and discrimination. The algorithm is to find some

J.-S. Pan et al. (eds.), *Advances in Intelligent Information Hiding and Multimedia*
Signal Processing, Smart Innovation, Systems and Technologies 63,
DOI 10.1007/978-3-319-50209-0_18

"stable point" in the image. These points are robust to perspective, light and noise disturbance.

2   Feature Information Database

This method applies SIFT algorithm to extract the main features of images to protect the copyright. Instead of store images, our method stores the main feature information of each image in database. It can save occupied space and accelerate matching identification. In order to offer more convenient, we build the user interface for extracting, storing and matching the main information of image.

3   Copyright Protection Method Based on the Main Feature of Digital Images

3.1   SIFT Algorithm

The main steps of the SIFT algorithm are as follows: (a) Generate scale-space; (b) Scale-space extreme detection; (c) Feature points localization; (d) Specified direction parameters for each feature point; (e) Generate the feature point descriptor .
(a) Generate scale-space
The scale-space theory [4] is designed to simulate the multi-scale features of the image data, and the Gauss convolution kernel has been proved to be the only one that can realize the scale transformation. It can define the scale-space $L(x,y,\sigma)$ as a variable size Gauss function $G(x,y,\sigma)$ with the original image $I(x,y)$ to do convolution:

$$L(x, y, \sigma) = G(x, y, \sigma) \times I(x, y). \tag{1}$$

Subtract two pictures which in the adjacent tower to get the Gauss difference pyramid $I(x,y)$:

$$D(x, y, \sigma) = L(x, y, k\sigma) - L(x, y, \sigma). \tag{2}$$

(b) Scale-space extreme detection;
Find the extreme points in scale-space. If the middle of the detection point is the minimum or the maximum point compared with its same sizes' 8 adjacent points and the adjacent sizes' 18 points. This point can be considered to be the feature point of the image at this scale.
(c) Feature points localization
Using the Taylor expansion to specified the precise location of extreme points in DoG (Difference of Gaussian) space.
(d) Specified direction parameters for each feature point

The magnitude and direction calculations for the gradient are done for every pixel in a neighboring region around the feature point in the Gaussian-blurred image L. An orientation histogram with 36 bins is formed, with each bin covering 10 degrees. Each sample in the neighboring window added to a histogram bin is weighted by its gradient magnitude and by a Gaussian-weighted circular window with δ that is 1.5 times that of the scale of the feature point. The peaks in this histogram correspond to dominant orientations. Once the histogram is filled, the orientations corresponding to the highest peak and local peaks that are within 80% of the highest peaks are assigned to the feature point. In the case of multiple orientations being assigned, an additional feature point is created having the same location and scale as the original feature point for each additional orientation.

(e) Generate the feature point descriptors

Previous steps found feature point locations at particular scales and assigned orientations to them. This ensured invariance to image location, scale and rotation. Now we want to compute a descriptor vector for each feature point such that the descriptor is highly distinctive and partially invariant to the remaining variations such as illumination, etc. This step is performed on the image closest in scale to the feature point's scale.

First a set of orientation histograms is created on 4×4 pixel neighborhoods with 8 bins each. These histograms are computed from magnitude and orientation values of samples in a 16×16 region around the feature point such that each histogram contains samples from a 4×4 sub-region of the original neighborhood region. The descriptor then becomes a vector of all the values of these histograms. Since there are 16 histograms each with 8 bins the vector has 128 elements. This vector is then normalized to unit length in order to enhance invariance to affine changes in illumination.

3.2   Feature Matching

The feature points are assumed to be $S=\{s_1, s_2, s_3, \ldots s_n\}$ and each item in S_i represents 128-dimension vector of corresponding points. In high dimensional space, the similarity of feature vector is usually measured by the Euclidean distance of feature vector. That is to say for one feature $s_i=\{f_1, f_2, f_3, \ldots f_n\}$, calculate it and all the rest of the features of the Euclidean distance. The 2NN (Neighbor 2-Nearest) method is used in matching. The threshold is set to be 0.6 in this paper. If the result less than 0.6, these feature points are considered matched. Then we use RANSAC [5-6] (Random Sample Consensus) to eliminate the wrong match points.

4   Experimental Results and Analysis

Find 100 different types of images from the Internet for the experiment. We extract their feature information and build the image feature database. The image match threshold is set to be adaptive [7] in order to find the same or similar feature points between two pictures [8]. If the number of image feature points in database more than 1000, select threshold to be 0.005; If it is between 500-1000, then select threshold to be

0.02; if it is less than 500, then select threshold to be 0.1. If the test results exceed the threshold, that means, their main feature information is matched. It can be preliminarily considered that this picture infringes copyright. (The experiment is done under the environment of Windows 7 and Matlab 2015a in this paper).

4.1 Anti-tampering Test

4.1.1 Partial Tampering on the Same Image

Take image Lena (299× 295, JPEG format) for example. This image adds sunglasses and scarf for tampering (Fig.1(a)).The user interface can detect whether there is a matching image in database automatically.

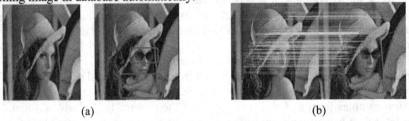

(a)   (b)

Fig. 1. Tampering on the same image. (a) The picture of Lena and its part tampering picture; (b) The results of feature matching.

The similar image from the database can be easily found automatically. The number of the matching feature points is 385(Fig.1(b)).

4.1.2 Partial Tampering On the Different Image

Take image (500× 610, JPEG format, Fig.2(a)) for example. Its main part is preserved but changed its background environment (Fig.2(b)). We aim to find the original picture from 100 different kinds of image in database.

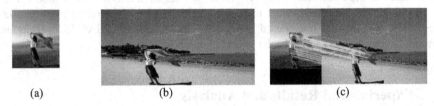

(a)   (b)   (c)

Fig. 2. Tampering between the different image: (a) Original image; (b)Tampered image (different background); (c)The result of feature matching from database.

Experimental results and analysis: This method can also solve the problem of copyright infringement in such situation. The number of the matching feature points is 230 (Fig.2(c)).

4.1.3 Large-scale Tampering

Take Lena's part tampering image (299×295, JPEG format) for example. We continue tamper it in large scale. The test result shows that the original image from database can't be detected in this case. This is mainly because the main information in this picture has been doctored beyond recognition. There is no need to detect out the original image in such situation.

4.2 Anti-JPEG Compression Test

Table 1. The test results of Anti-JPEG compression.

JPEG Compression	30%	50%	70%	90%
Test Result	√	√	√	√

100 images in database are tampered with different percentages of JPEG compression by Photoshop for this experiment. These data illustrate its resistance effect on JPEG compression is very good. All original images can be detected. (' √ ' indicates all image can be detected. This is the same meaning in the following experimental results).

4.3 Anti-noise Test

Table 2. The test results of Anti-noise.

Gaussian Noise (percent number)	15%	30%	50%	70%
Test result	√	√	83%	53%
Gaussian Blur (radius)	3.0	5.0	8.0	10.0
Test result	√	92%	64%	36%

100 images in database are tampered with different degrees of Gaussian noise and different radius of Gaussian blur for this experiment. Since different image has different pixels and content, so their resistance ability for noise and blur are also different. Experimental results show that this method can resist noise tamper to some extent.

4.4   Anti-geometric Distortion Test

4.4.1   Rotation

Table 3.   The test results of rotation.

Anticlockwise	10°	30°	50°	70°	90°
Test result	√	√	√	√	√

100 images in database are tampered with different degrees of rotation for this experiment. The experimental results show that this method can resist rotate tamper in general. In this experiment all original image are detected out.

4.4.2   Scaling

Table 4.   The test results of scaling.

Test result	10%	30%	50%	70%
Enlarge	√	√	√	√
reduce	√	√	86%	62%

100 images in database are tampered with different degrees of scaling for this experiment. The experimental results show that we can detect all enlarge tamper image, and can detect reduce tamper to some extent.

4.4.3   Tampering and Scaling

In this experiment, the main feature size of Fig. 2(a) becomes 70%. Our aim is to detect out the original picture from database.

Fig. 3. The test result of this experiment.

Experimental results and analysis: if we enlarge the main feature of Fig. 2(a), this method can detect out the original image in most case. Because of the threshold setting, the original image can't be detected out when the main feature reduced to 70% or less. Therefore, this method can resist geometric distortion tamper in some extent.

5 Conclusion

This paper proposes a copyright protection method based on the main feature of digital images by using SIFT algorithm. We store the main feature information of images instead of storing the entire image. It can save a lot of memory space and accelerate matching speed by this method. Since the SIFT algorithm has good robustness, so this method can resist tampering such as JPEG compression, noise, geometric distortion to some extent. Therefore the copyright of the image can be effectively protected.

Acknowledgment

This work is supported by National Natural Science Foundation of China (No. 61370195，U1536121).

References

1. Podilchuk, C., Zeng, W.: Image-adaptive Watermarking Using Visual Models. Selected Areas in Communications IEEE Journal on. vol. 16, no. 4, pp. 525--539 (1998)
2. Lowe, D.G.: Object Recognition from Local Scale-invariant Features. The Proceedings of the Seventh IEEE International Conference on Computer Vision. IEEE. vol. 2, pp. 1150--1157 (1999)
3. Lowe, D.G.: Distinctive Image Features from Scale-Invariant Keypoints. International Journal of Computer Vision. vol. 60, no. 2, pp:91--110 (2004)
4. Bas, P., Chassery, J.M., Macq, B.: Geometrically Invariant Watermarking Using Feature Points. IEEE Transactions on Image Processing. vol. 11, no. 9, pp.1014--1028 (2002)
5. Mair, E., Hager, G.D., Burschka. D., et al.: Adaptive and Generic Corner Detection Based on the Accelerated Segment Test. LNCS, vol. 6312, no. PART 2, pp. 183--196. Computer Vision - ECCV 2010. European Conference on Computer Vision. Heraklion, Crete, Greece, September 5-11, Proceedings (2010)
6. Lu, C.S., Liao, H.Y.M.: Multipurpose Watermarking for Image Authentication and Protection. IEEE Transactions on Image Processing. vol. 10, no.10, pp. 1579--1592 (2001)
7. Fischler, M.A., Bolles, R.C.: Random Sample Consensus: a Paradigm for Model Fitting with Application to Image Analysis and Automated Cartography. Communications of the ACM. vol. 24, no. 6, pp. 381--395 (1981)
8. Torr, P.H.S., Zisserman. A.: MLESAC: A New Robust Estimator with Application to Estimating Image Geometry. Computer Vision and Image Understanding. vol. 78, no. 1, pp.138--156 (2010)

Part II
Audio and Speech Signal Processing

Cross-Similarity Measurement of Music Sections: A Framework for Large-scale Cover Song Identification

Kang Cai, Deshun Yang, and Xiaoou Chen

Institute of Computer Science & Technology,
Peking University. Beijing, China
{caikang,yangdeshun,chenxiaoou}@pku.edu.cn

Abstract. For large-scale cover song identification, most previous works take a single feature vector as the representation of a song. Although this approach ensures structure invariance, it may cause overcorrection since it totally neglects the structure feature of the song. To address this problem, we put forward a novel framework for large-scale cover song identification based on music structure segmentation, aiming at matching the irrelevant sections and ignoring the irrelevant ones. In our implementation, we apply the average and weighted average methods to integrating similarities of section pairs. We evaluate the proposed framework based on three representative previous methods, including 2D Fourier magnitude coefficients, chord profiles, and cognition-inspired descriptors. The experimental results show that the all the three methods in our framework significantly outperform those in their original works.

Keywords: cross-similarity measurement, music structure segmentation, large-scale cover song identification

1 Introduction

It is urgent to develop effective music management due to the rapid growth of music in forms of video and audio on the Internet. Audio copy identification is a well-explored field and methods like audio fingerprinting achieve satisfying performances [1]. For the actual applications such as Youtube and Spotify, there exist tremendous songs covered by common people and professional singers. Compared with the original works, these covers vary in tempo, key, structure and so on [2]. Given this, cover song identification is more difficult than copy identification and most methods of audio copy identification cannot work well on the cover song identification.

Cover song identification has been a subtask of the Music Information Retrieval Evaluation eXchange[1] (MIREX) since 2006. The cover song dataset of MIREX contains 1000 songs. Methods using the feature series as the representation of a song achieve high mean average precision (AP) on the MIREX dataset.

---

[1] http://www.music-ir.org/mirex/wiki/MIREX_HOME

© Springer International Publishing AG 2017

J.-S. Pan et al. (eds.), *Advances in Intelligent Information Hiding and Multimedia Signal Processing*, Smart Innovation, Systems and Technologies 63,
DOI 10.1007/978-3-319-50209-0_19

In particular, the state-of-the-art method Q_{max} [3] achieves mean AP of 0.66. However, these methods cannot be directly applied to large-scale cover song identification due to the high computational complexity.

Large-scale cover song identification has a higher demand for efficiency, which means the representation of a song should be a fixed-and-low-dimension feature or even in the form of hash code instead of a feature time series. Thierry et al. [4] first published a large cover song dataset containing 18,196 songs in 2011 and applied the 'jumpcode' method to identifying the cover song and achieved the average rank of 4,361 from 12,960 on SHSD training dataset. After that, Ellis et al. [5] used 2D Fourier magnitude coefficient (2D-FMC) as the feature and proved that 2D-FMC is the state-of-the-art unsupervised representation in large-scale cover song identification. Humphrey et al. [6] put forward a supervised method by applying linear discriminant analysis (LDA) to 2D-FMC and this method achieves a higher mean AP. Apart from 2D-FMC, features including chord profiles [7], and cognition-inspired descriptors (CID) [8] were also proposed as the representations, but none of them outperforms the 2D-FMC.

Regarding representations of the songs applied in large-scale cover song identification, including 2D-FMC, chord profiles, CID and so on, they all neglect the local similarity between two songs. However, the local similarity is an important factor for cover song identification. First, cross-recurrent plot (CRP) significantly outperforms (dynamic time wrapping) DTW [3], which means the length of a local matching segment is a better metric than the similarity of two entire songs to identify covers. Second, when humans are detecting a cover, a certain melody and a high similarity of timbre in very short snippet can help to do the job [2]. The second fact means that a matching of salient segment pair determines the song pair to some extent. Based on above two facts, we put forward cross-similarity measurement of sections as a framework for cover song identification, which is in line with both previous experimental results and music cognition. The main idea is to segment a song according to its structure, and then determine the similarity of a song pair by integrating the cross-similarities of sections between two songs of the song pair. The proposed framework can be easily applied to the methods using a single feature vector to represent a song, which covers most of the previous methods used in large-scale cover song identification.

2 Our proposed framework

We proposed the framework based on music structure segmentation containing the following four sequential steps:

A. Music structure segmentation.
B. Feature representation of each section.
C. Similarity measurement of each section pair.
D. Integrating cross-similarities of sections.

The overall diagram is shown in Fig. 1. First, we apply music structure segmentation algorithm to segment each song into several sections, of which

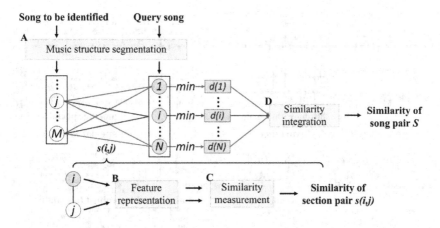

Fig. 1. Diagram of proposed framework. The execution order is A-B-C-D.

the time length is usually larger than 16 seconds [9]. Then we measure the cross-similarities between sections of two songs since the matching section pair cannot be directly determined. (Note that in this work, we calculate the distance or dissimilarity instead of similarity in actual practice.) To measure the cross-similarities, we extract the fixed-and-low-dimension feature vector as the feature representation of each section and then calculate the distances between feature vectors. Then comes to the vital step of our framework: how to integrate cross-similarities of sections into the similarity value of the song pair? A certain number of sections in the reference song means a fair comparison and we choose the query song as the reference song, i.e., for each section in the query song, we search the candidate song for the most matching section. In details, for each section i in the query song, we calculate the distances between section i and each the sections in the candidate song respectively and then keep the minimum distance as the dissimilarity label $d(i)$ for the section i. This process is formulated by

$$d(i) = \min(s(i, j)), j = 1, 2, ..., M \tag{1}$$

where $s(i, j)$ is the dissimilarity of section i in the query song and section j in the candidate song and M is the number of sections of the candidate song. Last, we integrate the $d(i), i = 1, 2, ..., N$ into a single value to represent the dissimilarity of song pair.

Our proposed framework has four advantages over previous methods using a single vector as the representation of a song, which is shown in Fig. 2. As Fig. 2(a) shows, our proposed framework can handle with the case of shifting key within song, but previous methods cannot. The framework also ignores the sections of the covers which do not have a matching section pair as shown in Fig. 2(b). Besides, each section in the query song only matches the most matching section and cannot be affected by other sections as Fig. 2(c) presents. Last, the framework also has a strong robust to the scalability of section by taking Fig. 2(d) as a example.

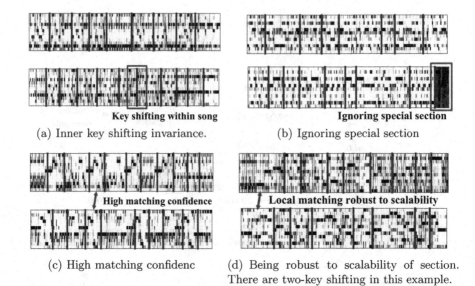

(a) Inner key shifting invariance.

(b) Ignoring special section

(c) High matching confidenc

(d) Being robust to scalability of section. There are two-key shifting in this example.

Fig. 2. Illustration of the advantages of our proposed framework. Each upper and lower picture pair in 2(a), 2(b), 2(c), 2(d) shows the beat-tracking chroma of a cover pair. Blue lines are the segmentation points for the songs. The segmentation algorithm used here is the structure features method proposed by [10].

3 Implementation of evaluated systems based on our proposed framework

In order to evaluate our framework, we implement several systems based on our framework. The details of implementation are as follows:

Chroma extraction. First of all, we use the chroma as the frame-level feature in our implementation. Although SHSD does not provide the audio files, the Echo Nest analyze API[2] can return the beat-tracking chroma by given a specific song. We tried the method of revising percussive beat by following the method of [11], but we finally drop it since it does not work well on some feature representations.

A. Music structure segmentation. Music structure segmentation methods can be summarized into three conceptually different approaches, which we refer to as repetition-based, novelty-based, and homogeneity-based approaches [12]. Besides, some methods use a mixed approach by combining above three approaches. We implement our proposed framework by applying two segmentation methods: the state-of-the-art mixed approach using the so-called structure feature [10] and a classic novelty-based method [13]. In this work, we name the first method as **'SF'** method and the second method as **'Novelty'** method for simplification. Note that the original low-level feature used in 'Novelty' method is

---

[2] http://developer.echonest.com/

Mel-frequency cepstral coefficients (MFCC) for general audio, we replace it with chroma to address the properties of the musical signals recommended by [14]. In this work, we use the beat-tracking chroma returned by the Echo Nest analyze API as the low-level features for both 'SF' and 'Novelty' methods. We set the parameters of 'SF' and 'Novelty' methods by reference to [10] and [13] respectively. We also evaluate the segmentation method by simply segmenting the song into sections with equal time lengths and we name it **'Naive'** method, which is used to further verify the effectiveness of 'SF' and 'Novelty' methods. We set the number of sections segmented by using 'Naive' method as 10 equalling to the average number of sections segmented by 'SF' and 'Novelty' methods for a fair comparison.

B. Feature representation. We apply three representative feature representations of large-scale cover song identification to the framework and compare their performances in our framework with the performances in the original works. The feature representations include **2D-FMC** [5], **chord profiles** [7], and **CID** [8]. 2D-FMC is the state-of-the-art fixed-dimensional representation although what it really means is unclear. Chord profiles method calculates the histogram of chords, which is used as the first step of the system in [7]. CID represents the pitch and pitch bigram from the respective of music cognition. It should be noted that 2D-FMC has no need to conduct the key transposition, while the chord profiles and CID need key transposition. In our work, we perform a circular permutation of a queried song chord profile 12 times instead of key estimation as their corresponding original works did for a fair comparison. We also set the parameters of three feature representations all by reference to their original works.

C. Similarity measurement of each section pair. For a fair comparison, we use the same distance measurement used in its original work for each feature representation: Euclidean distance for 2D-FMC by following [5], L_1 distance for chord profiles by following [7], and cosine distance for CID by following [8].

D. Integrating cross-similarities of sections. We apply two methods at the final step: the average method and weighted average method. The average S_a means that we regard each section of the query song as equal weight regardless of its length and can be represented as

$$S_a = \frac{1}{N} \sum_{i=1}^{N} d(i), \tag{2}$$

where $d(i)$ is the dissimilarity label of the section i in the query song given in Equation (1). The weighted average method attaches the time length of the section to each section as the weight and then calculate the weighted average value S_w given by

$$S_w = \frac{1}{\sum_{i=1}^{N} l(i)^\alpha} \sum_{i=1}^{N} (l(i)^\alpha * d(i)), \tag{3}$$

where $l(i)$ is the time length of section i in the query song and α is the parameter to be tuned in the experiment part. We use the number of beats as the time length.

We also tried the following three methods but none of them shows a better performance:

– Minimum value of d.
– Median value of d.
– Integrating similar sections within a song. We apply the clustering method by using the Q_{max} measure proposed in [10] to integrating similar sections.

4 Evaluations of the framework

We evaluate our framework on SecondHandSongs dataset[3] (SHSD), which is the largest cover song dataset ever released for academic research. SHSD lists 12,960 training cover songs and 5,236 testing cover songs, all part of the Million Song Dataset (MSD). SHSD contains only western popular songs and brings greater challenges since the cover of popular songs can differ much from the original work. We also use the subset of SHSD training set created by [4]. The subset consists of 1500 tracks and forms a 500 binary task, i.e., for each query song, our goal is to find its cover from two candidate songs.

We first tune the parameters on the dataset of 500 binary task and then evaluate the proposed framework on SHSD training set for a sanity check. The results on 500 binary task are given in Table 1.

Table 1. Accuracy on 500 binary tasks

Approaches	2D-FMC	Chord	CID
Original work	**82.2%**	79.4%	73.2%
SF-S_a	81.8%	80.2%	**77.2%**
SF-S_w	82.0%	**81.6%**	**77.2%**
Novelty-S_a	81.8%	80.6%	74.8%
Novelty-S_w	81.6%	81.0%	75.6%
Naive	78.6%	78.8%	76.4%

Table 2. mean AP on SHSD training set

Approaches	2D-FMC	Chord	CID
Original work	0.0948	0.0489	0.0701
SF-S_a	**0.1214**	**0.0704**	**0.0974**
SF-S_w	0.1172	0.0591	0.0838
Novelty-S_a	0.1133	0.0631	0.0894
Novelty-S_w	0.1011	0.0523	0.0812
Naive	0.0537	0.0505	0.0743

As Table 1 shows, both SF and Novelty methods outperform the original methods for chord profiles and CID, while Naive method only out outperforms the original work for CID. By varying the parameter α from 0.1 to 3.0, we find 1.8 for SF-S_w and 0.6 for Novelty-S_w are the best choices on the 500 binary task. Under this parameter setting, weighted average methods basically outperform average methods for both SF and Novelty methods. We also apply above parameter settings in the following experiment. The results on SHSD training set are shown in Table 2. We observe that both SF and Novelty methods achieve good performances for all the three feature representations. In particular, SF-S_a slightly outperform Novelty method. Based on these observations, we conclude

---

[3] http://labrosa.ee.columbia.edu/millionsong/

that a good music structure segmentation algorithm helps to improve the performance of our framework. On this larger dataset, however, weighted average methods become not that effective compared with the average methods. The reason may be that the value of α overfits the 500 binary task. We conclude that average method is more stable than the weighted average method for our framework, although the performance of weighted average methods may be improved by further adjusting the parameter α. This conclusion also means sections with various time lengths are all important to identify the cover song. In addition, It should be noted that SF-S_a of 2D-FMC achieves higher mean AP of 0.1214 than all the published unsupervised methods on SHSD training set.

Finally, we report the performances of SF-S_a on the SHSD testing set as presented in Table 3. As Table 3 shows, SF-S_a also significantly outperform the original methods on the testing set.

Table 3. mean AP on SHSD testing set containing 5,854 songs

Approaches	2D-FMC	Chord profiles	CID
Original work	0.1065	0.0661	0.0858
SF-S_a	**0.1349**	**0.0993**	**0.1145**

For a fine-grained observation, we present the recall of query-cover pairs by varying the position of cutoff in Fig. 3. The total number of query-cover pairs is 17772. As Fig. 3 shows, our proposed framework has a significantly higher recall at top rankings for each feature representation, which is a good property for practical applications.

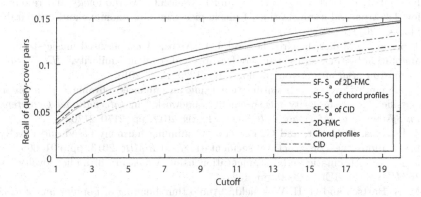

Fig. 3. The recall of query-cover pairs on SHSD testing set

5   Conclusion

In this work, we put forward a novel framework for large-scale cover song identification by measuring cross-similarities of music sections. By evaluating our
framework based on three representative previous methods, it shows that our
framework significantly improves their performances on SHSD dataset and also
confirms that local similarity is more proper than global similarity for cover song
identification. Most importantly, our proposed framework can be easily applied
to any methods using a single feature vector to represent a song, which covers
most of the previous methods used in large-scale cover song identification.

References

1. R. Typke, F. Wiering, R. C. Veltkamp *et al.*, "A survey of music information
   retrieval systems." in *ISMIR*, 2005, pp. 153–160.
2. J. Serra, E. Gómez, and P. Herrera, "Audio cover song identification and similarity: background, approaches, evaluation, and beyond," in *Advances in Music
   Information Retrieval.* Springer, 2010, pp. 307–332.
3. J. Serra, E. Gómez, P. Herrera, and X. Serra, "Chroma binary similarity and local
   alignment applied to cover song identification," *ICASSP*, 2008.
4. T. Bertin-Mahieux, D. P. Ellis, B. Whitman, and P. Lamere, "The million song
   dataset," in *ISMIR*. University of Miami, 2011, pp. 591–596.
5. D. P. Ellis and B.-M. Thierry, "Large-scale cover song recognition using the 2d
   fourier transform magnitude," in *ISMIR*, 2012, pp. 241–246.
6. E. J. Humphrey, O. Nieto, and J. P. Bello, "Data driven and discriminative projections for large-scale cover song identification." in *ISMIR*, 2013, pp. 149–154.
7. M. Khadkevich and M. Omologo, "Large-scale cover song identification using chord
   profiles." in *ISMIR*, 2013, pp. 233–238.
8. J. van Balen, D. Bountouridis, F. Wiering, R. C. Veltkamp *et al.*, "Cognition-
   inspired descriptors for scalable cover song retrieval," in *ISMIR*, 2014.
9. F. Bimbot, E. Deruty, G. Sargent, and E. Vincent, "Methodology and resources
   for the structural segmentation of music pieces into autonomous and comparable
   blocks," 2011.
10. J. Serra, M. Muller, P. Grosche, and J. L. Arcos, "Unsupervised music structure
    annotation by time series structure features and segment similarity," *IEEE Trans-
    actions on Multimedia*, vol. 16, no. 5, pp. 1229–1240, 2014.
11. X. Chuan, "Cover song identification using an enhanced chroma over a binary
    classifier based similarity measurement framework," in *International Conference
    on Systems and Informatics (ICSAI)*. IEEE, 2012, pp. 2170–2176.
12. J. Pauwels, F. Kaiser, and G. Peeters, "Combining harmony-based and novelty-
    based approaches for structural segmentation." in *ISMIR*, 2013, pp. 601–606.
13. J. Foote, "Automatic audio segmentation using a measure of audio novelty," in
    ICME, vol. 1. IEEE, 2000, pp. 452–455.
14. M. A. Bartsch and G. H. Wakefield, "Audio thumbnailing of popular music using
    chroma-based representations," *IEEE Transactions on Multimedia*, vol. 7, no. 1,
    pp. 96–104, 2005.

A Study on Tailor-Made Speech Synthesis Based on Deep Neural Networks

Shuhei Yamada, Takashi Nose, and Akinori Ito

Graduate School of Engineering, Tohoku University
Aramaki Aza Aoba 6-6-05, Aoba-ku, Sendai-shi, Miyagi, 980-8579 Japan
{s.yamada@dc,tnose@m,aito@spcom.ecei}.tohoku.ac.jp

Abstract. We propose "tailor-made speech synthesis," the speech synthesis technique which enables users to control the synthetic speech naturally and intuitively. As a first step to realizing tailor-made speech synthesis, we introduce F0 context into speaker model training of speech synthesis based on deep neural networks (DNNs). F0 context represents relative log F0 at the mora or the accent-phrase level of training data. It allows users to control the F0 of synthetic speech steplessly on the contrary to conventional F0 context in HMM-based technique. Experiments showed that F0 context was effective to control the F0 because the F0 of synthetic voice followed the value of F0 context.

Keywords: DNN-based speech synthesis, Prosody control, F0 context, Context label, Model training, Unsupervised labeling

1 Introduction

Recently, speech synthesizers are used in various situations such as smartphone applications such as Google Now [2] and Siri [1], announcement in public transport, and narration of video works. In near future, there will be needs to synthesize speech that represents an appropriate intention, especially in narration and games. We propose "tailor-made speech synthesis," a speech synthesis technique which enables users to control the synthetic speech naturally and intuitively. We aim to make the synthetic speech that is suitable for various situations.

Nose et al. [7] proposed a technique to control the speaking style with style vectors and multiple-regression HSMM in HMM-based speech synthesis. Also, Watts et al. [8] proposed a similar technique in DNN-based speech synthesis. With these techniques, we can control the speaking styles intuitively. However, we cannot synthesize speech with the speaking style which is not included in the corpus.

There are some studies on arbitrary prosody control of speaking style in HMM-based speech synthesis. Nishigaki et al. [6] proposed a technique to modify the prosody by replacing F0 of synthetic speech to that of user's speech. However, there are some disadvantages in this technique. The degradation of naturalness may be caused by the mismatch between F0 and spectrum parameters when users designate low F0 where predicted F0 is high and vice versa. Moreover, we cannot

© Springer International Publishing AG 2017   159
J.-S. Pan et al. (eds.), *Advances in Intelligent Information Hiding and Multimedia*
Signal Processing, Smart Innovation, Systems and Technologies 63,
DOI 10.1007/978-3-319-50209-0_20

modify only F0 or phoneme duration and users always have to designate the F0 by their voice to synthesize speech. On the other hand, in the technique proposed by Maeno et al. [5], first we synthesize speech without the F0 information as typical text-to-speech, then we can control the F0 of synthetic speech if we need. In particular, at first, we synthesize speech by typical HMM-based speech synthesis in a closed condition, then calculate the difference between log F0 of natural speech and the synthetic speech. We calculate the mean value of difference for each accent phrase and make the context by classifying the value into high, neutral and low. The context is named F0 context. Finally, we train models again with typical context and the F0 context as explicit information of the relative F0. This allows users to control the F0 of synthetic speech relatively.

As a first step to developing tailor-made speech synthesis, we focus on controlling the F0 of synthetic speech at the accent-phrase and the mora level. Considering usability of controlling the F0 in tailor-made speech synthesis, it is desirable that we do not always have to control the F0 and can control the F0 if we need as the technique by Maeno et al. [5]. In this study, we use DNNs for speaker models to improve the quality of synthetic speech compared with that of HMM-based speech synthesis. Also, the stepless control of F0 beyond 3-level control [5] is available. Besides, we aim to control the F0 not only at the accent-phrase level but also at the mora level.

2   Overview of Speech Synthesis Based on DNNs

In DNN-based speech synthesis [9], we train DNNs to represent the relationship between context at the frame level and acoustic features of training speech. Context usually includes the position in a phoneme, the type of phoneme and accent position and so on. In synthesis phase, first we create the context at the frame level by text analysis and duration of phonemes that was predicted separately. After that, we input the context to DNNs to predict acoustic features, and acoustic features are converted to the output waveform by a vocoder.

3   F0 Control Using F0 Context for Tailor-Made Speech Synthesis

3.1   Concept of Tailor-Made Speech Synthesis

We can vary synthetic speech at the sentence level by modifying F0, speaking speed of the whole sentence in parametric speech synthesis. Also, we can locally modify them heuristically, however, it is hard for us to modify them keeping speech natural. Because speech parameters like F0 and power of a phoneme relate that of surrounding phoneme, it is necessary for users to gain experience in modifying these parameters considering those of surrounding phoneme. Also, it is not reasonable to modify only F0 because mismatch may occur between F0 and spectrum parameters as we mentioned in section 1. We named "tailor-made speech synthesis" the framework that solves these problems and inexperienced users can modify synthetic speech naturally and intuitively.

As a first step to developing tailor-made speech synthesis, we propose a technique to allow users to control F0 of synthetic speech in DNN-based speech synthesis by designating relative F0 at accent-phrase and phoneme level. We introduce F0 context that represents relative F0 at the accent-phrase and mora level into the model training. F0 context allows users to modify F0 at the accent-phrase or the mora level and to synthesize F0 trajectories that naturally connects between phonemes.

3.2 F0 Context for F0 Modeling in DNN-Based Speech Synthesis

In this research, we created F0 context that represents relative log F0 for training data by the same idea of Maenos' technique [5]. Namely, we calculate the mean difference of log F0 in an accent phrase or a mora between the natural and synthetic speech in a closed condition. Using the F0 context along with the typical context in model training, we are able to control the F0 of synthetic speech relatively. In our tailor-made speech synthesis, stepless F0 control, unlike 3-step F0 control [5], is available because our F0 context has a continuous value.

3.3 F0 Control at Mora and Accent-Phrase Levels in Speech Synthesis Phase

We compared two methods to make F0 context: **ACCENT+MORA** and **MORA-ONLY** in this study. When we modify F0 of synthetic speech by tailor-made speech synthesis, users will often modify F0 at the accent-phrase level. We compared whether we should prepare F0 context at the accent-phrase level or mora-level context can be used to modify F0 at the accent-phrase level. In the method **ACCENT+MORA**, we supposed that first users modify the F0 by the F0 context at the accent-phrase level, and then users tune it by the F0 context at the mora level. In the method **MORA-ONLY**, users modify the F0 by the F0 context at the mora level only. In this method, users should modify F0 context of every mora in an accent phrase if they want to modify the F0 at an accent-phrase level.

In the method **ACCENT+MORA**, first we trained a model with typical context and predicted acoustic feature for training sentences by the model. Next, we normalized the difference between log F0 of predicted and natural speech to zero-mean and unit-variance. And then F0 context at the accent-phrase level was made by calculating the mean of the difference in each accent phrase.

We trained the second model with typical context and F0 context at the accent-phrase level to get the normalized difference of log F0 again. F0 context at the mora level was made by calculating the mean of the difference in each mora. The final model was trained with conventional full context and F0 context at the accent-phrase and mora level. In the method **MORA-ONLY**, we got the F0 context at the mora level by the first model. The final model was trained with typical context and F0 context at the mora level.

4 Experiments

4.1 Experimental Conditions

In this study, we used 10 speakers of ATR Japanese speech database set B as training and test data. The number of training sentences was 450, and the number of test sentences was 53. The context consists relative frame position (number from 0 to 1), 6- (**ACCENT+MORA**) or 3- (**MORA-ONLY**) dimensional F0 context, and answers (Yes: 1, No: 0) for 412 questions . The questions were about the type of phonemes, accent positions, breath group and the length of sentence. They were used in the decision tree based context clustering in HMM-based speech synthesis. We used not only F0 context for current accent phrase (mora) but also previous and next accent phrases (moras) by the same idea that we used triphone as linguistic context. For the first or the last accent phrase (mora), we copied the value for current accent phrase (mora) to the value for previous or next accent phrase (mora) respectively.

The acoustic features were extracted by STRAIGHT [3] and consisted of 40 mel-cepstral coefficients, log F0 , 5-band aperiodicity coefficients, and their delta and delta-delta features. The sampling frequency of speech was 16 kHz. Frame-shift length was 5 ms. Log F0 of the unvoiced region was linearly interpolated. We copied the value of log F0 at the head and the end of voiced region to unvoiced region at the head and the end of speech, respectively. Band aperiodicity parameters were the mean value of 0–1, 1–2, 2–4, 4–6, 6–8 kHz of aperiodicity. The topology of DNNs was determined by preliminary experiments. DNNs had 3 hidden layers. Each hidden layer had 1024 units. We used sigmoid function as activation function and set dropout probability to 0.5. We trained the model of V/U flag for 10 epochs and the model of other features for 30 epochs. Fig. 1 shows the architecture of DNNs in this study. We trained models for voiced/unvoiced flag and other features separately. The weights of DNN was initialized by samples of gaussian distribution that have zero-mean and deviation $\sqrt{1/n}$ (n means the dimension of input vectors). After that, the weights were optimized by Adam algorithm [4] to minimize squared error. The size of mini-batch was 100. Acoustic features were scaled to be within a 0 to 1 range for training data.

When we synthesized speech, we obtained parameter sequences considering delta and delta-delta features as with the experiment conducted by Zen et al. [9]. In this study, we used the phoneme duration of natural speech for synthesis.

4.2 Objective Evaluations

First, we conducted an objective evaluation to compare **ACCENT+MORA, MORA-ONLY, CONVENTIONAL** (typical DNN-based speech synthesis without F0 context). We created ideal F0 contexts for the test sentences by training data in this experiment to examine whether proposed methods can control F0 by the F0 context. Also, we compared which method can control F0 in detail.

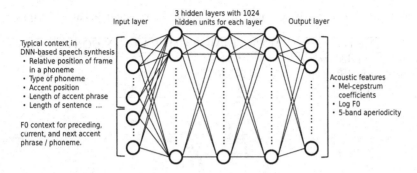

Fig. 1. Architecture of DNNs in this study.

Table 1. Result of objective evaluation with F0 context from original sample (mean of 10 speakers).

Method	ACCENT+MORA	MORA-ONLY	CONVENTIONAL
Mel-cepstral distance [dB]	4.55	4.52	4.57
Log F0 RMSE [cent]	139.7	144.5	224.4
V/U error rate [%]	5.23	5.20	5.61

Table 1 shows the mean result of 10 speakers. From the result, we found that the appropriate F0 context reduced the error of log F0 and slightly reduced the error of V/U error. The error of log F0 in **ACCENT+MORA** was slightly less than the one in **MORA-ONLY**. In **ACCENT+MORA**, F0 context included not only the variation of F0 at the mora level but also the gradual variation of F0 at the accent-phrase level. We considered that the RMSE of log F0 in **ACCENT+MORA** was smaller than that in **MORA-ONLY** because **AC-CENT+MORA** could control the gradual variation of F0.

Next, we conducted another objective evaluation without controlling F0, that is when we set F0 context for all accent phrases and moras zero. We conducted this experiment because at first users synthesize speech with all-zero F0 context and users may adopt it as final synthetic speech in tailor-made speech synthesis. We compared the method **ACCENT+MORA**, the method **MORA-ONLY**, and **CONVENTIONAL**. Table 2 shows the mean result of 10 speakers. As a result, we found that typical method had least error of log F0, followed in order by the method **MORA-ONLY** and the method **ACCENT+MORA**. The mel-cepstral distance and V/U error rate of all method were similar. F0 context corresponds to mean difference of log F0 between natural and synthetic speech at the accent-phrase or mora level, however, F0 context does not always correspond to log F0 at the frame level. It seemed that incorrespondence caused the increase of the error of log F0. We considered that the method **ACCENT+MORA** was affected by the mismatch at the frame level more than **MORA-ONLY** because we used higher dimensional F0 context in the method **ACCENT+MORA**.

Table 2. Result of objective evaluation with all-zero F0 context (mean of 10 speakers).

Method	ACCENT+MORA	MORA-ONLY	CONVENTIONAL
Mel-cepstral distance [dB]	4.56	4.55	4.57
Log F0 RMSE [cent]	245.5	239.6	224.4
V/U error rate [%]	5.65	5.58	5.61

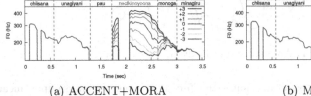

(a) ACCENT+MORA   (b) MORA-ONLY

Fig. 2. Example of F0 trajectories of synthetic speech when F0 context set at the accent-phrase level.

4.3 F0 Control Examples

We synthesized speech with some variation of F0 context for an accent phrase or a mora. First, we selected one accent phrase for each sentence randomly (excluding pause and silence) and set the F0 context (from -3 to $+3$) of the accent phrase. "1" in F0 context means the standard deviation of difference of log F0 between natural speech and synthetic speech because F0 context is the standardized mean difference of log F0. We set F0 context of unselected region to zero.

Example F0 trajectories of synthetic speech with F0 context are shown in Fig. 2(a) (**ACCENT+MORA**) and Fig. 2(b) (**MORA-ONLY**). Speech content was "Chiisana unagiyani nekkinoyoona monoga minagiru. (in Japanese)" Space means boundaries of accent phrases. The selected accent phrase was "nekkinoyoona." Focusing on only the selected accent phrase, we compared mean log F0 in the accent phrase of speech synthesized with non-zero F0 context to that with zero F0 context. We examined how F0 context affects log F0 of synthetic speech by this experiment. Fig. 3(a) for the method **ACCENT+MORA** and Fig. 3(b) for the method **MORA-ONLY** show the correspondence between F0 context and mean difference for all speakers and all evaluation sentences.

In both methods, we found that F0 context was effective to control log F0 at the accent-phrase level because F0 context almost corresponded linearly to the difference of log F0. The method **ACCENT+MORA** generated smoother F0 trajectories over the previous and next accent phrases than the method **MORA-ONLY** because we included previous and next accent phrase into F0 context.

As well as an experiment of F0 control at the accent-phrase level, we selected one mora for each sentence randomly (excluding pause, silence and moras include unvoiced vowel or phoneme /N/) and set the F0 context of the mora. In the

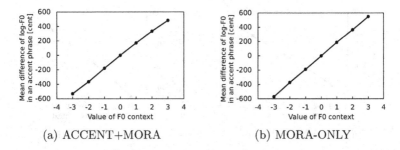

Fig. 3. Correspondence between the value of F0 context and the difference of log F0 when F0 context set at the accent-phrase level.

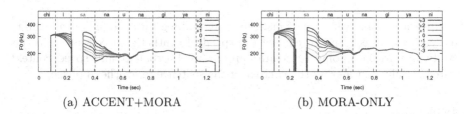

Fig. 4. Example of F0 trajectories of synthetic speech when F0 context set at the mora level.

method **ACCENT+MORA**, we used the answer F0 context at accent-phrase level and set the F0 context for the selected mora. In the method **MORA-ONLY**, we set the F0 context only for the selected mora and set zero for the other moras. Example F0 trajectories of synthetic speech with F0 context (from -3 to $+3$) are shown in Fig. 4(a) (**ACCENT+MORA**) and Fig. 4(b) (**MORA-ONLY**). Speech content was "Chiisana (in Japanese)." The selected mora was "sa." Similarly focusing on only to the selected mora, Fig. 5(a) for the method **ACCENT+MORA** and Fig. 5(b) for the method **MORA-ONLY** show the correspondence between F0 context and mean difference for all speakers and all evaluation sentences.

As well as controlling the F0 at the accent-phrase level, in both methods, we found that F0 context was effective to control log F0 at the accent-phrase level because F0 context almost corresponded linearly to the difference of log F0.

5 Conclusions

In this study, we proposed "tailor-made speech synthesis," the speech synthesis technique which enables users to control prosody of synthetic speech naturally and intuitively. As a first step to developing tailor-made speech synthesis, we introduced F0 context into DNN-based speech synthesis, and we found that F0 context was effective for stepless control of F0. Future work includes subjective evaluation of naturalness of synthetic speech. Also, we are planning to

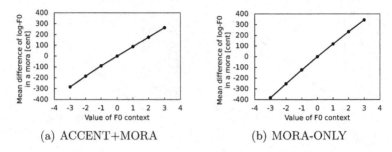

(a) ACCENT+MORA   (b) MORA-ONLY

Fig. 5. Correspondence between the value of F0 context and the difference of log F0 when F0 context set at the mora level.

compensate global variance of mel-cepstrum and log F0 to improve the quality of synthetic speech and to control the speed of synthetic speech as F0 in this study. We are also going to create appropriate F0 context and speed context automatically considering the context of the sentence in synthesis phase.

References

1. Apple Inc.: iOS - Siri - Apple, `http://www.apple.com/ios/siri/`
2. Google Inc.: Google Now,
   `https://www.google.com/search/about/learn-more/now/`
3. Kawahara, H., Masuda-Katsuse, I., de Cheveigné, A.: Restructuring speech representations using a pitch-adaptive time–frequency smoothing and an instantaneous-frequency-based F0 extraction: Possible role of a repetitive structure in sounds. Speech Communication 27(3–4), 187–207 (1999)
4. Kingma, D., Ba, J.: Adam: A method for stochastic optimization. arXiv preprint arXiv:1412.6980 (2014)
5. Maeno, Y., Nose, T., Kobayashi, T., Koriyama, T., Ijima, Y., Nakajima, H., Mizuno, H., Yoshioka, O.: Prosodic variation enhancement using unsupervised context labeling for HMM-based expressive speech synthesis. Speech Communication 57, 144–154 (2014)
6. Nishigaki, Y., Takamichi, S., Toda, T., Neubig, G., Sakti, S., Nakamura, S.: Prosody-controllable HMM-based speech synthesis using speech input. In: Proc. MLSLP (2015)
7. Nose, T., Yamagishi, J., Masuko, T., Kobayashi, T.: A style control technique for HMM-based expressive speech synthesis. IEICE Trans. Inf. & Syst. E90-D(9), 1406–1413 (2007)
8. Watts, O., Wu, Z., King, S.: Sentence-level control vectors for deep neural network speech synthesis. In: Proc. Interspeech. pp. 2217–2221 (2015)
9. Zen, H., Senior, A., Schuster, M.: Statistical parametric speech synthesis using deep neural networks. In: Proc. ICASSP. pp. 7962–7966 (2013)

Introduction and Comparison of Machine Learning Techniques to the Estimation of Binaural Speech Intelligibility

Kazuhiro Kondo and Kazuya Taira

Graduate School of Science and Engineering, Yamagata University,
4-3-16 Jonan, Yonezawa, Yamagata 9928510, Japan
kkondo@yz.yamagata-u.ac.jp

Abstract. We proposed and evaluated a speech intelligibility estimation method for binaural signals. The assumption here was that both the speech and competing noise are directional sources. We trained a mapping function between the subjective intelligibility and some objective measures. We attempted SNR calculation on a simple binaural to monaural mix-down, better SNR selection from left and right channels (better-ear), and a sub-band wise better-ear selection (band-wise better-ear). For the mapping function training, we tried neural networks (NN), support vector regression (SVR), and random forests (RF). A combination of better-ear and RF gave the best results, with root mean square error (RMSE) of about 4% and correlation of 0.99 in a closed set test.

Keywords: Speech Intelligibility, Binaural Speech, Objective Estimation, Machine Learning, Diagnostic Rhyme Test

1   Introduction

Information communication using speech may potentially be conducted in all sorts of ambient noise conditions. For example, a lecture might be conducted either in a large classroom, or a small room with varying degree of reverberation. A conversation might be conducted with significant surrounding noise in a busy shopping mall. Accordingly, techniques for efficient and accurate speech communication quality assessment is necessary in order to conduct regular quality measurement to assure stable and sufficient speech communication over these various environments.

There are basically two types of speech quality measures, i.e., the overall perceptual listening quality, and the speech intelligibility, of which the latter measures the accuracy of the perceived speech signals over a transmission channel, and thus is a crucial measure of the communication quality [5, 7].

Speech intelligibility is measured using human subjects. The subjects listen to read speech samples, and identify the content of this speech. The content of the read speech may be syllables, words, or sentences. The subjects typically listen to each sample, and write down or select what they heard. The number of

© Springer International Publishing AG 2017   167
J.-S. Pan et al. (eds.), *Advances in Intelligent Information Hiding and Multimedia Signal Processing*, Smart Innovation, Systems and Technologies 63,
DOI 10.1007/978-3-319-50209-0_21

stimuli that needs to be evaluated needs to be large enough to cover all aspects of the language that is being tested, such as the phonetic context. The test also needs to include enough number of subjects so that the variation in the responses by subject are averaged out. Thus, intelligibility tests are generally time-consuming, and expensive.

Accordingly, numerous efforts to estimate the intelligibility without using human subjects have been conducted. One of the earliest examples of such estimation is the Articulation Index (AI) [2], which estimates the intelligibility from Signal-to-Noise Ratio (SNR) measurements within a number of frequency bands combined using a perceptual model. In a later effort, Steeneken and Houtgast proposed the Speech Transmission Index (STI) [8], which uses artificial speech signals communicated over the test channel to estimate the intelligibility of the received signal by measuring the weighted average modulation depth over frequency sub-bands.

However, most of these estimation methods estimate the monaural speech intelligibility using monaural signals. In the real world, however, the human subjects listen to speech signals using both ears, i.e., binaural signals. This can potentially improve the speech intelligibility since the human auditory system can potentially discriminate sources traveling from different directions. However, it is often the case that speech intelligibility estimation is conducted using monaural signals. This can lead to significant underestimation of speech intelligibility, especially when one can expect distinct noise sources to be located away from the target speech source.

Thus, there has been efforts to estimate the speech intelligibility from binaural signals. For example, Wijngarrden et al. attempted to improve the accuracy of STI on binaural signals [10]. They employed the inter-aural cross-correlogram, which is a plot of cross-correlation vs. the lag time, to adjust the contribution of each channel signal in each of the bands to the final Modulation Transfer Function (MTF), and showed that they can estimate the binaural intelligibility from binaural signals with their model at comparable accuracy that a conventional STI can predict the intelligibility on monaural signals.

We have also attempted to estimate the binaural speech intelligibility using binaural signals [9]. We calculated the frequency-weighted SNR for each channel, and applied the better-ear model [1] to this measure, and mapped this to intelligibility using a pre-trained logistic regression function. This seems to give a relatively accurate intelligibility estimation, with Root Mean Square Error (RMSE) and Pearson correlation of about 10% and 0.79, respectively. However, obviously there was still room for improvement.

In this paper, we expanded the better-ear model to conduct the objective measure selection by sub-bands. We also introduce more sophisticated machine learning techniques to model the relation between subjective intelligibility and the objective measures. As we will see, the selection of objective measure by sub-bands do not seem to be advantageous over selecting the channel signal as a whole. However, the use of Random Forest to map the objective measure to intelligibility significantly improves the accuracy of the estimated intelligibility,

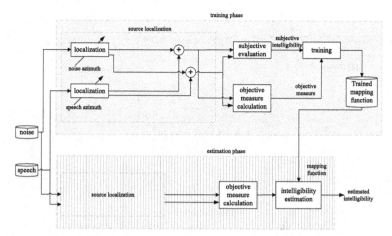

Fig. 1. Block Diagram of Speech Intelligibility Estimation

to a practical level. A Random Forest is a relatively efficient machine learning method that uses multiple regression trees trained on data subsets, and outputs prediction values as the mean of the output of these trees.

2   Estimation of Binaural Speech Intelligibility

Fig. 1 shows a block diagram of the proposed binaural speech intelligibility estimation method. In this method, we try to estimate the binaural intelligibility of a mixture of speech and noise source coming from various directions. We assume that not only the target speech, but also the noise is a directive source, such as a group of bystanders talking loudly from a specified direction, or an automobile or a train passing by from one direction to another.

We first train a mapping function between an objective measure calculated using the binaural signal to the subjective intelligibility. To train this mapping function, we compile a database of target speech traveling from various directions by convolving monaural target speech samples with the corresponding Head Related Transfer Functions (HRTFs). We also prepare noise sources from different directions by convolving this with the same HRTFs. Then, these two sources are mixed to compile a database of localized speech and noise with various azimuth combinations.

We conducted subjective intelligibility evaluations using the above database to compile a database of subjective intelligibility to use as supervisory signals in the training. The objective measure of each of the mixed signal is calculated, and the mapping function from this measure to the supervisory subjective intelligibility is trained. In our previous work [9], we used conventional logistic regression (LR) function for this mapping. However, we found that this function does not match the objective measure to the subjective intelligibility in the lower SNR range. Thus, in this paper, we attempted the use of some machine learning techniques, such as neural networks (NN), support vector regression (SVR), and random forests (RF) to improve the mapping accuracy at all SNR ranges.

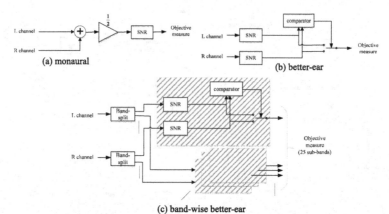

(c) band-wise better-ear

Fig. 2. Monaural and Binaural Objective Measures

The trained functions were then used to estimate the intelligibility of either a localized speech and noise combination used in the training (closed set testing), or a combination of unknown noise and speech (open set testing).

2.1 Objective Measures

Fig. 2 depicts the three methods to calculate objective measures to be mapped to intelligibility from binaural signals.

- The Monaural Model: Fig. 2 (a) depicts the monaural model, which is intended as the baseline. The monaural model simply averages the left and right channel signal into a single channel signal. This model corresponds to conventional monaural speech intelligibility estimation, and should give lower intelligibility than subjective intelligibility using binaural signals since the binaural unmasking is not taken into account.
- The Better-Ear (BE) Model: Fig. 2 (b) depicts the BE model, which selects either the left or the right channel based on the channel-wise SNR. The channel selection is conducted frame by frame.
- The Band-wise Better-Ear (BBE) Model: Fig. 2 (c) depicts the BBE model, which selects either channel for each of the sub-band based on the sub-band SNR of left and right channel. The selection is also conducted in each temporal frame.

2.2 Estimation from Objective Measures using Machine Learning Techniques

In a previous paper, we reported on using a logistic regression function trained with the maximum likelihood method to map the objective measure to speech intelligibility [9]. In this paper, we evaluated three popular machine training methods to train the mapping function. All these used the respective packages in the R environment.

- Neural Networks (NN): The nnet function in the nnet package was used to train an NN that maps the objective measures into intelligibility. The

number of units used in the input, the hidden, and the output layer was 25, 8, and 1, respectively. All other parameters were left at default values.
- Support Vector Regression (SVR): The svm function in the e1071 package was used. The Radial Basis Function was selected as the kernel function. The cost parameters and the kernel parameters (gamma) were set to values output by the tune function in the package.
- Random Forest (RF): The randomForest function in the randomForest package was used. The tuneRF function in the package was used to adjust the parameters in each tree.

3 Estimation Accuracy Evaluations

The speech intelligibility estimation accuracy was evaluated for localized speech mixed with localized competing noise at various azimuths. We tested both data within the training data (closed set testing), as well as speech mixed with unknown noise (not found in the training set) for open set testing.

3.1 Experimental Conditions

We selected 60 words out of the Japanese Diagnostic Rhyme Test (DRT) word list [3]. The words were read by one female speaker. Three noise samples were selected from the JEIDA noise database [4]; babble, A/C fan coil, and local train. For closed set testing, the mapping functions were trained and evaluated using all samples. No cross-validation was conducted in this case. For open set testing, the functions were trained using samples mixed with babble and A/C fan coil, and tested on samples with the local train noise. Other training and testing noise combinations have been tested, and will be reported in a coming paper.

Both the speech and noise samples were localized by convolving with the KEMAR HRTF, available from MIT. The azimuths for either the speech and noise sources were 0 (directly in front of the listener), ± 45, and ± 90 degrees (positive degrees to the right, negative degrees to the left of the listener). We also included a diotic noise source to simulate a non-directional noise source, which is common since noise may reverberate and travel at almost equal level from all directions. The localized sources for all possible azimuth combinations were then scaled and mixed. The noise level was adjusted so that the SNR resulted in -6, -12, and -18 dB.

In the objective measure calculation, the SNR calculation for the monaural and the BE model were frequency-weighted SNR, in which the SNR was calculated in 25 sub-bands, and averaged with weighting corresponding to the sensitivity of the human auditory system [6]. The BBE model also used the same 25 sub-bands.

Table 1. RMSE Between Subjective and Estimated Intelligibility

Mapping Function	closed set			open set		
	mono	BE	BBE	mono	BE	BBE
LR	19.6	14.3	14.4	23.3	16.9	17.0
NN	15.7	7.7	7.4	31.4	54.3	57.9
SVR	15.9	6.1	6.1	41.7	24.5	25.7
RF	8.9	**4.1**	4.6	25.4	**16.1**	**16.0**

Table 2. Pearson's Correlation Between Subjective and Estimated Intelligibility

Mapping Function	closed set			open set		
	mono	BE	BBE	mono	BE	BBE
LR	0.594	0.812	0.808	0.540	0.802	0.799
NN	0.765	0.946	0.951	0.318	0.316	0.222
SVR	0.763	0.969	0.969	0.542	0.670	0.647
RF	0.937	**0.986**	0.982	0.419	**0.863**	**0.861**

3.2 Subjective Speech Intelligibility Evaluation

The speech intelligibility of all of the localized speech mixed with localized noise samples was measured using four subjects. All subjects were in their early twenties, and reported normal hearing in their annual audiology tests.

3.3 Results and Discussions

Tables 1 and 2 shows the RMSE and Pearson's correlation between the subjective and the estimated intelligibility for both closed and open set testing, with all of the combinations of binaural objective measure calculation and mapping function training methods. As can be seen, RF combined with BE seems to give the most accurate estimation in both closed and open test. NN and SVR seem to give fairly accurate results for closed set test, but the accuracy decreases significantly for open set, indicating that these models do not generalize well to unseen data. It may be that these models were over-trained to the training data, and we still may need to tweak the parameters.

Fig. 3 compares the subjective to estimated intelligibility using LR, in combination with mono, BE and BBE, respectively. The diagonal solid line in the middle is the ideal, where the measured and the estimated intelligibility completely match, and the upper and lower dotted lines are ±20% from this line,

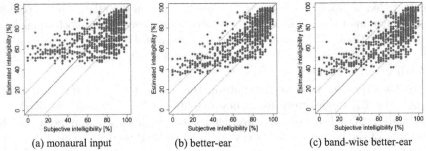

(a) monaural input   (b) better-ear   (c) band-wise better-ear

Fig. 3. Distribution of Subjective vs. Estimated Intelligibility by Objective Measure Calculation Method using LR (closed set)

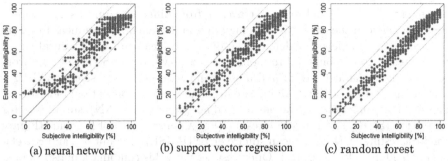

Fig. 4. Distribution of Subjective vs. Estimated Intelligibility by Machine Learning Method using the BE Model (closed set)

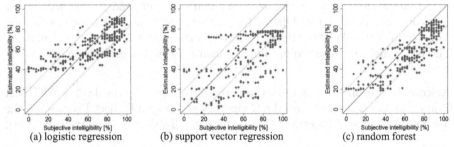

Fig. 5. Distribution of Subjective vs. Estimated Intelligibility by Machine Learning Method using the BE Model (open set)

which we consider acceptable deviation from the ideal. As can be seen, the mono model seems to give slightly more concentrated plots in the upper half, showing that this model cannot estimate intelligibility of samples with low subjective intelligibility. BE and BBE shows some improvement, but not significantly, and indicates that LR is not a very good match to this set of data.

Fig. 4 compares the distribution with NN, SVR, and RF combined with BE in a closed set test. As can be seen, all three give plots that are relatively concentrated on the diagonal axis, showing a very accurate estimation. However, the plots with RF give a very tight fit on the diagonal, significantly better than the other two, indicating significantly higher estimation accuracy.

Fig. 5 shows the distribution for LR, SVR, and RF combined with the BE model for the open set test. Plots for NN is not shown since this model was not able to estimate intelligibility at all. Again, RF shows a higher concentration of plots along the diagonal line compared to others, although much more scattered compared to the closed set. It is also apparent from plots for NN and SVR that these models cannot estimate the intelligibility for this set, i.e., a significantly lower generalization capability to unseen data.

4   Conclusion

A speech intelligibility estimation method for binaural signals was proposed and evaluated for its estimation accuracy. Both the speech and competing noise were assumed to be directional sources, traveling from different azimuths. A mapping function between the subjective intelligibility and some objective measures were trained. Three models to calculate the objective measure from binaural signal was tested; a simple binaural to monaural mix-down, better SNR selection from left and right channels (BE), and a sub-band wise better-ear selection (BBE). For the mapping function training, we originally used a conventional logistic regression (LR) function, but we newly tried neural networks (NN), support vector regression (SVR), and random forests (RF). A combination of BE and RF gave the best results, with root mean square error (RMSE) of about 4% and correlation of 0.99 in a closed set test. Open test set with this combination also showed better accuracy than others, with RMSE and correlation of approximately 16% and 0.86, respectively.

So far, we have only used three noise sources, and need to test with other types of noise. We also would like to test other distance measures, such as the Log Area Ratio or the Weighted Spectral Slope, in the objective measure calculation. The BE model may also require improvements for higher estimation accuracy.

Acknowledgments. This work was supported in part by the JSPS KAKENHI Grant Number 25330182, and the Cooperative Research Project Program of the Research Institute of Electrical Communication, Tohoku University (H26/A14).

References

1. Edmonds, B.A., Culling, J.F.: The spatial unmasking of speech: Evidence for better-ear listening. J. Acoust. Soc. Am. 120(3), 1539–1545 (Sept 2006)
2. French, N.R., Steinberg, J.C.: Factors governing the intelligibility of speech sounds. J. Acoust. Soc. Am. 19(1), 90–119 (1947)
3. Fujimori, M., Kondo, K., Takano, K., Nakagawa, K.: On a revised word-pair list for the Japanese intelligibility test. In: Proc. Int. Symp. on Frontiers in Sp. and Hearing Res. Tokyo, Japan (Mar 2006)
4. Itahashi, S.: A noise database and japanese common speech data corpus. J. Acoust. Soc. Japan 47(12), 951–953 (Dec 1991), in Japanese
5. Kondo, K.: Subjective Quality Measurement of Speech. Springer-Verlag, Heidelberg, Germany (2012)
6. Ma, J., Hu, Y., Loizou, P.C.: Objective measures for predicting speech intelligibility in noisy conditions based on new band-importance functions. J. Acoust. Soc. Am. 125(5), 3387–3405 (May 2009)
7. Quackenbush, S.R., III, T.P.B., Clements, M.A.: Objective Measures of Speech Quality. Prentice-Hall, Englewood Cliffs, NJ, USA (1988)
8. Steeneken, H.J.M., Houtgast, T.: A physical method for measuring speech transmission quality. J. Acoust. Soc. Am. 67(1), 318–326 (1980)
9. Taira, K., Kondo, K.: Estimation of binaural intelligibility using the frequency-weighted segmental SNR of stereo channel signals. In: Proc. APSIPA-ASC. pp. 101–104. Hong Kong (Dec 2015)
10. Wijngaarden, S.J., Drullman, R.: Binaural intelligibility prediction based on the speech transmission index. J. Acoust. Soc. Am. 123(6), 4514–4523 (June 2008)

Impact of variation in interaural level and time differences due to head rotation in localization of spatially segregated sound

Daisuke Morikawa

Graduate School of Advanced Science and Technology, Japan Advanced Institute of Science and Technology, 1-1 Asahidai, Nomi, Ishikawa, Japan
morikawa@jaist.ac.jp
http://researchmap.jp/morikawa/?lang=english

Abstract. In this paper, we clarify the role of variation in interaural level difference (ILD) and interaural time difference (ITD) due to head rotation in localization of spatially segregated sounds. Listeners were asked to distinguish between two sources of white noises having various ILDs/ITDs under head rotation. In ILD condition, the segregation rate reached 80% when the ILD between two sources at an angular difference of 36° corresponded to different sides, i.e., left and right hemisphere. However, the sound image was integrated into one when the sources corresponded to the same side. In ITD condition, two or three images were perceived regardless of the ITDs. This was because when only one source was used, it was perceived as separate lower- and higher-frequency images. In confirmations with low- and high-pass noises, the lower-frequency image was contained for lower than 1.2 kHz and the higher-frequency one was contained for higher than 1.7 kHz.

Keywords: Interaural level difference; Interaural time difference; Spatially segregated sound; Sound localization; Head rotation

1 Introduction

Attention plays an important role in separating target sound stream from other sound streams. The phenomenon is well-known as "cocktail party effect" [1]. Listeners have ability to segregate the sound streams, i.e., separate the target stream from others and group them into a stream. Previous research have investigated effects of acoustic feature of sounds, such as sound direction, timber, and temporal structure, on the process. It is well known that the interaural level difference (ILD) and interaural time difference (ITD) are cues for sound localization [2]. The temporal variation of these cues greatly contribute to sound localization [3]. However, it is not clear as to which acoustic features play the role of important cues when the sound is segregated. Some researchers have reported that differences in direction are particularly important in the cocktail party effect. In addition, signals are easy to hear when the signals and maskers have different interaural cues or are spatially separated. These are also referred as to

© Springer International Publishing AG 2017
J.-S. Pan et al. (eds.), *Advances in Intelligent Information Hiding and Multimedia Signal Processing*, Smart Innovation, Systems and Technologies 63,
DOI 10.1007/978-3-319-50209-0_22

the binaural masking level difference [4] and spatial release from masking [5][6]. However, paired stimuli (a signal and masker) have different types of timber such as pairs of voice and random noise. Listeners perceived two sounds separately for different ILDs or ITDs when two sounds had the same timber [7]. However, the effectiveness of the temporal variation of ILD and ITD due to head rotation is sill not clear.

This paper presents the results of a comprehensive study on understanding the effect of ILD and ITD due to head rotation on sound segregation.

2   Experimentals

2.1   Experimental system

Figure 1 outlines the experimental system, which consisted of a Windows-based personal computer (PC), digital-to-analog converters (DACs) (RME, Fireface UCX), a headphone amplifier (audio-technica, AT-HA21), headphones (Sennheiser, HDA-200) and motion sensor (Logical Product, LP–WS1105). The motion sensor was connected to the PC via a USB interface, and it was fastened to the band of headphones. The sampling frequency of the DACs was 192 kHz and that of the motion sensor was 1 kHz. The apparent sampling frequency of the motion sensor was 200 Hz, because, the sensor sends 5 samples at once. The angular resolution capability was 1°. The experiment was carried out in a soundproof chamber. The background A-weighted sound pressure level of the room was less than 21 dB.

Four listeners with normal hearing from 23 to 30 years of age participated for each of the experiment.

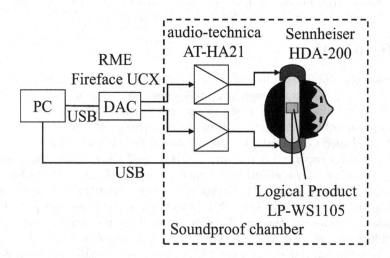

Fig. 1. Experimental system.

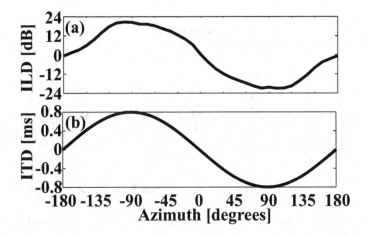

Fig. 2. Interaural difference (a) mean measured ILD, (b) ITD model of sphere

2.2 Stimuli

Two uncorrelated white noise sources (WN1, WN2) were used. The duration of the white noise was 3 s. A 30 ms linear taper window was applied at the beginning and end of the white noise sources.

The PC received the angle of motion sensor, and switched the ILD or the ITD in response to them in real time. Binaural signals were synthesized from each of the white noise sources. ILD or ITD were generated using the overlap-add method like Otani's method [8]. The ILD or the ITD were switched frame by frame. The frames of left and right channel stimuli are given by Eq. (1,2)

$$S_l = A_l(\theta_1 + \theta_s) \otimes \text{WN1} + A_l(\theta_2 + \theta_s) \otimes \text{WN2} \tag{1}$$

$$S_r = A_r(\theta_1 + \theta_s) \otimes \text{WN1} + A_r(\theta_2 + \theta_s) \otimes \text{WN2} \tag{2}$$

where A_l and A_r are impulse responses which are estimated from delta function $\delta(t)$, WN1 and WN2 are one frame segments of WN1 and WN2, θ_1 and θ_2 are configuration angle of WN1 and WN2, θ_s is head-rotation angle, and, $\otimes$ denotes the convolution. One frame consisted of 4092 points (about 21 ms). The initial value of θ_s was 0° in all the stimuli. The sound pressure level was 70 dB when $\theta_1 + \theta_s$ and $\theta_2 + \theta_s$ was 0°.

In ILD condition, ILD was modified by amplitude of A_l and A_r based on the mean measured ILD. The mean ILD was calculated from the measured head-related transfer functions (HRTFs) [7]. Figure 2(a) plots the mean ILD.

In ITD condition, ITD was modified by delay of A_l and A_r based on the ITD model. The ITD model of sphere for each azimuth proposed by Kuhn [9] was used for modification. These ITDs are given by Eq.(3), where r is the radius of the sphere, c is the sound speed and θ is the stimulus angle.

$$\text{ITD}(\theta) = 3\frac{r}{c}\sin(\theta) \tag{3}$$

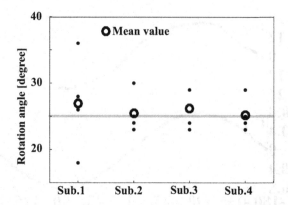

Fig. 3. Rotation angle of listeners when they stop their head.

Figure 2(b) plots the ITD model when r is 90 mm and c is 340 m/s. r

3   ILD Experiments

3.1   Preliminary experiment

Listeners were instructed to freely rotate their head in a horizontal direction and were asked to number the sound images and answer either one or two. All the listeners perceived two separate sounds under a head-still condition [7]. However, when the listeners rotated their head largely, the sound images were sometimes perceived as one sound image.

3.2   Experiment 1

Here, we clarify when sound images were segregated and integrated. In this experiment, θ_1 and θ_2 were configured 25° and −25°. Listeners were instructed to turn their heads left or right and stop when two sound images became one sound image.

Figure 3 plots the rotation angle $|\theta_s|$ when listeners stopped turning their heads. $|\theta_s|$ was about 25°. In this angle, either of the sound sources corresponded right in front of the listener. In other words, angle of A_l and A_r were $|50°|$ and 0°.

3.3   Experiment 2

In this experiment, we clarify the difference of head-rotation and head-still in sound segregation by ILD when the ILDs corresponded to different sides. Hear, θ_1 and θ_2 were ±10, 12, 14, 16, 18, 20, 22, 24, 26 and 0°. Where 0° was single sound image stimulus. The experiments consisted of four sessions, and each session consisted of 60 trials. The stimuli were presented in random order from

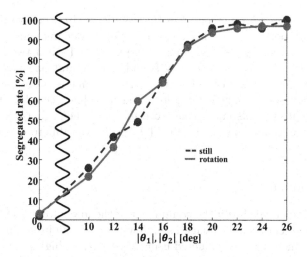

Fig. 4. Sound segregating rate of ILD condition

10 types of stimuli. Since all experiments consisted of four sessions, 24 answers were obtained for each type stimulus. The inter-stimulus-interval of each stimuli was 3 s.

In head-rotation condition, listeners were instructed to freely to rotate their head in a horizontal direction while the stimulus was being reproduced. In head-still condition, they were instructed to hold their head still while the stimulus was being reproduced. When motion sensor was turned off, θ_s was always 0°. They were asked to number the sound images and to mark either one or two on an answer sheet. They were also asked to answer their perception about the number of sound image when their heads were towards front, and if the number changed when their head rotated.

Figure 4 plots the mean segregation rate for each condition for four listeners. The mean segregation rate reached 80% when $|\theta_1|$ and $|\theta_2|$ were over 18°. Over 36° angular difference between A_l and A_r was necessary for sound segregation. This tendency was similar to previous work of head-still condition [7]. There was no significant difference in segregation rate between head-rotation and head-still conditions. All the listeners were reported that judging segregation at head-rotation condition was easier than head-still condition. Listeners were separated into two group. Two listeners rarely gave one sound image as their response, while another two rarely gave two sound images as their response.

4 ITD Experiments

4.1 Preliminary experiment

Listeners were instructed to freely rotate their head in a horizontal direction and were asked to number the sound images and answer either one or two. Two or three sound images were perceived in this condition. In other words, one sound image was not perceived in this condition. One lower-frequency sound image and one higher-frequency sound image were perceived when two sound images were perceived. Two lower-frequency sound images and one higher-frequency sound image were perceived when three sound images were perceived. The lower-frequency sound images were localized an angle according to θ_1 and θ_2 of which synchronous listeners head rotation, and the higher-frequency sound image was localized center of head. This result was different from head-still condition [7].

4.2   Experiment 3

In this experiment we clarify the component of the lower-frequency sound image and higher-frequency sound image using low-pass and high-pass noises.

The experiments were conducted using low-pass and high-pass filtered noises in addition to the WN1. WN2 and θ_2 did not affect the S_l and S_r. The cut-off frequencies of low-pass and high-pass filter were 1.0, 1.2, 1.4, 1.7, 2.0, 2.4, 2.8, 3.4 and 4.0 kHz. The noise conditions (e.g., low-pass filtered condition, and high-pass filtered condition) were included of the respective 9 filtered-noises and WN1. The other procedures were same as that for head-rotation condition of experiment 2.

Figure 5, 6 plots the mean segregating rate for each condition for the four listeners. Fig. 5indicates that the mean segregation rate increased with cut-off frequency. The mean segregation rates were smaller than the other mean segregation rates and not greatly differed with cut-off frequency when the cut-off frequency was lower than 1.2 kHz. Fig. 6 shows that the mean segregation rate dropped with cut-off frequency. The mean segregation rates were smaller than the other mean segregation rates and not greatly differed with cut-off frequency when the cut-off frequency was over 1.7 kHz.

5   Discussion

In experiment 1, sound images were integrated when either of the sound source corresponded to right in front of the listener. The two sound images were perceived separately when sound sources corresponded to different sides. This means that, the two sound images were integrated into one, when the ILDs corresponded to the same side. This integration does not happen in modification of ITD condition.

In experiment 2, the two sound images were perceived when the ILD between the two sources at an angular difference of 36°. There was no significant difference between head-rotation and head-still conditions. However, listeners easily judged segregation by head rotated. This means that the various ILDs had some effect for sound segregation.

In experiment 3, the lower-frequency sound image was contained for lower than 1.2 kHz and the higher-frequency sound image was contained for higher

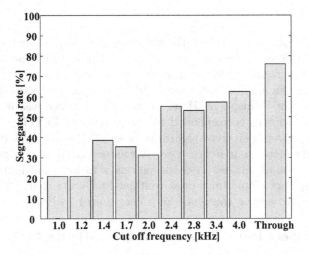

Fig. 5. Sound segregating rate of Low-pass condition

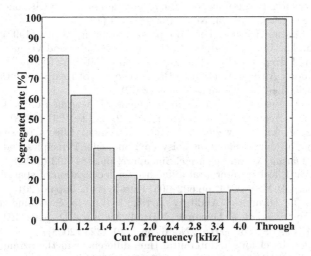

Fig. 6. Sound segregating rate of High-pass condition

than 1.7 kHz. Therefore, the ITD was mainly calculated from lower than 1.5 kHz [2], and the ILD calculated from higher than 1.5 kHz [2]. In addition, lower-frequency sound image was localized from ITD, and higher-frequency sound image was localized from other cues. Higher frequency sound image does not synchronize listeners head rotation. This segregation does not happen in modification of ILD condition.

These results suggest that, in the sound localization strategy, the ITD is more dominant in the perception of sound image segregation, while the ILD is more dominant in the perception of sound image integration.

6   Conclusion

Sound segregation experiments with two white noise sources under various ILD
and ITD conditions were conducted under the head-rotated condition. In ILD
condition, the segregation rate reached 80% when the ILD between the two white
noise sources at an angular difference of 36°, corresponded to the different side,
i.e., left and right hemisphere. However, the sound image was integrated into
one, when the sources corresponded to the same side. In ITD condition, two
or three images were perceived regardless of the ITDs. This was because when
only one white noise source was used, it was perceived as separate high and
low frequency sound images. In confirmations with low- and high-pass noises,
the lower-frequency sound image was contained for lower than 1.2 kHz and the
higher-frequency one was contained for higher than 1.7 kHz.

Acknowledgment Part of this work was supported by KAKENHI (15K16022).

References

1. E. C. Cherry: Some experiments on the recognition of speech with one and two ears,
   J. Acoust. Soc. Am., Vol. 25, No. 5, pp. 975–979 (1953)
2. J. Blauert: Spatial Hearing, The MIT press, Cambridge, MA., (1997)
3. H. Wallach: The role of head movements and vestibular and vis-ual cues in sound
   localization, J. Exp. Psychol., Vol. 27, No. 4, pp. 339–368 (1940)
4. B. C. J. Moore: An Introduction to the Psychology of Hearing: Sixth Edition, pp.
   271–275, Brill, Leiden, The netherlands (2012)
5. M. Ebata, T. Sone, T.Nimura: improvement of hearing ability by directional
   information, J. Acoust. Soc. Am., Vol.43, No. 2, pp. 289–297 (1968)
6. N. Kuroda, J. Li, Y. Iwaya, M.Unoki, M. Akagi,: Effects of spatial cues on
   detectability of alarm signals in noisy environments, Principles and Applications
   of Spatial Hearing, World Scientific, Singapore, pp. 484–493 (2011)
7. D. Morikawa: Effect of interaural difference for localization of spatially segregated
   sound, Proc. IIH-MSP 2014, pp.602–605, Kita-kyusyu, Japan (2014)
8. M. Otani, T. Hirahara: Auditory Artifacts due to Switching Head-Related
   Transfer Functions of a Dynamic Virtual Auditory Display, IEICE TRANS.
   FUNDAMENTALS, Vol. E91-A, No. 6, pp.1320–1328 (2008)
9. G. F. Kuhn: Model for the interaural time differences in the azimuthal plane, J.
   Acoust. Soc. Am., Vol.62, No. 1, pp. 157–167 (1977)

An improved 5-2 channel downmix algorithm for 3D audio reproduction

Xingwei Sun[1], Risheng Xia[1], Lei Yang[2], Lizhong Wang[2], and Junfeng Li[1]

[1]Institute of Acoustics, Chinese Academy of Sciences
[2]Samsung R&D Institute China-Beijing (SRC-B)

Abstract. With the continuous development of virtual reality (VR), 3D audio is widely used particularly for headphones. Therefore, a multi-channel to two-channel downmixing approach is in great demand. Many traditional methods have been proposed, but these methods cannot meet the requirements for higher perceptual quality in VR scenes. In this paper, we propose an improved five-two channel downmixing approach based on coherence suppression and spatial correction. This method can reduce the timbre change and spatial inaccuracy problems which traditional method commonly encountered. Its effectiveness is validated through the preliminary experiments.

Keywords: Virtual reality, 3D audio, Downmix

1 Introduction

3D audio technologies have already attracted more and more research interest, which is highly motivated by the strong and promising applications of VR in recent days. Regardless of the 3D audio creation methods (e.g., multi-channel approach and object-based approach), the present commonly-used audio reproduction device in VR is headphone. To reproduce 3D audio content over the present VR devices, therefore, a multi-channel to two-channel downmixing approach is in great demand.

Many downmix methods have been reported in the literature [1], which includes the method suggested by ITU [2], the one by Dolby [3] and the matrix-encoding approach [4], to name a few. In these traditional methods, the fixed downmixing equations are usually exploited by blindly summing up the corresponding signals, which may results in the timbre changes and spatial inaccuracies in the downmixed sound. To solve this problem, in this paper, we propose an improved five-two channel downmixing approach based on coherence suppression and spatial correction. Its effectiveness is examined by the spectrograms of the downmixed signals and the informal listening test.

2 Traditional Downmixing Method

In this section, the traditional downmixing method on which the proposed downmixing method is presented will be briefly introduced. As an example, the

J.-S. Pan et al. (eds.), *Advances in Intelligent Information Hiding and Multimedia
Signal Processing*, Smart Innovation, Systems and Technologies 63,
DOI 10.1007/978-3-319-50209-0_23

matrix-encoding five-two channel downmixing method is used and its implementation can be given by

$$L = FL + 0.71 * C + j * (\cos\alpha * LS + \sin\alpha * RS) . \tag{1}$$
$$R = FR + 0.71 * C + j * (\sin\alpha * LS + \cos\alpha * RS) . \tag{2}$$

where j implies a 90 degree phase shift and α is typically 30-35 degrees. And FL, C, FR, LS and RS are the abbreviations for the channels of front left, center, front right, left surround and right surround in 5.1 channel audio. Likewise, L and R are the abbreviations for the channels of left and right in 2 channel audio.

In the traditional downmixing method, the spectral and timbre changes are often introduced, which are mainly from the following reasons. The first two terms (i.e., FL and C in $Eq.(1)$, FR and C in $Eq.(2)$) might have similar but phase shift components, which leads to the comb-filter artifacts in the downmixed signals. And as the traditional downmixing methods use fixed coefficients and mixing equations to blindly combine five input channels into two output channels, the location information is encoded into the amplitude and phase characteristics and may be impaired due to the neglect of the pre-existing spatial relationships.

3   Improved Downmixing Method

To solve the problems of timbre changes and spatial inaccuracies in the traditional downmixing methods, in this section, we present an improved downmixing algorithm incorporating the coherence suppression and spatial correction techniques.

3.1   Coherence Suppression

The Coherence Suppression component is used to minimize the spectral and timbre distortion. As the downmix processings for the left and right output channels are similar, the following description is only for the left output channel. Since the first two terms in $Eq.(1)$ might be correlated at least partially, their sum should be further processed to alleviate the spectral and timbre distortion. Their corresponding representations in the short time Fourier transform (STFT) are denoted as $FL(k, m)$ and $C(k, m)$, where k and m are the discrete frequency and time indices. k and m will be omitted for brevity later.

The essential idea is to suppress the coherent signal components between FL and C by formulating C signal as

$$C = W * FL + U . \tag{3}$$

where W describes the correlation between FL and C, and the uncorrelated signal component is given by U. It is clear that we can separate the correlated component and uncorrelated component by using a suppression gain G designed on the coherence-function. To preserve the unchange of overall energy of the correlated signal, a scaling factor G_F is further introduced. The sum of the first two terms is finally given by

$$L' = G_F * FL + G * C .\tag{4}$$

The detail implementation can be referred to [5].

Finally, the downmixed output of left channel is given by

$$L = L' + j * (\cos\alpha * LS + \sin\alpha * RS) .\tag{5}$$

After this process, we obtain a downmixed two channel signal in which the timbre changes have been minimized. In the next step, the spatial inaccuracies will be corrected.

3.2 Spatial Correction

The Spatial Correction component is used to detect frequency bands where s-patial errors in the downmix processing have occurred. By modifying the inter-channel level difference (ICLD) and inter-channel phase difference (ICPD) prop-erties of the downmixed signal per frequency band, the spatial imaging of the downmixed signal can be made consistent with the source signal. For comparing the spatial cues of the original 5.1 channel signal and those of the downmixed two channel signal, both signals are mapped to a normalized 2-dimensional plane.

The 5.1 channel signal is mapped to a normalized position vector as shown with exemplary locations on a 2-dimensional plane comprised of a lateral axis which indicates the left and right direction and a depth axis which indicates the front and rear direction. As shown in Fig.1, the value of the location for (X_{LS}, Y_{LS}), (X_{RS}, Y_{RS}), (X_R, Y_R), (X_L, Y_L) and (X_C, Y_C) are assigned to $(0, 0)$, $(0, 1)$, $(1, 1)$, $(0, 1)$ and $(0.5, 1)$, respectively. These coordinates are determined by the locations of speakers in the physical space.

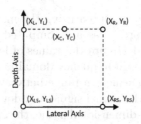

Fig. 1. The 2-dimensional plane where the 5.1 channel signal is mapped .

The estimated image position vector $P(F)$ can be calculated per sub-band as set forth in the following vector equation:

$$P(F) = M_L(F) * (X_L, Y_L) + M_R(F) * (X_R, Y_R) + M_C(F) * (X_C, Y_C) + M_{LS}(F) * (X_{LS}, Y_{LS}) + M_{RS}(F) * (X_{RS}, Y_{RS}) .\tag{6}$$

where $M(F)$ can be viewed as normalized sub-band magnitude estimates for each channel. The $M(F)$ for each channel is calculated as the ratio of the magnitude of each channel and the sum of the five channels in frequency domain.Thus, for each frequency band, a position vector $P(F)$ is provided and that is used to define the position of the apparent frequency source for that frequency band.

For the downmixed two channel signal, the ICLD and ICPD characteristics are used to generate a representation of the sound field. And as shown in Fig.2, the downmixed sound field is mapped to a normalized 2-dimensional plane where the ICLD characteristic is mapped to the X-dimension and the ICPD characteristic is mapped to the Y-dimension.

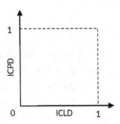

Fig. 2. The 2-dimensional plane where the downmixed two channel signal is mapped .

The coordinates of the downmixed two channel signal can be calculated as:

$$X_D(F) = \frac{|D_R(F)|}{|D_L(F)| + |D_R(F)|} . \tag{7}$$

$$Y_D(F) = \frac{\pi + (\arg(D_R(F)) - \arg(D_L(F)))}{\pi} . \tag{8}$$

where the X-coordinates are computed using the magnitude values of each frequency band, and the Y-coordinates are computed using the phase angles of each frequency band. D_L and D_R are the values of left and right channel of the downmixed two channel audio in frequency domain. Note that the numerator of Y_D may need to add $\pm 2\pi$ to achieve a range between $-\pi$ and π.

With both the original 5.1 channel signal and downmixed two channel signal mapped onto a normalized 2-dimensional plane, the spatial characters of the signals can be compared. Relevant spatial cues of ICLD and ICPD are modified per frequency band in the downmixed signal. Correction of the ICLD is performed through multiplication of the magnitude value by multiplying the value generated by the deviation of $X_D(F)$ and the X-coordinate of $P(F)$ in each frequency band. Correction of the ICPD is performed through addition of the phase angle by adding the value generated by the deviation of $Y_D(F)$ and the Y-coordinate of $P(F)$ in each frequency band. The corrected magnitude and phase angle are recombined to form the complex value of the downmixed two channel signal for each frequency band and are then converted into time domain signal.

4   Performance Evaluation

The effectiveness of the proposed algorithm is demonstrated by the following examples in terms of spectrogram and informal listening test.

In order to verify the validity of the coherence suppression component, a 5.1 channel white noise whose center channel is similar but phase shifted to front left and front right channels was selected. The spectrograms of the left-channel output signals by the traditional downmixing algorithm and by the improved downmixing algorithm are plotted in Fig.3. It is clear that the traditional downmixing algorithm introduced the comb-filter effect characterized by the clear visible notches in Fig.3, and that the improved downmixing algorithm alleviate this undesired comb-filter effect to a large degree.

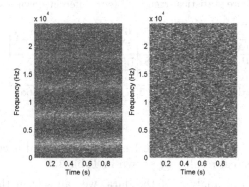

Fig. 3. Spectrogram of the downmixed signals by the traditional downmixing algorithm (left) and by the improved downmixing algorithm (right).

For a quantitative evaluation, we computed the distortion of each downmixed binaural audio with respect to the original audio. The distortion measure called log-spectral distance was used which is given by

$$D = \sqrt{\frac{1}{k} \sum^{k} (\log(P_5) - \log(P_2))^2} \ . \tag{9}$$

where P_5 is the total power of the original five channels, P_2 is the total power of the downmixed two channels and k is the discrete frequency indices. We used four input audio signals for the distortion analysis. In Fig.4, it is obvious that the improved downmixing algorithm introduced less distortion than the passive method in the comparison of all the four audio signals.

In the informal listening test, we selected four downmixing method, namely, (a) ITU recommended method, (b) matrix downmixing method mentioned in section 2, (c) downmix with HRTF processing and (d) the proposed improved downmixing method in this paper, to process the same audio. Ten students with normal hearing were invited to attend this test.They were asked to score the

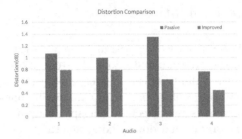

Fig. 4. Mean log-spectral distances of four audio signals downmixed by two different methods.

Table 1. The score statistical results of four different downmixing methods.

	1	2	3	4	5	6	7	8	9	10	average
(a)	7.5	7	7	6.5	7	7	8.5	6	6	7.5	7
(b)	8	7.5	7.5	8	8	7.5	7	7	8	7	7.55
(c)	5	5	7	4	5	6	7	6	7	5	5.7
(d)	7.5	7	7.5	7	7.5	7	8.5	8	8.5	8	7.65

four methods in accordance with the feeling of space and accuracy of imaging location. The statistical results are shown in Table 1.

Through statistical analysis of the data, we can see: in the current downmixing algorithm, the matrix downmixing method has a good effect. In the downmixing with HRTF processing, the MIT HRTF database was used. As it brings a serious timbre distortion, the result of the listening test didn't perform well. The improved downmixing algorithm has the best result in the overall scores, which shows its validity in improving the feeling of space and accuracy of imaging location.

5   Conclusion

An improved five-two channel downmix algorithm based on coherence suppression and spatial correction has been described. The Coherence Suppression component minimizes the spectral and timbre distortion caused by comb-filter effects. And the Spatial Correction component ensures that the spatial cues of the downmixed signal accurately reflect the source through modification to the ICLD and ICPD characteristics. The analysis of spectrogram and informal listening test were carried out, which demonstrate the improvements of timbre consistency and spatial imaging accuracy.

6 Acknowledgements

This work is partially supported by the National Natural Science Foundation of China (Nos. 11461141004, 61271426, 11504406, 11590770, 11590771, 11590772, 11590773, 11590774), the Strategic Priority Research Program of the Chinese Academy of Sciences (Nos. XDA06030100, XDA06030500, XDA06040603), National 863 Program(No. 2015AA016306), National 973 Program (No. 2013CB329 302) and the Key Scienceand Technology Project of the Xinjiang Uygur Autonomous Region (No. 201230118-3).

References

1. Kuba Łopatka, et al.: Novel 5.1 downmix algorithm with improved dialogue intelligibility. Audio Engineering Society. (2013)
2. ITU: ITU-R Recommendation BS.775-1, Multichannel stereophonic sound system with and without accompanying picture. (1994)
3. Digital Audio Compression Standard (AC-3,E-AC-3).(2010)
4. C. Faller and P. Schillebeeckx: Improved ITU and Matrix Surround Downmixing. Audio Engineering Society. (2011)
5. A. Adami, et al.: Down-Mixing Using Coherence Suppression. IEEE International Conference on Acoustic, Speech and Signal Processing. (2014)

Investigation on the head-related modulation transfer function for monaural DOA

Nguyen Khanh Bui, Daisuke Morikawa, and Masashi Unoki

School of Information Science, Japan Advanced Institute of Science and Technology
1-1 Asahidai, Nomi, Ishikawa, Japan
{khanhsbn,morikawa,unoki}@jaist.ac.jp

Abstract. This paper investigates the head-related modulation transfer function (HR-MTF) to find out important trends and features for estimating the monaural direction of arrival (DOA). Previous studies suggested that the HR-MTF contains significant trends and features for estimating monaural DOA. However it is still unclear where the range of meaningful modulation frequencies is and what the exact trends of the features are. To resolve these issues, the reported study carried out a geometrical analysis of the human head and a modulation spectrum analysis using three different datasets of. The results indicated that the HR-MTF in a specific modulation frequency range showed significant trends from 110 to 350 azimuth degree, -30 to 100 degree and 120 to 220 elevation degree. It was also found that there are notable features such as a significant peak from 80 to 110 azimuth degree and dips at 210 azimuth and 110 elevation degree.

Keywords: head-related modulation transfer function, monaural direction of arrival, monaural modulation spectrum

1 Introduction

It has been reported that human beings can recognize an incoming sound sources location by using monaural cues [1]. Thus, the ability of estimating the monaural direction of arrival (DOA) has been considered an interesting and challenging problem whose solution would yield a better understanding of human hearing. Available monaural cues for sound localization can be regarded as spectral cues in the head-related transfer function (HRTF), which is a transfer function between a sound source position and the eardrum position in an ear, such as peaks and dips in the monaural spectral envelope. However, it is still unclear how these peaks and dips in the spectral envelope are relevant to the estimation of monaural DOA. These spectral cues alone, which often change with the signal, would not be enough for estimating the sound source locations.

It has been revealed that modulation cues are important for estimating monaural DOA [2]. Ando et al. reported the feasibility of estimating monaural DOA on the basis of a monaural modulation spectrum (MMS) with a few types of source signal [3][4]. The method is based on observing the MMS shapes, which are approximately drawn as arcs with azimuth variations in the same

© Springer International Publishing AG 2017

J.-S. Pan et al. (eds.), *Advances in Intelligent Information Hiding and Multimedia Signal Processing*, Smart Innovation, Systems and Technologies 63,
DOI 10.1007/978-3-319-50209-0_24

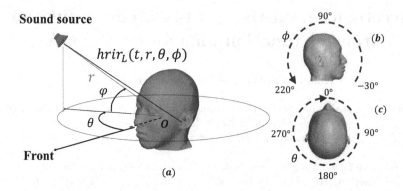

Fig. 1: Definitions of (a) head-related impulse response of left ear as a function of horizontal plane θ, median plane ϕ, and distance r, (b) elevation measurement convention, and (c) azimuth measurement convention.

direction as the ear (left or right). However, the head-related modulation transfer function (HR-MTF), which retains the DOA-related characteristics, was not rigorously investigated. As it is based on the concept of the MTF, good understanding of the HR-MTF is crucial for estimating monaural DOA. Moreover, the above studies simulated a DOA model with a limited dataset and only observed the MMS with an unexplained upper modulation frequency limit of 200 Hz.

This paper investigates the HR-MTF to identify important trends and features for monaural DOA. HR-MTFs with different sound source positions on the horizontal and median planes in three different datasets were analyzed by using modulation spectrum analysis in the range of meaningful modulation frequency.

2   Head-related modulation transfer function

Before reaching the eardrum, the original source sound is reflected and diffracted by the human body structures. These modifications contain the source location, which can be captured by an impulse response that represents the relationship between the sound source and eardrum location. That impulse response is termed the head-related impulse response (HRIR). Figure 1 shows the definitions of HRIR at a specific ear as a function of time t, horizontal plane θ, median plane ϕ, and distance r.

The MTF was proposed to account for a relation between the transfer function of frequency in an enclosure in terms of the envelopes of input and output signals and characteristics of the enclosure [5]. Regarding HRIR and monaural DOA, the complex modulation transfer function is defined as:

$$E_h(f_m, \theta, \phi) = \frac{\int_0^\infty h(t, \theta, \phi)^2 \exp(jwt)dt}{\int_0^\infty h(t, \theta, \phi)^2 dt} \tag{1}$$

where $h(t, \theta, \phi)$ is the HRIR and $E_h(f_m, \theta, \phi)$ is the HR-MTF. Equation (1) is the normalized Fourier transform of the squared HRIR divided by its total energy.

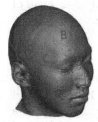

Measurements	Half wavelength(cm)	Frequency(Hz)
A (bitragion arc)	36.8	466
B (sagittal arc)	36.4	471
C (half of head circumference)	28.5	601

Fig. 2: Human head geometry and meaningful modulation frequency.

3 Modulation spectrum analysis

3.1 Materials

Three different datasets of HRIRs were used to investigate the important trends and features for monaural DOA in the HR-MTF. The first dataset was recorded by the Research Institute of Electrical Communication, Tohoku University (RIEC) [6][7]. The dataset included 104 subjects (208 ears) at 936 positions. The source distance r was 1 meter, and the sampling frequency was 48 kHz. The second one was an HRIR dataset that was recorded by ourselves in the RIEC environments. There were four subjects (8 ears) at 1296 positions. The source distance r was 1 meter, and the sampling frequency was 48 kHz. The third database was calculated using computer models of human heads based on the boundary element method (BEM) [8]. It consisted of seven heads (14 ears), each at 1278 positions. The source distance r was 1.2 meter, and the sampling frequency was 44.1 kHz.

3.2 Geometrical analysis

Figure 2 shows three different measurements of the head shape, including (A) bitragion arc, (B) sagittal arc, and (C) half of the head circumference [9]. To determine the range of meaningful modulation frequency for monaural DOA, we assessed the human head geometry and the frequency of an arbitrary simple signal as shown in Fig. 2. The frequencies corresponding to labels A, B, and C in Fig. 2 were derived from measurements related to the head and the sound speed (343 m/s). Suppose that there is a sound source moving in the median and horizontal planes. The head shape in the bitragion arc has effects on both of the median and horizontal planes. The sagittal arc and half of the head circumference might mainly affect the elevation and azimuth estimation.

From the above considerations, it was found that the minimum meaningful modulation frequency with regard to the head shape should be 500 Hz. At a higher modulation frequency, smaller parts of the head will have more of an effect on the HR-MTF. In addition, the upper limit of the modulation frequency should be related to the amplitude modulation with the carriers, so 1,000 Hz was used as the upper limit. Therefore, in this study, the meaningful range of modulation frequency was chosen to be from 500 Hz to 1,000 Hz.

3.3 Method

There were three steps involved in the method of investigation. Firstly, HR-MTFs were derived from three different HRIR datasets on the horizontal and median planes. This provided overviews of HR-MTFs for all subjects of the datasets, so that trends and features could be extracted. The similarities and differences between datasets were also indicated. Secondly, the trends and features were subjected to a more detailed observation. The results were analyzed regarding to the relationship between the HR-MTF and the different sound source locations. Finally, an assessment of the trends and features with different modulation frequency variations was conducted. The purpose of this step was to see how HR-MTF changes within the meaningful modulation frequency range.

4   Results and discussion

Figures 3 and 4 provide overviews of averaged azimuth and elevation HR-MTFs of left ears in the three different datasets. The abscissas represent the azimuth or elevation, and the ordinates show the modulation depth in decibels with the standard deviation in the left panel and modulation depth/degree with derivative in the right panel. The red solid bars show the standard deviations at different degrees, while the solid blue curve shows the mean value. It was found that the standard deviations with regard to the derivatives were smaller than those of the modulation depth. This means there were considerable differences in modulation depth level among the subjects in a database.

An obvious trend is one in which the HR-MTFs approximately formed curves in three ranges. The first range was from 110 to 350 degree of azimuth, while the other ones were from −30 to 100 degree and 120 to 220 degree of elevation. In the elevation HR-MTF, the first range had a larger modulation depth level compared with the latter one. The derivative standard deviations were small in these three ranges, which showed that the subjects in the three different datasets showed similar trends.

The results show other notable features. A significant peak and a small dip existed in the azimuth HR-MTF, from 80 to 110 degree and at about 210 degree. In addition, a big dip at about 110 degree often occurred in the elevation HR-MTF. The standard deviations were larger in the elevation degree and showed that the modulation depth level of the features varied from subject to subject.

The three different datasets showed similar trends and features. However, there were also notable differences. In Fig. 3(e), from −30 to 140 degree, the azimuth HR-MTF of the BEM dataset had a significant lower modulation depth level. However, the derivative standard deviation showed that this trend was similar among the subjects in the BEM dataset. The reason was likely the absence of the shoulder reflection from the BEM dataset. Unlike the others, the BEM dataset only used the head for the HRIR calculation.

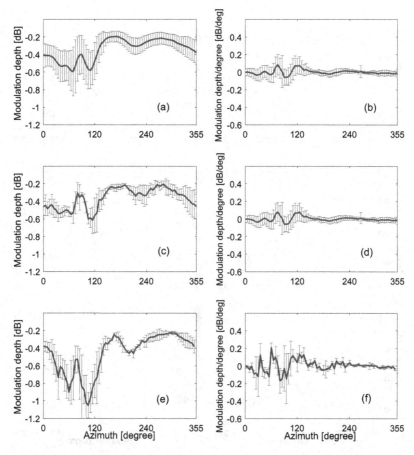

Fig. 3: Analysis results for azimuth. (a)(c)(e): mean and standard deviation, (b)(d)(f): derivative mean and derivative standard deviation. (a)(b): RIEC database, (c)(d): Self-recorded database, (e)(f): BEM database (modulation frequency $f_m = 500$ Hz).

Figures 5 and 6 show detailed observations in order to see the above trends and features in different subjects from the RIEC dataset and the averaged RIEC data with modulation frequencies. The trends and features are discussed below. In Fig. 5, the abscissas represent the azimuth or elevation, while the ordinates display the modulation depth of the HR-MTF in decibels. The inner four lines on each panel show the HR-MTFs of different subjects.

It can be seen from Fig. 5(a) that there are three regions, corresponding to three degree ranges of the azimuth HR-MTF. The first region starts from around 0 to 80 degree, containing out-of-trend data with different trends. Accordant to the left ear, this region was additionally affected by HR-MTF passive effects;

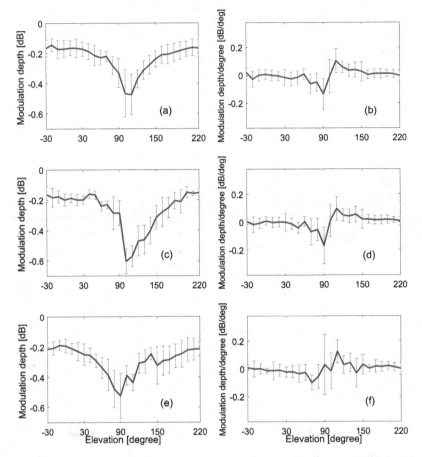

Fig. 4: Analysis results for elevation. (a)(c)(e): mean and standard deviation, (b)(d)(f): derivative mean and derivative standard deviation. (a)(b): RIEC database, (c)(d): Self-recorded database, (e)(f): BEM database (modulation frequency $f_m = 500$ Hz).

therefore, there were differences from subject to subject. The second region has a significant peak from 80 to 110 degree. This degree range was directly opposite the ear, so that the sound signal had various equal-length paths to get to the ear on other side. This might be an explanation for the common peak. The third region covers the rest of the azimuth range, in which the HR-MTF approximately formed a curve. The fact that this region was mostly in the front of the ear might be the reason for this trend.

Figure 5(b) indicates the three notable regions in the elevation HR-MTF. In the first and third regions, the HR-MTFs could be approximately drawn as arcs. The incoming sounds at these positions could easily reach the ear; this

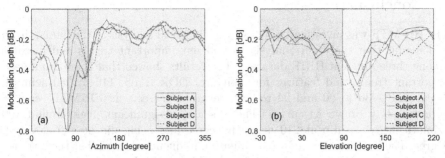

Fig. 5: Trends and features of HRMTF (different subjects).

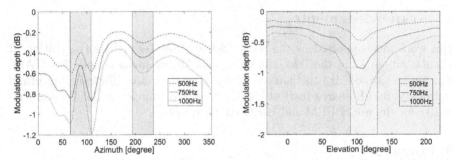

Fig. 6: Trends and features of HRMTF (different modulation frequencies).

might be the reason for the trend. Moreover, the same reason could explain why HR-MTF at the front of the head had a larger modulation depth compared with the one at the back. Additionally, there is a large dip in the second region of the elevation HR-MTF. The modulation depth level also varies more among subjects. This elevation range is around the top of the head, where there are not many differences in body structure from one person to another. Therefore, the modulation depth levels tend to be unalike between subjects.

Figure 6 shows the relationship between HR-MTF and the range of meaningful modulation frequency. The abscissas show the azimuth or elevation, while the ordinates display the modulation depth of the HR-MTF. The inner three curves on each side show the HR-MTFs with different frequency variations.

It can be seen that there are similar trends and features in the HR-MTF with modulation frequency variations on both the horizontal and median plane. However, with increasing modulation frequency, the significant peak at 80 to 110 azimuth degree seems to become larger, in comparison with the overall trend. The other features such as the dip at 210 azimuth degree or the big dip at 100 elevation degree, however, do not share the same tendency.

As with different sides of the ear, although the results have been omitted from this paper, both HR-MTFs show similar trends and features.

5   Conclusion

The HR-MTF was investigated by using geometrical analysis of the human head and modulation spectrum analysis to find out important trends and features among three different HRIR datasets. The results showed that there are indeed important trends and features for monaural DOA. From 110 to 350 azimuth degree and −30 to 90 and 90 to 130 elevation degree, the HR-MTFs can be approximately drawn as curves. There is also a significant peak around 100 azimuth degree, a dip at 210 azimuth degree and a big dip at 100 elevation degree. The investigations of the modulation frequency showed that these trends and features are common among the meaningful range. In the future, the results will be incorporated in a monaural DOA method.

6   Acknowledgments

This work was supported by the Secom Science and Technology Foundation. It was also supported by the Telecommunications Advancement Foundation (TAF). The authors would like to thank Prof. Tatsuya Hirahara (Toyama Prefectural University) and Research Institute of Electrical Communication of Tohoku University for the use of BEM and RIEC HRTF dataset.

References

1. K. Strelnikov, M. Rosito, and P. Barone, "Effect of Audiovisual Training on Monaural Spatial Hearing in Horizontal Plane," PloS one, vol. 6, no. 3, pp. 1-9 (2011)
2. E. R. Thompson and T. Dau, "Binaural processing of modulation interaural level difference," J. Acoust. Soc. Am., vol. 123, no. 2, pp. 1017-1029 (2008)
3. M. Ando, D. Morikawa, and M. Unoki, "Study on Method of Estimating Direction of Arrival Using Monaural Modulation Spectrum." Journal of Signal Processing, vol. 18, no. 4, pp. 197-200 (2014)
4. D. Morikawa, M. Ando, and M. Unoki, "Feasibility of Estimating Direction of Arrival Based on Monaural Modulation Spectrum." 2015 International Conference on Intelligent Information Hiding and Multimedia Signal Processing (IIH-MSP2015), pp. 384-387, Adelaide, Australia (2015)
5. M. Unoki, "Speech Signal Processing Based on the Concept of Modulation Transfer Function -Basis of Power Envelope Inverse Filtering and Its Applications-," Journal of Signal Processing, vol. 12, no. 5, pp. 339-348, 2008.
6. K. Watanabe, Y. Iwaya, Y. Suzuki, S. Takane, and S. Sato, "Dataset of Head-related Transfer Functions Measured with a Circular Loudspeaker Array." Acoust. Sci. and Tech. Acoustical Science and Technology vol. 35, no. 3, pp. 159-165 (2014)
7. "The RIEC HRTF Dataset," http://www.riec.tohoku.ac.jp/pub/hrtf/index.html
8. Makoto Otani, and Shiro Ise, "Fast Calculation System Specialized for Head-related Transfer Function Based on Boundary Element Method." The Journal of the Acoustical Society of America J. Acoust. Soc. Am., vol.119, no. 5, pp. 2589-2598 (2006)
9. ISO 7250-3, "Basic human body measurements for technological design - Part 3: Worldwide and regional design ranges for use in ISO product standards." (2015)

Temporal characteristics of perceived reality of multimodal contents

Shuichi Sakamoto†, Hiroyuki Yagyu†, Zhenglie Cui†, Tomoko Ohtani†,
Yôiti Suzuki†, and Jiro Gyoba‡

†Research Institute of Electrical Communication and
Graduate School of Information Science, Tohoku University
2-1-1 Katahira, Aoba-ku, Sendai, Miyagi, 980-8577, Japan
‡Graduate School of Arts and Letters, Tohoku University,
27-1 Kawauchi, Aoba-ku, Sendai, Miyagi, 980-8576, Japan
{saka, yagyu, sai}@ais.riec.tohoku.ac.jp, ohtani.tomoko@noc.geidai.ac.jp,
yoh@riec.tohoku.ac.jp, gyoba@m.tohoku.ac.jp

Abstract. Understanding the mechanism related to how people perceive reality is crucially important for the development of high-definition communication systems in multimedia contents. Various sensory information is well known to affect perceived reality from multimodal contents. For this purpose, it is important to present multimodal information with proper timing. This study investigated temporal characteristics of perceived reality induced by multimodal contents including vibration information. Results indicate that temporal characteristics of perceived reality depended on the timing of the events in the content, e.g. the timing with which the train just passes through an observation point. Moreover, the sense of presence was more sensitive to the timing of the presentation of sensory information with long contents.

Keywords: sense of presence, sense of verisimilitude, multimodal integration, stimulus onset asynchrony, reality perception

1 Introduction

Developing advanced multimedia systems requires knowledge of how humans process multimodal information. Audio-visual information has been regarded mainly as crucially important sensory information to induce a high sense of reality [4]. However, other sensory information aside from audio-visual information also plays an important role in enhancing a user's immersive experience from the system. Vibration information is one of the important sensory information. Actually, some entertainment systems, such as those used in motion rides, driving simulators, and movie theaters, use rich vibration information effectively to increase user's perceived reality.

Some researchers have reported the effects of vibration information on perceptual reality from audio-visual contents. The results indicate clearly that full-body vibration enhances perceived reality [3, 11, 5]. These studies only addressed the

© Springer International Publishing AG 2017   199
J.-S. Pan et al. (eds.), *Advances in Intelligent Information Hiding and Multimedia
Signal Processing*, Smart Innovation, Systems and Technologies 63,
DOI 10.1007/978-3-319-50209-0_25

sense of reality which a user perceived immediately after experiencing the contents. Perceived reality depends on the events presented to the user. Therefore, how users' perceived reality is enhanced or reduced according to the events is a matter that demands investigation. These temporal characteristics of perceived reality can be used effectively to create high-definition multimedia contents. For example, if perceived reality is strongly related to the certain key events, then multimedia content developers can introduce events to the contents when they intend to enhance users' perceived reality.

This study investigated how people perceived the sense of reality as they experienced the multimodal contents. As the multimodal contents, audio, visual, and vibration information were presented to users. To analyze the temporal characteristics of the perceived reality more precisely, stimulus onset asynchrony (SOA) of audio-visual information and visual-vibration information are changed as parameters. The temporal characteristics of the perceived reality of multimodal contents were investigated by analyzing the effects of the SOAs on the intensity of perceived reality.

2   Index of perceived reality [5]

The sense of reality includes many meanings [4, 6, 7]. Therefore, it is difficult to ask observers directly about their amount of perceived reality. We hypothesized that perceived reality is divisible mainly into two parts: the reality obtained from the object or target and that obtained by the environment around the people. The former and latter might be respectively designated as the "figure" and the "background" of a scene. If the former reality is increased, then people might perceive the object or target as the real object, even if it is an imitation. If the latter reality is increased, then people might perceive it as though they were inside the environment. The importance of each reality is therefore dependent on the contents presented by the multimodal display. Based on this hypothesis, we selected two indicators to evaluate the perceived reality.

2.1   Sense of Presence

The sense of presence is defined as the subjective experience of being in one place or environment even when one is physically situated in another place [10]. This sense is the most popular indicator used to evaluate virtual reality (VR) systems and environment. When people see a movie with a very large screen and rich spatial audio information, they would feel as if they were in the scene. Under such circumstances, they feel high sense of presence. One might consider an extraterrestrial alien being. Although the alien does not exist in the world, one might easily imagine the shape and characteristics of the alien. In this case, some key characteristics make it an alien. When these key components are well presented, people feel a high sense of verisimilitude. Many researchers have examined cues for the sense of presence. Teramoto *et al.* reported that the sense of presence is important not only for evaluating the VR environment, but also for the expression of the existing "real" experience [8]. Moreover, they showed that the sense of presence is dominated mainly by the absolute amount of stimuli and that it is related dominantly to background components in a scene [7, 2].

2.2 Sense of Verisimilitude

The sense of verisimilitude is the appearance of being true or real. We extend the meaning of this indicator as the existence of essence for the target object in a scene, although the scene may have incomplete information overall.

We have investigated the mechanisms of perceiving the sense of presence and the sense of verisimilitude using audio-visual contents [7, 2]. Results suggest that the sense of verisimilitude corresponds to appreciation of foreground components in multimodal content.

3 Experiment [12]

3.1 Stimulus

To create multimodal content including vibration, we recorded sound, moving pictures, and ground vibration when a train passed near the measurement point. Figure 1 shows the recording setup. A dummy head (SAMRAI; Koken Co. Ltd.) with binaural microphones (4101; Brüel & Kjær) inside its ears and a DV camera (AG-DVX100A; Panasonic Inc.) were set on the ground. The dummy head and the DV camera were set as close as possible. Moreover, we tried to fix the direction of the DV camera at the same direction of the dummy head as if an observer had stood and seen the view at the point of the dummy head. Binaural microphones were connected to audio inputs of DV camera via an amplifier (2639; Brüel & Kjær). Binaural signals were recorded at the sampling frequency of 48 kHz. Two acceleration pickups (VM-80; RION Co. Ltd.) were fixed on both sides of the wooden board on the ground. When the vibration information was recorded, an approximately 60-kg weight was put on the board. These recording systems were set beside the JR local line (Tōhoku Honsen) near Iwakiri Station. The total duration of the stimulus was 35 s. The maximum amplitude of the recorded vibration was around 0.2 cm. Audio-visual and vestibular information were recorded simultaneously at the scene. Figure 2 presents the scenario of the multimodal contents that were recorded. The train is approaching and passing just beside the observer. A horn is sounded three times in the contents.

3.2 Experimental procedure

Ten men and six women, all with normal hearing acuity and normal/corrected-to-normal vision, participated in the experiment. As explained in the previous section, two indicators were used to evaluate the sense of reality. For each indicator, eight observers (five male, three female) were assigned. Therefore, observers only evaluated the senses of presence or verisimilitude. The mean age of the

Fig. 1. Recording setup.

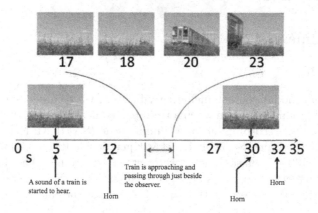

Fig. 2. Scenario of the contents.

observers who evaluated the sense of presence was 22.0 (SD=1.8), whereas that of the observers who evaluated the sense of verisimilitude was 22.1 (SD=3.1).

Figure 3 portrays the experimental setup. They stood directly on the motion platform during experiments. Stimulus onset asynchrony (SOA) of audio-visual information and visual-vibration information was set from −800, −400, −200, −100, 0, 100, 200, 400, and 800 ms. Here, 0 ms signifies that audio, visual and vestibular information was presented at the timing when the three types of information were recorded. A positive value signifies that auditory or vibration information is started to present before visual information.

Before the experiment, the following instructions, which were almost the same as those given to observers in earlier studies [7, 2], were given to observers to clarify differences between the two indicators:

- **Sense of presence**: Please rate the degree to which they felt they were now just in the presented place.

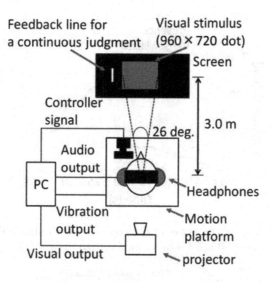

Fig. 3. Experimental setup.

- **Sense of verisimilitude**: Please rate the degree to which they felt as if presented object were real.

During experiments, observers were asked to report the degree of the sense of presence/verisimilitude in real time using a throttle lever (Throttle Quadrant; SAITEK). A small bar is shown just beside the content to indicate how much the lever was moved. The bar length corresponds to the amount of the movement of the lever according to continuous cross-modality matching using line length [9]. The maximum bar length (400 dot) was corresponded to the maximum value of the sense of presence/verisimilitude that observers experienced in daily life, whereas the minimum length (2 dot) corresponded to the condition in which observers felt no sense of presence/verisimilitude at all.

After each trial, observers were also requested to rate the perceived subjective senses of presence/verisimilitude from 0 (low) to 6 (high). The obtained data were normalized based on the observers' average and were converted to a z-score for comparison of scores assigned to each perceived subjective reality.

Each session consisted of 81 trials (9 audio-visual SOAs × 9 visual-vestibular SOAs). Observers experienced one SOA condition three times. Therefore, each observer examined three sessions in total.

4   Results and Discussion [12]

Figure 4 presents the temporal change of perceived reality at 0 SOA condition, in which all sensory information was presented synchronously. The figure shows clearly that the highest score is observed when the train is passing among observers in both senses. However, the sense of verisimilitude maintains a high

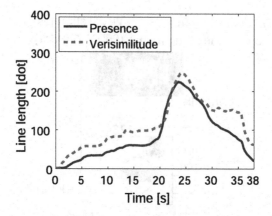

Fig. 4. Averaged value of the senses of presence and verisimilitude during the experiment (audio-visual SOA, 0 ms; visual-vestibular SOA, 0 ms).

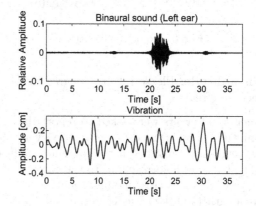

Fig. 5. The amplitude of presented auditory and vestibular information.

score even after the train disappeared, whereas the score of the sense of presence is decreasing rapidly. When people perceive the sense of verisimilitude, the prototype of the scene which people retain in memory is expected to play an important role. Abe *et al.* reported that the image intrinsic to a sound source, evoked by verbal and visual information, affects evaluation of the sound [1]. Even when the train is passing, an observer would keep the image of the train in mind. Perhaps this image induces an observed high score of the sense of verisimilitude.

Figure 6 shows the averaged seven-scale scores of the senses of presence and verisimilitude as a function of the audio-visual SOA and visual-vestibular SOA. The obtained data were normalized based on the observers' average and were converted to a z-score for comparison of scores assigned to each perceived subjective reality.

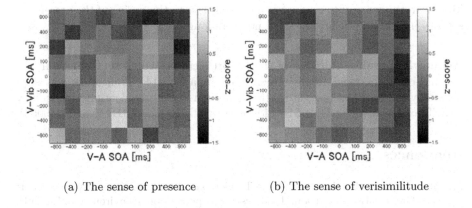

(a) The sense of presence   (b) The sense of verisimilitude

Fig. 6. Averaged seven-scale scores of the senses of presence and verisimilitude (z-score).

The results of three-way (2 indicators $\times$ 9 audio-visual SOA $\times$ 9 visual-vestibular SOA) ANOVA revealed that a three-way interaction was statistically significant ($F_{64,896} = 1.454$, $p < .05$). In each indicator, the analysis of a simple main effects of audio-visual SOA and visual-vestibular SOA shows that the simple interaction of both SOAs was statistically significant ($F_{64,896} = 1.713$, $p < .001$) for the sense of presence, although no significant difference was found for the sense of verisimilitude ($F_{64,896} = 1.037$, $n.s.$). When the visual-vestibular SOA were -400 ms, -100 ms, and 200 ms, the simple simple main effects of audio-visual SOA were statistically significant (-400 ms: $F_{8,1008} = 1.984$, $p < .05$, -100 ms: $F_{8,1008} = 3.786$, $p < .001$, 200 ms: $F_{8,1008} = 2.604$, $p < .01$).

These results suggest that the sense of presence is sensitive to the timing of the presented sensory information, whereas the sense of verisimilitude is robust to asynchronicity. As explained previously, the sense of verisimilitude is related closely to the prototype that people keep in their memory. Therefore, the effects of asynchrony among the presented sensory information are expected to be small for the sense of verisimilitude, especially experiencing very long contents.

5   Conclusions

Various sensory information is well known to enhance perceived reality from multimodal contents. To increase such effect, it is important to present this information with proper timing. This study investigated the temporal characteristics of perceived reality induced by multimodal contents including vibration information. To analyze the effect of the presentation timing of all sensory information, audio-visual SOA and visual-vestibular SOA were changed as parameters. Results demonstrate that the sense of verisimilitude maintained a high value even when the event was finished, although the sense of presence decreased rapidly.

Moreover, the sense of presence was more sensitive to the timing of the presenting sensory information when the contents were long.

Acknowledgments Portions of this report were presented in Japanese in the Journal of the Virtual Reality Society of Japan (2014). This work was supported by the National Institute of Information and Communications Technology (NICT) of Japan and a Grant-in-Aid for Scientific Research (B) (No. 26280067).

References

1. Abe K., Ozawa K., Suzuki Y., Sone T.: Comparison of the effects of verbal versus visual information about sound sources on the perception of environmental sounds. Acta Acustica united with Acustica 92, 51–61 (2006)
2. Honda A., Kanda T., Shibata H., Asai N., Teramoto W., Sakamoto S., Iwaya Y., Suzuki Y., Gyoba J.: Determinants of senses of presence and verisimilitude in audio-visual contents. Journal of the Virtual Reality Society of Japan 18(1), 93–101 (2013) (in Japanese)
3. Kim S., Sakanashi H., Martens W.L.: Multimodal representation of electronic piano sound for pianoplayers. In: INTER-NOISE 2011, 6-page manuscript (2011)
4. Meyer G.F., Wang L.T., Timson E., Perfect P.: White MD Objective fidelity evaluation in multisensory virtual environments: Auditory cue fidelity in flight simulation. PLoS One 7(9), e44381 (2012)
5. Sakamoto S., Hasegawa G., Honda A., Iwaya Y., Suzuki Y., Gyoba J.: Body vibration effects on perceived reality with multi-modal contents. ITE Trans. on MTA 2(1), 46–50 (2014)
6. Slater M., Steed A., McCarthy J., Maringelli F.: The influence of body movement on subjective presence in virtual environments. Human Factors 40(3), 469–477 (1998)
7. Teramoto W., Yoshida K., Hidaka S., Asai N., Gyoba J., Sakamoto S., Iwaya Y., Suzuki Y.: Spatiotemporal characteristics responsible for high vraisemblance. Journal of the Virtual Reality Society of Japan 15(3), 483–486 (2010) (in Japanese)
8. Teramoto W., Yoshida K., Asai N., Hidaka S., Gyoba J., Suzuki Y.: What is "sense of presence"? A nonresearcher's understanding of sense of presence. Journal of the Virtual Reality Society of Japan 15(1), 7–16 (2010) (in Japanese)
9. Tato T., Namba S., Kuwano S.: Continuous judgment of level-fluctuating noise, In: INTERNOISE 94, pp. 1081–1084 (1994)
10. Witmer B.G., Singer M.J.: Measuring presence in virtual environments: A presence questionnaire. Presence 7(3), 225–240 (1998)
11. Woszczyk W., Cooperstock J., Roston J., Martens W.L.: Shake, rattle, and roll: Getting immersed in multisensory, interactive music via broadband networks. J. Audio Eng. Soc. 53, 336–344 (2005)
12. Yagyu H., Cui Z., Sakamoto S., Ohtani T., Suzuki Y., Gyoba J.: Effects of multimodal contents on the temporal shifts in reality perception. Journal of the Virtual Reality Society of Japan 20(3), pp. 199–208 (2015) (in Japanese)

Part III
Communication Protocols, Techniques, and Methods

An MTSA Algorithm for Unknown Protocol Format Reverse

Fanghui Sun*, Shen Wang, and Hongli Zhang

Department of Computer Science and Technology
Harbin Institute of Technology
Harbin, 150001, China

Abstract. In the process of unknown protocol specification reverse, a crucial procedure is clustering messages into different classes, from which we can infer the formats of each class and thus construct the grammar and syntax of that protocol. This paper proposed a Message Token Sequence Alignment algorithm(MTSA) to measure the similarity of message token sequences of unknown Internet protocol by applying a similarity score strategy. According to the result of a verification experiment, we found that this strategy is able to measure the message token sequence reasonably.

Keywords: Protocol reverse · Sequence alignment · Evaluation strategy

1   Introduction

With the Internet environment becoming increasingly complicated, numerous novel private protocols continually come into sight. This promotes the importance of protocol reverse engineering in intrusion detection and prevention, Trojan tracking, and Internet traffic management. In the process of reversing an unknown protocol specification, clustering messages into different classes is a crucial procedure, from which we can further infer the format (grammar and syntax) of that unknown protocol from each class.

Before clustering the messages from an unknown protocol, we need to measure the difference between those messages. Several sequence alignment algorithms are employed to handle this process since the Protocol Information Project launched by Beddoe in 2004 [1]. In the work of [2], Needleman Wunsch algorithm (N/W algorithm) [3] is introduced to get the structure information of messages. ScriptGen [4], and Roleplayer [5] introduced byte-wise sequence alignment techniques to discover message formats needed for session replay. In Discover [6], aimed to find out the integral protocol format, a type-based sequence alignment is leveraged to merge clusters with similar formats to avoid over classification. In the work of ProDecoder [7], the N/W algorithm [3] is again used to find the common byte sequences to represent the stable part of protocol messages among

---

* Email: sunfanghui@hit.edu.cn

© Springer International Publishing AG 2017   209
J.-S. Pan et al. (eds.), *Advances in Intelligent Information Hiding and Multimedia*
Signal Processing, Smart Innovation, Systems and Technologies 63,
DOI 10.1007/978-3-319-50209-0_26

each message cluster and thus build the protocol's format in form of regular expression.

Those methods mentioned above mostly launch the N/W algorithm to find the longest common substring. But that is not very much suitable when evaluating the message token sequence similarity. Considering the target of inferring the format of an unknown protocol, the similarity measuring process would need to take some format traits into account.

We summarize two format alignment traits: *position advantage* and *distance weaken*. After inspecting the formats of some mainstream Internet protocols, we found that most important format tokens are usually centred on two ends of a message, especially the head. Therefore, the similarity appears near the two ends should have a higher proportion. We call that position advantage. The formats of messages usually have a vertical correspondence, which means that the messages of same format should show similarity at akin offsets. And we call that distance weaken.

In this paper, a Message Token Sequence Alignment algorithm is proposed to measure the similarity of unknown protocol messages considering the aforementioned two format aligning traits, and is based on the basic Levenshtein Distance Algorithm (LD algorithm) [7].

This paper is divided into four sections. We explain the structure and scheme of MTSA algorithm in Section 2. A verification experiment is then given in Section 3. We conclude the work of this paper and discuss about the future work in Section 4.

2 Structure and Scheme of MTSA

The Message Token Sequence Alignment is composed of two parts: token similarity score and sequence alignment, as in Fig. 1. The similarity score of a token pair from two message token sequences is measured in part 1, and the second part uses those scores to calculate the message similarity by designing a dynamic programming algorithm: MTSA.

Prior to the alignment process, the messages of an unknown protocol are captured and fragmented in to token sequences, with each token representing a field of the protocol message format.

Here we assume that the partition process is highly precise, so that quality of the fragment would not affect the score and alignment result.

2.1 Token Similarity Score

This part has three steps: basic score, position and distance amendment. The first step matches the attributes of two tokens and gets a basic score. The other two part will revision that score considering the position advantage and distance weaken traits discussed in section 1. At the end of this part, a score matrix of token similarity is built.

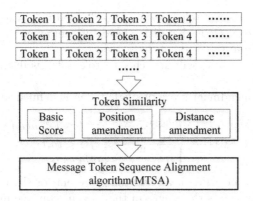

Fig. 1. Structure of the Strategy

Basic Score As mentioned before, tokens in a concrete Internet communication packet are corresponding to the fields of a protocol message format. Thus the attributes are basically same between them, and are matched to be the Basic Score of token similarity, as in (1):

$$TS_{M,N}(i,j) = \sum_k \mathscr{I}_{(M_i.k=N_j.k)} \tag{1}$$

where k represents for token attributes, and $\mathscr{I}(.)$ is an indicator function, as in (2):

$$\mathscr{I}_{(x=y)} = \begin{cases} 1, \, if \, x = y \\ 0, \, if \, x \neq y \end{cases} \tag{2}$$

In (1), $TS_{M,N}(i,j)$ is the Basic Score of the token similarity of M_i and N_j which are from two message token sequences M and N.

Position Amendment According to the position advantage trait of protocol format, a corresponding amendment is designed to amplify the weight of tokens at certain position when they are matched, that is, to enlarge the matching score of tokens at the beginning or end of a message token sequence by adding a position amendment score, as shown in Fig. 2.

		HEAD				MIDDLE	TAIL	
Message	Token 1	Token 2	Token 3	Token 4	Token 5	······	Token n-1	Token n
	+0.5	+0.4	+0.3	+0.2	+0.1	······	+0.1	+0.2

Fig. 2. Position Amendment Sketch

In this step, a message token sequence is divided into three positions: head, tail, and middle. Tokens in each position have different amendment score, according to their distance from two ends of the sequence.

For short sequences, we will first check a token if it is in the head, then tail, and middle at last.

The position amendment score of single token is defined in (3):

$$PA_M(i) = \begin{cases} \frac{head-i+1}{10} & , if \ 1 \leq i < head \\ \frac{tail-(len-i)}{10} & , if \ i < tail \\ 0 & , otherwise \end{cases} \tag{3}$$

where i is the token position, head and tail represent for the length of head and tail part separately, len is the length of that token sequence measured by token numbers.

Sometimes the two tokens are located in different positions, so a priority rule is set up for this: if one of them is in middle, there will be no amendment; if they are in the same position, the amendment will be the average of their scores; if one of them is in tail, the other is in head, then the amendment will be the score of that one in tail, as is expressed in (4):

$$PA_{M,N}(i,j) = \begin{cases} \frac{PA(i)+PA(j)}{2} & , if \ 1 \leq i,j < head \\ \frac{(PA(i)(i \in tail)+PA(j)(j \in tail)}{2} & , if \ i < tail \ or \ j < tail \\ 0 & , otherwise \end{cases} \tag{4}$$

The token similarity score with position amendment ($TSPA_{M,N}$) is designed to be (5):

$$TSPA_{M,N}(i,j) = TS_{M,N}(i,j) * (1 + PA_{M,N}(i,j)). \tag{5}$$

Distance Amendment According to the distance weaken trait of protocol format when aligning messages, similar tokens at close places should have high authority, as interpreted in Fig. 3.

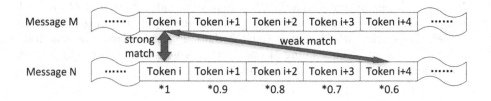

Fig. 3. Distance Amendment Sketch

When we align the token i in message M and token $i+4$ in message N, their matching score will be reduced to 60% because of their relative distance. For the tokens that are considerably far, their matching will become almost meaningless, and their similarity score will be zero.

This distance amendment step is explicated in (6):

$$TSPDA_{M,N}(i,j) = \begin{cases} TSPA_{M,N}(i,j) * (1 - \frac{|i-j|}{max\_dis}) & , |i-j| \in [0, max\_dis] \\ 0 & , |i-j| > max\_dis \end{cases}$$

(6)

In (6), $|i-j|$ is the relative distance of the two tokens, and max_dis is the maximum countable distance. $TSPDSA_{M,N}$ is the token similarity score with position and distance amendment.

2.2 The MTSA Algorithm

The MTSA algorithm constructs a score matrix for two message token sequences with length M and N separately. Score $MTSA_{M,N}[i,j]$ in matrix represents the similarity between prefix $m_1, m_2, ..., m_i$ of M and prefix $n_1, n_2, ..., n_j$ of N. The similarity of message token sequences M and N can be achieved by calculating the scores iteratively according to (7), where subscript M, N of $MTSA_{M,N}[i,j]$ is omitted for short.

$$MTSA[i,j] = \begin{cases} 0 & , min(i,j) = 0 \\ max \begin{cases} MTSA[i-1,j-1] + TSPDA_{M,N}(i,j) \\ MTSA[i-1,j] \\ MTSA[i,j-1] \end{cases} & , i,j > 0 \end{cases}$$

(7)

The detail of MTSA algorithm is spelled out as follows:

```
[Algorithm-MTSA]
    Input:  M, N, L_M, L_N
    Output: MTSA[L_M, L_N]
    1. for i=0 to L_M do
    2.     for j=0 to L_N do
    3.         if i=0 or j=0 do MTSA[i,j]=0
    4.         else MTSA[i,j]=max{MSTA[i-1,j-1]+
                   TSPDA(i,j),MTSA[i-1,j],MTSA[i,j-1]}
    5.         end if
    6.     end for
    7. end for
```

Let M and N be the length of two messages measured by token series, and A be the number of attributes each token have.

Computing Time In the Basic Score step, each token pair needs to match A attributes, and there are $M \times N$ token pairs. So step 1 needs $O(AMN)$ computing time. Step 2 and 3 need to make amendment on $M \times N$ pairs of tokens separately, so the computing time in the later two steps is $O(MN)$. The MTSA algorithm is a dynamic programming process, so it has $O(MN)$ computing time. To sum up, the whole MTSA process has a computing time of $O(AMN)$.

Computing Space The token similarity measuring part builds and maintains a $M \times N$ score matrix. Based on that, the MTSA algorithm compute the sequence similarity using a matrix of $(M + 1) \times (N + 1)$ to store intermediate results. Therefore the computing space of MTSA is $O(M \times N + (M + 1) \times (N + 1)) = O(MN)$.

3 Verification Experiment of MTSA

We choose SMTP protocol to verify MTSA by measuring the similarity among the messages and comparing the result with knowledge from the RRC document.

3.1 Data Preparation

The chosen SMTP messages are shown in Fig. 4. The SP symbol stands for the space character, and 'seq' stands for the serial number of tokens. Attributes for Message B and C are likewise.

Message A	MAIL	SP	FROM	:	<Alice@gmail.com>
seq	1	2	3	4	5
offset	1	5	6	10	11
width	4	1	4	1	17
data	MAIL	SP	FROM	:	<Alice@gmail.com>
Message B	RCPT	SP	TO	:	<Bob@sina.com.cn>
Message C	RCPT	SP	TO	:	<Alice@gmail.com>

Fig. 4. Input Data for MTSA

We use three attributes in this experiment: offset(distance of bytes from start of the message), width(number of bytes this token has), and data(the content of this token).

3.2 MTSA Process and Result

The computing process of MTSA is shown in Fig. 5, including the intermediate results of basic score of Token Similarity in Fig. 5(a), scores after position and distance amendment in Fig. 5(b), Fig. 5(c) separately, and the sequence alignment process in Fig. 5(d).

B\A	1	2	3	4	5
1	2	0	1	0	0
2	0	3	0	1	0
3	0	0	1	0	0
4	0	1	0	2	0
5	0	0	0	0	1

(a) $TS_{A,B}$

B\A	1	2	3	4	5
1	3	0	1.4	0	0
2	0	4.2	0	1.3	0
3	0	0	1.3	0	0
4	0	1.3	0	2.4	0
5	0	0	0	0	1.1

(b) $TSPA_{A,B}$

B\A	1	2	3	4	5
1	3	0	1.12	0	0
2	0	4.2	0	1.04	0
3	0	0	1.3	0	0
4	0	1.04	0	2.4	0
5	0	0	0	0	1.1

(c) $TSPDA_{A,B}$

B\A	0	1	2	3	4	5
0	0	0	0	0	0	0
1	0	3	3	3	3	3
2	0	3	7.2	7.2	7.2	7.2
3	0	3	7.2	8.5	8.5	8.5
4	0	3	7.2	8.5	10.9	10.9
5	0	3	7.2	8.5	10.9	12

(d) $MTSA_{A,B}$

Fig. 5. Computing Process of MTSA

In this experiment, we assign the head to be 5, and tail to be 0, as the messages are relatively short.

According to Fig.5(d), the similarity of token sequences A and B is measured to be 12. Using the same method, similarity of message B, C and message A, B is measured to be 18.4 and 13.1 respectively.

It is fit with the prior knowledge of RFC 821 that message B and C follow the same kind of protocol format, and message A follows another format.

4   Conclusion and Discussions

An Message Token Sequence Alignment algorithm is proposed in this paper based on a token similarity measure strategy. This strategy takes token attributes and protocol format alignment traits(*Position Advantage* and *Distance Weaken*) into account, to make a suitable measurement of the similarity between message token sequences. The verification experiment checked the availability of this method.

As introduced in section 1, messages captured from the Internet need to be partitioned into token sequences before MTSA. Therefore the experiment in this paper is relatively brief and hoping for a more scaled test on an integrated implementation in the future.

The relative measure rules designed in this paper are coarse gained to some extent. Our main idea is considering the unique attributes of protocol format to measure message similarity. The similarity evaluation strategy will be revised to become fine-gained in an entire unknown protocol format reverse system.

Acknowledgement This work is supported by the National Natural Science Foundation of China (Grant Number: 61471141, 61361166006, 61301099), and Basic Research Project of Shenzhen, China (grant Number: JCYJ20150513151706561).

References

1. Protocl Information project, http://4tphi.net/~awalters/PI/PI.html
2. Beddoe, M.A.:Network protocol analysis using bioinformatics algorithms. http://www.baselineresearch.net/PI/
3. Needleman, S.B., Wunsch, C.D.: A general method applicable to the search for similarities in the amino acid sequence of two proteins. J. Mol. Biol. 48:444-453 (1970)
4. Leita, C., Mermoud, K., Dacier, M.:ScriptGen: An Automated Script Generation Tool for Honeyd. In: Proceedings of the 21st Annual Computer Security Applications Conference (ACSAC 2005),
5. Cui, W., Paxson, V., Weaver, N.C., Katz, R.H.:Protocol-Independent Adatpive Replay of Application Dialog. In: Proceedings of the 13th Symposium on Network and Distributed System Security (NDSS 2006)
6. Cui, W., Kannan, J., Wang, H.J.:Discoverer: automatic protocol reverse engineering from network traces. In: Proceedings of the 16th USENIX Security Symposium (2007)
7. Wang, Y., Yun, X., Shafiq, M. Z., Wang, Y., Liu, A., Zhang, Z., et al.:A semantics aware approach to automated reverse engineering unknown protocols. In: Proc. IEEE ICNP, pp. 110 (2012)
8. Levenshtein, V.I.:Binary codes capable of correcting deletions, insertions, and reversals. Soviet Physics Doklady 10 (8): 707710 (1966)

High Precision 3D Point Cloud with Modulated Pulses for LiDAR System

Kai-Jiun Yang and Chi-Tien Sun

Industrial Technology Research Institute,
195, Sec. 4, Chung Hsing Rd., Chutung, Hsinchu, Taiwan 31040, R.O.C.
{ykj,ctsun}@itri.org.tw

Abstract. The LiDAR system uses laser pulses to delineate the 3D point cloud. Conventional LiDAR equipment is bulky and expensive because it contains multiple sets of laser guns and photo diodes with mechanical motors. If there are multiple LiDAR devices in the same filed or other laser beams that are of the same wavelength, the measurements can be interfered by one another. In this paper, we proposed the architecture to differentiate the desired signal from the ambient interference. The detecting laser pulses are encoded while the reflected laser pulses are analysed in both time and frequency domain. Additionally, the accuracy is further improved by phase equalization of the reflected laser pulses. The FPGA platform with MEMs mirror was built to validate the proposed architecture, and the dimension of the platform is greatly reduced so that the prototype is portable.

Keywords: LiDAR, ToF, Multi-user, Fast Fourier Transform, Time Delay and Phase Rotate Conversion.

1 Introduction

The modern real-time object detection uses the methods such as radar, infra-red, sonar, night vision, and LiDAR (Light Detection And Ranging). These approaches have various physical limitations and the cost concerns. The detecting range of radar is hundreds of meters, and the detection is based on single point scanning. It only can tell whether the object exists within the effective range. The provided result cannot fully depict the contour of the objects, and the horizontal resolution is less desirable. The costs of infra-red and sonar are low, but the effective ranges are within few tens of meters. As to the resolution, it is poor and the contour or the objects cannot be fully illustrated as well. The night vision technology captures the image in dark environment through amplifying the dimly reflected light. Its effective range is within hundreds of meters, and the acquired images are in two dimension without depth information. All of the above techniques are susceptible to the ambient lights, the surrounding climates, or the semi-blockage such as foliage or vapour. LiDAR uses laser pulses which the energy is penetrative and concentrative, such that the reflected signal can be discerned after hundreds of meters travelling. The depth map can be

© Springer International Publishing AG 2017   217
J.-S. Pan et al. (eds.), *Advances in Intelligent Information Hiding and Multimedia*
Signal Processing, Smart Innovation, Systems and Technologies 63,
DOI 10.1007/978-3-319-50209-0_27

acquired through analysing the timing, the amplitude, and the width of the received pulses. Additionally the generated depth map includes layer information which is selective so that the underneath contour can be told. Moreover, the effective range can be above hundreds of meters depending on the laser type. However, the dimension of the LiDAR system is bulky and expensive, and the calibration is needed for high precision measurement.

As the progress of development in self-driving cars, the features such as long-range and the high-precision in LiDAR are applied for the navigation. The self-driving cars developed by Google have applied the LiDAR system designed by Velodyne [1]. In the actual field, the LiDAR system needs to counteract various interferences caused by the weather and the blockage. Another important issue is the interference caused by the other light sources with the same wavelength in the co-existed field. The LiDAR systems actively generate the laser pulses, which can be regarded as the interference when the other LiDAR systems pick up the detecting signals. Therefore, the signal differentiation and encryption also play important roles in the design of LiDAR system. The design of the LiDAR system can be categorized into five agendas: the laser signal sources, the laser modulation techniques, the detection of the laser signals, the estimation of the distance, and the reconstruction of the depth map. This paper reviews the fundamentals of the LiDAR system, and the focuses on laser modulation techniques and the estimation of the distance. The algorithms are proposed to suppress those non-ideal interference, which refines the estimation in the frequency domain and compensates the results in time domain. Also an FPGA prototype was built to verify the feasibility and the performance, which was meant for portable LiDAR system that shall be applied in self-driving vehicles or artificial visual enhancement.

2   Fundamentals of LiDAR and the Applications

2.1   Laser Ranging Principles

The LiDAR system uses densely distributed laser pulses to scan the field and renders the 3D depth map based on those reflected signal. Since the laser is a kind of non-dispersive light which can penetrate water or gaseous medium, it is also suitable for all kinds of non-contact ranging. Based on the types of the carriers, there are three kinds of LiDAR system: airborne LiDAR, bathymetric LiDAR, and terrestrial LiDAR. Since the LiDAR system uses active light source, it is less susceptible to ambient interference, such that the generated depth map is far more accurate than those converted from 2D images which can also be compromised by the angles during photographing. According to the forms of the laser signals, there are two kinds of detection used in LiDAR systems: Tof and phase detection.

2.2   Time of Flight (ToF) Detection

The ToF LiDAR estimates the distance based on the elapsed time of the round trip from the laser gun that generates the pulse at the transmitter, reflected by the object, to the photo diode that captures the pulse at the receiver [2]. Currently this is the major method used in long range LiDAR. At the transmitter, the laser gun emits the pulses that are of high instantaneous power within the width less than one nano second as shown in Fig.1. To avoid damaging human

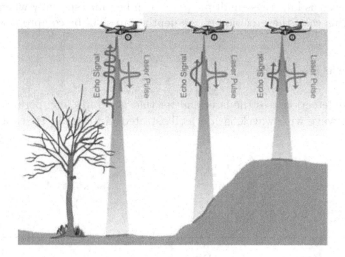

Fig. 1. The principle of ToF LiDAR.

eyes, the laser emitter must comply to eye-safe class one laser. The laser pulses travels through the medium and bounces back when hitting impenetrable objects. As a result, the target distance can be estimated by calculating the time t that the pulses have travelled

$$R = C \times t/2 \tag{1}$$

where C is the speed of the light. This means that one nano second difference is 15cm apart. Therefore, one of the import key factors in ToF estimation is the timing resolution of the receiver.

The reflected laser pulses carry additional information other than the distance. As the pulse travels through the air, the delay spread transforms the pulse into the shape that can be approximated by Gaussian distribution. Based on the amplitude and the width of the return pulses, the material of the targets can also be delineated as shown in Fig.1. The larger the intensity in pulse, the harder the object. For example, the intensity shall be different between the pulses that hit the steel and those hit the mud. On the other hand, the width of the pulses can tell whether the surface is tilted.

The crucial feature that makes LiDAR unique is selectable layer rendering. As mentioned earlier, the received time, the pulse width, and the intensity can be interpreted to specific physical characteristics. Therefore, the users can choose the representation based on the features, for example, the reflectivity, of the objects. Additionally, the laser pulses can penetrate multiple layers of certain objects as shown in Fig.1. The single laser pulse not only bounced back when hitting the ground. On its way back, the laser pulse also sweeps through several branches and generates additional small pulses. Hence, the reconstructed depth map of the branches can be selected independently. However, when the resolution of the receiver is low, these multiple pulses can be blur especially when they are close. In this case, the precision of the depth map shall be compromised.

2.3 Phase Detection

The phase detection uses the laser emitter that generates long period of continuous laser wave with wavelength λ as illustrated in Fig.2 [3]. By capturing the

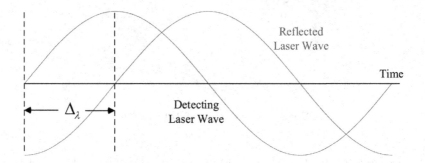

Fig. 2. The principle of phase detection used in LiDAR.

relevant amplitude of the reflected laser wave, the receiver can tell the phase delay of the emitted detecting laser wave by the delayed wavelength Δ_λ . The estimated distance is then derived by

$$R = (M\lambda + \Delta_\lambda)/2 \tag{2}$$

where M is the number of the periodic laser wave. If the distance to be measured is longer than the wavelength of the periodic laser wave, a timing mark should be added into the laser wave and it is logged at the receiver.

The distance estimation that uses phase detection usually is applied in short range measurement within 100 meters. Due to the class-one laser power limitation, the laser emitter cannot generate periodic laser wave with high energy. However, this kind of devices can be made in compact size and the precision is within centimetres, such as the hand-held laser range finder Bosch GLM series. During the measurement, if there is other laser source with the same laser wavelength superimposed on the same target, the result shall be erroneous or the measurement fails due to such an interference.

2.4 Generation of 2D Detection Matrix

The 2D detection matrix is generated by one or more laser emitters with corresponding photo diodes and the mechanical module for movement control. In Velodyne HDL-64E[1], there are 64 laser emitters with the sensors that are mounted on the servomotor to capture the panorama. Hence, the size is large and the frame rate is restricted. MEMs mirror was applied to replace the servomotor for compact and economical solution [4]. MEMs mirror steers the reflective surface in 1-D or 2-D, and the round-trip frequency can be as high as thousands of hertz. Therefore, the scan provides hundreds of frames per second for better resolution. Yet the module is fragile. If there is dust that falls on the mirror, it can damage the device. Polygon mirror is another option to generate the detecting matrix [5]. However, the image refresh rate is reduced due to the mechanical RPM limit of the servomotor.

3 Signal Processing for User Differentiation

3.1 Encoded Pulses for User Differentiation

To differentiate the detecting signals from different LiDAR sources, we propose to modulate the source signals so that each LiDAR device can discern its own signal. As to the phase-detection based architecture, the symbols can be modulated as that are used in orthogonal frequency division multiplexing (OFDM). Since the transmitter and the receiver are both on the same platform, the timing and frequency offset can be well monitored. With these methods, the user identity can be hidden in the modulated symbols. Furthermore, the signal integrity can be preserved by DSP techniques such as tracking and equalization.

3.2 Rectification of Phases for Better Precision

The duality of fast Fourier transform (FFT) suggests that the time delay is the phase shift in frequency domain [6].

$$x[n - t_D] \leftrightarrow e^{-j\frac{2\pi}{N}kt_D} X[k] \tag{3}$$

The timing acquisition using only convolution depends highly on the sampling rate of ADC. For example, if the sampling rate is 1GHz, the resolution of such ToF LiDAR shall be limited to 15cm. However, the cost of high sampling rate ADC is expensive and the resolution is limited. With the help of FFT and the duality in (3), such a limitation can be improved by comparing the phase difference between the modulated symbols at the transmitter and the receiver. The proposed architecture is as shown in Fig.3. As the laser emitter is triggered, the transmitted encoded pulses is convoluted with the source encoding pattern, and the start point can be marked by the max scan of the convoluted outputs. Both the FIFOs at the transmitter and the receiver are enabled at the start of the TX encoded pulses, and the same amount of samples are logged for the

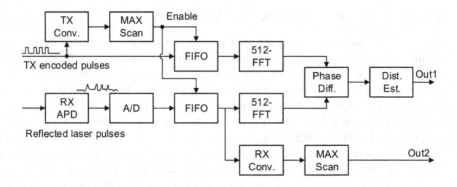

Fig. 3. The fundamental DSP blocks applied in multi-user LiDAR system.

following N-FFTs. For implementation convenience, not all the N outputs from FFTs are needed to be compared. In our case, only three sub-carriers that of of maximal energy are excerpted for comparing the phase difference. The phase differences between the transmitted and the received symbols are tracked by coordinate rotation digital Computer (CORDIC) [7]. For the kth sub-carrier, the phase difference is $e^{j\Delta_\theta[k]}$. With the duality in (3), the time delay of the kth subcarrier is

$$t_D[k] = \frac{Nk}{2\pi}\Delta_\theta[k] \tag{4}$$

By averaging the delay of the selected sub-carriers, the distance is therefore estimated as in (1).

4   Experiments

4.1   Testing Platform

The proposed multi-user differentiation was implemented on FPGA platform for real-time depth map rendering. The encoded laser pulse sequence was generated by Mitsubishi laser emitter which the wavelength was 658nm and the average power was 100mW, and it was driven by the encoded sequence modulated from FPGA. Meanwhile, the 2D laser matrix was reflected by the MEMs mirror OP-6111 that was produced by OPUS Microsystems. As to the receiver, the avalanche photo diode (APD) C5658 that was produced by Hamamatsu was hooked to the optical lenses for receiving the reflected laser pulses. The received power signal was then sampled by the ADC with sampling rate 1.25G per second, which the model is FMC126 from 4DSP. In connection with these external components, the DSP function as depicted in Fig.3 was implemented on the FPGA platform ML605 from Xilinx. Afterwards the processed estimation was fed back to PC, and the 3D depth map was rendered by Matlab in real time.

The ADC output of the received pulse sequence and the superimposed result are as drawn in Fig.4. The update of the depth map ran in real time, and

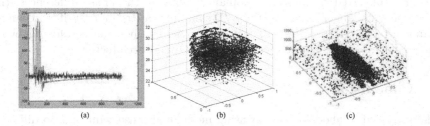

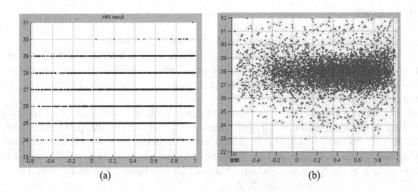

Fig. 4. The detected hand toy was placed 50cm in front of the laser gun, and (a) the transmitted (red) and the received (blue) encoded laser pulse sequence. (b) The point cloud generated from Out2 in Fig.3. (c) The point cloud generated from Out1 in Fig.3

the frame rate was one frame per second. Since the detecting laser pulses are encoded, the other laser sources with the same wavelength but different coding sequence did not affect the desired inputs to the receiver. Hence, the proposed system is robust against different LiDAR devices.

4.2 Analysis

Let us discuss the accuracy of the estimation with the proposed algorithm. The

Fig. 5. (a)The lateral view of Fig.4(b), and (b) the lateral view of Fig.4(c).

proposed modulated pulses detection can not only focus on its own source, but the accuracy can also be greatly improved. The results in Fig.5(a) uses only convolution for distance estimation. In such a case, the precision is bounded by the A/D sampling rate. In our experiment, the sampling rate of A/D is

1.25G samples per second. Hence, the precision is limited to 12cm. On the other hand, the precision can be enhanced to few centimetres as shown in Fig.5(b) if the proposed estimation is applied. Since the received samples in time domain are first converted to frequency domain to extract the phase differences, the precision is no longer bound by the A/D sampling rate. Moreover, if the channel gain can be properly estimated, each sub-carrier can be equalized for removing the channel response as the method used in OFDM architecture.

5 Conclusion

The proposed method focuses on using modulated pulse sequences to differentiate the LiDAR users in the same field, and the phase rectification helps to refine the distance estimation, such that the precision of the 3D depth map can be improved. Additionally, the hardware platform have been built as a proof of concept. MEMs mirror is applied to reduce the size of LiDAR system and the cost. The proposed algorithm can effectively counteract the interference in the optical channels, and the estimations are converged accordingly. Moreover, the proposed method, which uses encoded pulse sequence, is suitable for multiple LiDAR systems to work in the same filed at the same time. Further signal processing techniques such as timing and frequency tracking that are used in OFDM architecture shall be applied in the future study. To better counteract the channel imperfection and the variation in different devices, we will add in the gain control and adaptive filters to excavate the received samples that are buried in the background noise. As a result, the detecting range can be reliably extended.

References

1. Velodyne HDL-64e, http://velodynelidar.com/hdl-64e.html.
2. J. Kostamovaara, K. Maatta, and R. Myllyla, "Pulsed time-of-fight laser range finding techniques for industrial applications," *Proc. SPIE Conf. Intelligent Robotic Systems-Optics, Illumination and Image Sensing for Machine Vision*, Nov. 1991, vol. 1614, pp. 283–295.
3. M. Ou-Yang, "High-dynamic-range laser range finders based on a novel multimodulated frequency method", *Optical Engineering*, vol. 45, no. 12, p. 123603, 2006.
4. K. Ito, C. Niclass, I. Aoyagi, H. Matsubara, M. Soga, S. Kato, M. Maeda and M. Kagami, "System design and performance characterization of a MEMS-based laser scanning time-of-flight sensor based on a 256 × 64-pixel Single-Photon Imager", *IEEE Photonics J.*, vol. 5, no. 2, 2013.
5. C. Niclass, M. Soga, H. Matsubara, M. Ogawa and M. Kagami, "A 0.18-μ CMOS SoC for a 100-m-Range 10-Frame/s 200 × 96-Pixel Time-of-Flight Depth Sensor," *IEEE Journal of Solid-State Circuits*, vol. 49, no. 1, pp. 315–330, Jan. 2014.
6. A. Oppenheim and R. Schafer, Discrete-time signal processing. Englewood Cliffs, N.J.: Prentice Hall, 1999.
7. Y.H. Hu, "CORDIC-based VLSI architectures for digital signal processing," *Signal Processing Magazine, IEEE*, vol. 9, no. 3, pp.16–35, July 1992.

Research on Frequency Automatically Switching Technology for China Highway Traffic Radio

Lei CAI, Zhong-teng YU *, and Chun-lei MENG

National Center of ITS Engineering and Technology, Research Institute of Highway, M.O.T
Key Laboratory of Intelligent Transportation Systems Technologies, M.O.T
No.8 Xi Tucheng Road, Haidian District, Beijing, 100088, P.R.C
http://www.rioh.cn

Abstract. Based on the strictly investigate of worldwide researches on traffic radio, this paper presents the differences among China Highway Traffic Radio(CHTR), traditional traffic radio and network broadcast. Data transmitting technology over subcarrier is studied first. Since China Highway Traffic Radio is faced with multi-frequency problem in different region, the mainly research is to development frequency automatically switching technology. The FM receiver design and experiment are shown in the paper. The experiment result indicates that the FM receiver can automatically switch its frequency when necessary. At last, the conclusion present the significance of this research on promoting China Highway Traffic Radio and raising listeners' feeling and safety.

Keywords: Information service, China Highway Traffic Radio, Data Radio, Frequency automatically switch

1   Introduction

According to the China national economy and social development statistical bulletin, by the end of 2015, the number of private car in China had reached to 143.99 million, which was 14.4% more than last year. Meanwhile, the public transport industry and tourism developed rapidly. However, the bad weather such as fog, typhoon, heavy snow and freezing weather occurred more frequently these years. At the same time, the traffic accident and traffic blocking event happened more than before. The demand of the public to timely and reliable travel information is increasingly urgent. When the weather disaster in 2008 and "5.12" Wenchuan earthquake happened, the roads, communication and electric power were destroyed. Radio became the significant information channel. In addition, road network management and service are faced with higher requirements, because of bad weather and rising car ownership. How to serve the public better, improve the capacity of road emergency disposal is an urgent problem in the transportation industry.

---

* National Key Technology R&D Program(2014BAG01B02)

© Springer International Publishing AG 2017
J.-S. Pan et al. (eds.), *Advances in Intelligent Information Hiding and Multimedia Signal Processing*, Smart Innovation, Systems and Technologies 63,
DOI 10.1007/978-3-319-50209-0_28

Fig. 1. Rain, snow and freezing weather in China

Since strategic cooperation agreement between China Ministry of Transport and China National Radio was signed in 2009, the national highway traffic radio came into existence. Broadcast technology for highway traffic radio have been studied and explored[1–7] .Meanwhile China highway traffic radio Jing-Jin-Tang highway pilot project and China highway traffic radio demonstration project in Beijing, Tianjin, Hebei, Hunan and Sichuan five provinces have been established and implemented. In 2016, the construction of China highway traffic radio has been included into the 13th Five-Year Plan for Transportation Informatization. China highway traffic radio will be constructed in not only these demonstration project, but also other provinces in China.

Fig. 2. China's highway traffic radio premiere ceremony on June 26th,2012

During highway traffic radio demonstration project in five provinces was implemented, due to the shortage of frequency resource, there was a problem that the frequency have to be different in different provinces even in different cities of one province the radio frequency resources carry out the process, due to the lack of frequency resources in our country, many adjacent province or unity can't realize the frequency broadcast, driving on the highway of the public may need to manually switch across the province driving frequency, frequency uncertainty, unsafe, FM process shows the problem such as reception, how to realize automatic frequency switch, how to ensure China's highway traffic radio long-term operation, it is necessary to study frequency switching technology.

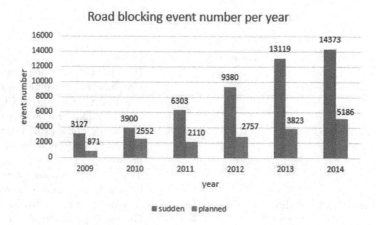

Fig. 3. Road blocking event number per year from 2009 to 2014

2   Global circumstance

2.1   Circumstance of foreign research development

Traffic broadcasting as the most effective and efficient method of information transmission while travelling are fully covered at other developed countries. In 1997, German was first to applicate TMC service while use to broadcast live traffic and weather information through FM radio, and then TMC was rapidly and widely covered at other European countries. In addition, TMC is based on the principle of serve the whole section of a highway; however, it is totally different in China, a highway might be divide into different sections as the highway normally across serval provinces or regions management may be an issue, as well as trans-industrial issue. Moreover, the transmission data length is comparably shorter, and do not support Chinese. In 1996, Japan has established VICS, information announced through FM with multicast to targets on a wide range of coverage in certain area, as the standard of VICS is only permitted within Japan, and transmit and receive devices are highly customized, so that build VICS seems an unable finished project in China.

2.2   Differences between CHTR and traditional traffic radio

At the end of 1980s, traffic radio or other specific traffic broadcast began to opening across the country. At present, most provinces or regions had established their own traffic radio station in China. With the largely increasing number of vehicles radio audiences has a rapid expansion during the last few years. In an increasingly congested city, especially giant city, traffic radio has becoming the most favorite program, with the largest number of audiences, and gain the highest advertising profit.[8, 9]

Table 1. current situation of traffic radio

Radio station	Frequency	Detail
Jiangsu Communication Broadcasting Station	FM101.1	Broadcast FM stereo in Jiangsu Province and surrounding regions
Shanghai Traffic Radio	FM105.7/AM648	The first traffic radio in China, two of the main programs are traffic and automobile, provide traffic and hospitality information in every 2 hours
Guangzhou Traffic Radio	FM105.2	Traffic information covered in Guangzhou and surrounding regions, and highways within Guangdong province, mostly in Cantonese

3   Frequency automatically switching technology research and equipment development

3.1   Data Radio sending and receiving technology

The traffic data information will be sent and received through FM subcarrier. The traffic data information system will be mainly composed of transmitting base station and FM reception, which are shown in fig.4. Encoding and modulation part is combined by a data channel encoding module and a modulation module. Decoding and demodulation part mainly realize the separation and demodulation of subcarrier frequency and data recovery.

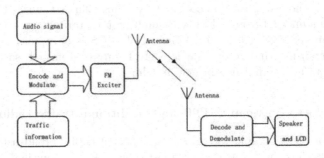

Fig. 4. Data transmit system structure

Table 2. Comparison among different broadcast

	China Highway Traffic Radio	Traditional Traffic Radio	Network Broadcast
Base station	Using traditional broadcast base stations and low power base stations which are set up along highway	Using only traditional high power broadcast base station	Using mobile communication base station and Internet access point
Service area	Highway as well as primary road and major city	Only City	City Signal is weak between cities.
Organization department	Transportation industry and radio and TV industry	Radio and TV industry	Radio and TV industry or Internet company
Framework	National unified framework	Planning in a province or a city	Some are set up by traditional traffic radio, others are pure network broadcast
Function	FM radio cover different provinces and different condition including tunnel.Set up as part of National emergency broadcast.	Traffic information service for city	pure network broadcast lack of traffic information and audience.
Traffic information broadcast frequency	At any time the need arises	In fixed time period	Barely
Information acquire and broadcast delay	Obtain information and broadcast without delay	Obtain information in fixed time period and broadcast without delay	Broadcast with about 1min delay

3.2 Frequency automatically switching technology

Frequency automatically changing technology is mainly for car and portable radio reception, when it left from one frequency area to another frequency area. For these two frequency belong to the same FM station, the programs are also the same. If the car radio select the better performance frequency automatically, both the audience listening experience and information services will be raised. Based on FM subcarrier, there are three modes of frequency automatically changing applications as shown in table.3.

Only a very few FM station in China transmit program type information. Therefore, the second mode is difficult to apply. In order to solve the different frequencies problem of China Highway Traffic Radio, the first mode is mainly studied.

Table 3. Frequency automatically switching mode

Mode	Function
FM station alterative frequencies	Some FM stations have different frequencies in different regions. When people drive to a new region, receiver will select the alterative frequency. Since this, people can listen the same program all the time.
Same program type	Receiver can choose another same type program when current program signal is weak or lost.
Traffic announcements	When some emergency traffic information arrive, receiver will stop current program and broadcast the traffic information.

The flow of alternative frequency switching is shown in fig.5. When signal is received and this function is turned on, first step is to judge whether Received Signal Strength Indication(RSSI) is below the threshold. If it is lower than the thresholdall the alternative frequencies will be tried one by one until some RSSI satisfy the condition.

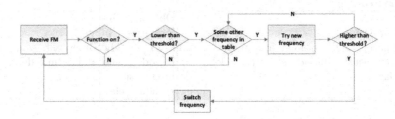

Fig. 5. Frequency automatically switching flow

3.3   Receiver development

Hardware structure is shown in the fig.6. As shown, SI4721 chip is used as the FM reception and decoding chip, which has plenty functions, such as stereo demodulating, subcarrier demodulating and signal decoding. ARM stm32f103 is used as MCU to control all the components.

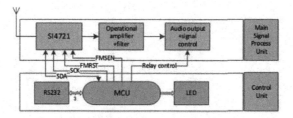

Fig. 6. Receiver hardware structure

4 Experiment and analysis

4.1 Experiment purpose

The experiment purpose is to find out whether the receiver can switch to the right frequency in FM station frequency table when current frequency strength becomes weak and the strength of some other frequency in the table meets the requirement. If it works, user can listen China Highway Traffic Radio all the way without manual changing frequency, no matter which frequency area he or she drives in.

4.2 Experiment plan

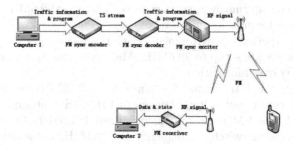

Fig. 7. experiment structure

As shown in fig.7 the experiment plan use computer 1 as traffic information source and program source. Computer 1 transmits both traffic information and program information to FM sync encoder, in which the two information are combined and encoded into a TS stream. After receiving the TS stream, the FM sync decoder extract traffic information and program signals and transmit them to the FM exciter. The exciter modulates them on a fundamental frequency and send out it by the antenna. At the receive part, receiver get the signal and process. Some results and states are shown at computer 2. The port connection diagram is shown in the fig.8 In order to provide Audio Engineer Society (AES)

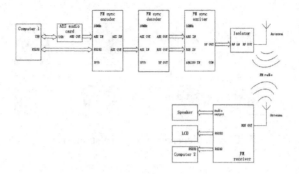

Fig. 8. experiment port connection diagram

standard signal, an audio card is linked between computer 1 and FM sync encoder. The TS stream is transmitted over Asynchronous Serial Interface (ASI). In addition, Radio Data System(RDS) means the FM subcarrier signal.

4.3 Experiment steps

step 1. In order to avoid interruption of other FM station, the experiment is conducted in the experiment tunnel of Research Institute of Highway. Test RSSI in the tunnel before experiment.
step 2. Setup the alternative frequency table. There are 99.6MHz,101.9MHz and 102.1MHz three frequencies in the table.
step 3. Setup experiment environment. Both the FM exciter frequency and the receiver frequency are setup to 99.6MHz. Make sure the receiver can present the program and traffic information.
step 4. Switch the FM exciter frequency from 99.6MHz to 101.9MHz. Test whether the receiver switch its frequency to 101.9MHz automatically.
step 5. Switch the FM exciter frequency from 101.9MHz to 102.1MHz. Test whether the receiver switch its frequency to 102.1MHz automatically.
step 6. Repeat step3-5 several times.

4.4 Experiment result

The fig.9 presents the experiment record of the FM exciter and the receiver. When the FM exciter frequency was changed, the receiver detected that the current RSSI was lower than threshold. After about 2 seconds, the receiver switched to the same frequency as the FM exciter. The program and traffic information receiving and performing would continue. The receiver had passed all the experiments above. The frequency automatically changing technology was proved to be effective.

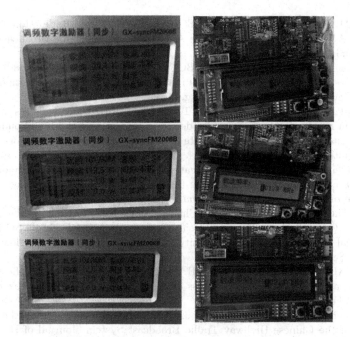

Fig. 9. experiment record

5   Conclusion

China Highway Traffic Radio was established by the Ministry of Transport of the People's Republic of China (MOT) and the China National Radio, a significant foundation of the National Emergency Broadcasting System. Currently, there are serval regions had been covered by CHTR, such as, Beijing, Tianjin, Chongqing, Hebei Province and Hunan Province this five regions, and Beijing-Tianjin-Tangshan Expressway. In this article, through strictly investigate of current researches worldwide, had been found that the differences between CHTR and traditional traffic radio, and analysis the technology of transmit and receive broadcasting data, and automatic frequency control system. In addition, automatic frequency control terminal devices were developed and had been taking into numerous tests at labs and outdoor. As result, the findings demonstrate that these devices were able to switch radio frequency automatically, transmit and receive audio and data properly. These results are create a significant foundation of promoting CHTR to the national wide, and CHTR plays an important role of leading Chinas highway traffic broadcasting to a healthy developing industry.

References

1. ZONG Feng, LI Kang, LI Wei-feng. Understanding of The Standard on Frequency Modulation (FM) sound sync-broadcasting [J]. Radio & Tv Broadcast Engineering, 2001,(1):68-72

2. HOFMANN F., HANSEN C., and SCHAFER W., Digital Radio Mondiale(DRM) Digital Sound Broadcasting In The AM Bands, IEEE Trans. Broadcasting, 2003 ,49(3):319-328.
3. Chun-lei MENG . Utilization of IBOC Technology to Provide Broadband Data Services via Highway FM Radio, Journal of Highway and Transportation Research and Development, Vol.29, No.8, Aug. 2012
4. Chun-lei MENG . Research on Key Technologies of Highway Seamless Information System in Mountainous Area, Academic Proceedings of the first Young Specialists Committee of Research Institute of Highway Ministry of Transport Academic Committee, Vol. 7, 2012
5. Chun-lei MENG , Lei CAI. Application on Auto Synchronous Technology Used in Travel Information Service along the Highway, Journal of Highway and Transportation Research and Development, Vol. 29, No. S1, Jul. 2012
6. CUI Tian-xiao, MENG Chun-lei, Chao-shi CAI, Xiao-ning FENG, Feng HU. An Active Constellation Extension Architecture for MIMO-OFDM STBC Decoding, Journal of Communication University of China (Science and Technology), Vol.19, No.3, Sep. 2012
7. Lei CAI, Chun-lei MENG, Liang HAO. The Chinese Highway Traffic Dedicated Broadcast. 20th ITS world congress Tokyo 2013, October14,2013/Tokyo International Forum
8. CAI Lei, MENG Chunlei, HAO Liang, WANG Xinke. Research on the Architecture of the Chinese Highway Traffic Broadcast System, Journal of Highway and Transportation Research and Development, Vol. 32, No.S1, Apr. 2015
9. Xin-ke WANG, Sheng-xi CAI, Lei CAI, Chun-lei MENG . Research of Highway Traffic Radio Information Safety Management, Journal of Highway and Transportation Research and Development, Vol. 6, 2015

An Automatic Decoding Method for Morse Signal based on Clustering Algorithm

Yaqi Wang [1,2,3*], Zhonghua Sun [1,2,3*] and Kebin Jia[1,2,3*]

1 Beijing Laboratory of Advanced Information Networks, Beijing, China
2 College of Electronic Information and Control Engineering, Beijing University of Technology,
Beijing, China
*[*kebinj@bjut.edu.cn](mailto:kebinj@bjut.edu.cn)*

Abstract. For technical problems of low accuracy of automatic decoding for Morse signal, an automatic decoding method for time-frequency spectrum of manual or mechanical Morse signal is put forward, which based on time-frequency analysis method and machine learning technology. It generates time-frequency image based on STFT, which used for extraction of Morse signal based on adaptive image enhancement later. K-means clustering algorithm have been introduced to identify the dots, dashes and interval between them. Error-correction algorithm put forward to improve the accuracy of decoding. Simulation experiment and engineering practice on Morse signal demonstrate the effectiveness and feasibility of this algorithm.

Key words: Morse Code; Image Enhancement; Automatic Decoding; Clustering Algorithm

1 Introduction

Short wave telegraph is an important part of the military wireless communication, and receiving telegrams is mostly achieved by radio operators to listen and transcribe. With the progress of science and technology development, this artificial way of receive increasingly exposed some disadvantages. This way of working requires radio operators with a strong ability to listen and write down. Artificial decoding work is repetitive and boring, and people's ability to respond is limited. Mistranslation and omissions situation is inevitable. Therefore, the study of an automatic identification and decoding method of Morse signal which used to instead of the traditional way is a very important issue [1] [2]. The study is the need for Automation construction.

Automatic decoding of Morse signal, aimed at extract information from noisy signals at the non-manual auxiliary case. Morse code is a non-uniform code, use different combinations of dots, dashes, and intervals to represent letters, numbers, punctuation and symbols. In this paper, characterization of Morse signal properties in the frequency domain used to extract graphic feature of Morse code in time-frequency domain. Machine learning methods used to restore Morse code message into meaningful numbers, letters or symbols, in order to achieve automatic identification and decoding of mechanical and manual Morse code.

© Springer International Publishing AG 2017
J.-S. Pan et al. (eds.), *Advances in Intelligent Information Hiding and Multimedia Signal Processing*, Smart Innovation, Systems and Technologies 63,
DOI 10.1007/978-3-319-50209-0_29

235

2   Pretreatment

2.1 Generate Time-frequency Image of Morse Signal

In both time domain and frequency domain, Morse signal has prominent features. Time-frequency analysis method, which taking into account the characteristics of both time domain and frequency domain analysis Morse signal, is the more popular approach and has been applied in many fields [3]. Firstly, generate time-frequency distribution of one-dimensional signal by time-frequency conversion. Then the time-frequency distribution is converted into time-frequency image by a mapping function. The typical feature of time-frequency image of Morse signal is it looks like a rectangular light bars.

Generate time-frequency distribution matrices f(x, y) of Morse signal by short-time Fourier transform (STFT) [4], where x represents the rows of the matrix coordinate and y coordinate represents the column matrix. Generate a time-frequency image g(x, y) by the gray mapping function, then g(x, y) used for analysis and object extraction. Gray mapping function is as follows,

$$g(\mathrm{x}, \mathrm{y}) = \frac{f(\mathrm{x},\mathrm{y})-\min\_f}{\max\_f-\min\_f} * 255. \tag{1}$$

Where the maximum value of Matrix $f(\mathrm{x}, \mathrm{y})$ is $\max\_f$, and the minimum value is $\min\_f$.

2.2   Image Enhancement

Image enhancement technique of Digital image processing is used to improve image visual effect [5]. Contrast enhancement is a common method of image enhancement technology. In this paper the method of contrast enhancement is used to enhance Morse - frequency image, so as to achieve the purpose that outstanding signal area and suppressing noise. Prior to enhance the contrast of time-frequency image, range of gray values of the region of interest (ROI) of the image will be estimated. Here, we calculate gray level distribution interval [low high] of ROI according to the distribution histogram.

For the input image g, where g(x, y) represents the gray scale value of the pixel of the x row and y column of the image. Equation (2), (3), respectively, give both inclusive grayscale distribution,

$$\mathrm{low} = \frac{(|\mathrm{peak\_gray}-\mathrm{mean\_gray}|+ \mathrm{peak\_gray})}{255} \tag{2}$$

$$\mathrm{high} = \frac{\mathrm{max\_gray}}{255} \tag{3}$$

Where peak_gray is the peak of the image gray values, max_gray represents mean grayscale of the image, and max_gray is the maximum grayscale of the image. According to the estimated range of grayscale, we enhance the contrast of the image g. The formula is,

$$g(x,y) = \begin{cases} 0 & , g(x,y) \leq low \\ \frac{g(x,y) - \text{min\_gray}}{\text{max\_gray} - \text{min\_gray}} * 255 & , g(x,y) > low \end{cases} \tag{4}$$

2.3 Binaryzation and Morphological Denoising

Image segmentation is a key step in the image processing to image analysis [6]. The existing image segmentation methods mainly include: threshold-based segmentation, region-based segmentation, edge-based segmentation. The proportion of the target area of Morse signal time-frequency image is relatively small and the contribution of its gray information to the whole image is small. So in the paper, Otsu segmentation method is used to obtain an adaptive threshold. And to separate target and background, we use the threshold-based segmentation method to obtain a binary image. Binary image still exist some small noise.

Outlier removal Method and the small fragments removal method of Morphological processing are used for further denoise.

3 Morse Code Decoding

3.1 Parameter Extraction

In the time-frequency image, Morse signal appears as a series of small rectangular area. The length of rectangular block on the time axis represents the code length. The distance between adjacent rectangular blocks is the length of the interval, there are three kinds of intervals: code interval, character interval and character interval. The key of the decoding is calculate dynamically two kinds of basal code and the three kinds of interval.

By sequentially recording the length of dot, dash and intervals of the dot-dash in the binary image, a two-dimensional matrix that represents the sequence of Morse code will be obtained. The matrix includes information that alternating code and interval, and the corresponding code length and interval length.

3.2 Decoding Algorithm

Decoding transform Morse signal which expressed in the form of data length into a Morse code message. Decoding algorithm consists of two parts: identify dot-dash based on K-means clustering method and decoding based on lookup code table method. The standard ratio of the time length between elements of Morse code (dots, dashes, code interval, character interval, word interval) is: 1: 3: 1: 5: 7. But in reality, radio operators have difficulties in maintaining a stable typing of Morse code and serious noise interference that would cause the length of the data obtained cannot strict satisfy the above proportional relationship. Hence, the automated decoding algorithm of Morse code is becoming more on demand. K-means clustering algorithm of Machine Learning can improve the recognition rate of dot and das.

Based on the matrix obtained in the stage of parameter extraction, we can get the distribution diagram of the five elements (Fig.1). The duration normalized for ease of analysis. In the figure, five elements are marked out respectively, it is not hard to find out that the distribution of the same type of elements are gathered.

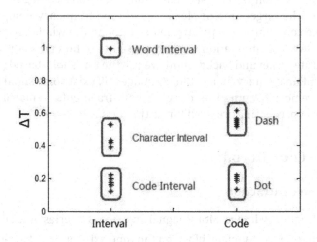

Fig.1. Distribution of five elements.

K-means clustering algorithm is the most classical clustering method based on classification [7]. Its basic idea is: cluster by centering on the k point in space, and classifies the object that most close to them. Through iterative method, successively update the value of the clustering center until the best clustering results are obtained. The biggest advantage of the algorithm is simple and rapid, and the key problem of the algorithm is the choice of initial center and distance formula. Since the element distribution of Morse signals has significant clustering feature, K-means clustering algorithm can effectively differentiate between the five elements. A Morse code decoding algorithm based on K-means clustering method is proposed:

Step1: Normalization the length values of codes and intervals. $T1i$ represents the length of code elements and $T0j$ represents the length of interval elements. Found out the maximum length max1 in $T1i$, and the maximum length max0 in $T0j$. Normalization by formula $T1i' = T1i/max1$ and $T0j' = T0j/max0$.

Step2: Choose five initial centers. Through observation and analysis, the 5 initial clustering center is set to: (1,0),(1,1),(0,0),(0,0.4),(0,1).

Step3: Calculate the distance between each sample to the clustering center; and update clustering centers.

Step4: Obtain the best clustering results through iterative method.

Step5: The five clustering results classified as dots and dashes, code interval, character interval and word interval.

Step6: Output identification results of dot and dash.

Step7: Lookup the Morse code table automatically and output the decoding result.

3.3 Error-correction Algorithm

The length of code or interval obtained by parameters extraction may be wrong, this is the cause of an error decoding result. There are two main causes that generates error codes. One is the noise interference, it resulted in shortening the length of the code, and code will be submerged by noise when serious interference exists; another is the incorrect estimation of the interval, that led to the incorrect combination of dot and dash, generates invalid code.

There lists the error correction algorithm under some specific error conditions as follows:

① If the length of a code is more than seven times the length of dot, consider that code is not even and split into two dash.

② If the length of dot-dash sequence of a character is more than six, split it at the maximum interval until get valid codes.

③ If there are two consecutive invalid codes, priority merging the two invalid codes, consider the merger of invalid code and valid code if the code cannot be effectively merged.

④ If an invalid code adjacent to valid code at each side, consider whether there is abnormal dash or dot in the invalid code sequence firstly, split or combine the sequence. Otherwise, consider the merger of invalid code and its adjacent valid code;

⑤ In addition to the above situation, which is absolutely invalid code.

4 The Experimental Results and Analysis

To verify the suitability of this algorithm, simulation experiments of decoding were conducted. Morse signals having the same content on different Signal Noise Ratio (SNR) are automatically decoded, and Table 1 shows the experimental results.

The results show that the algorithm can adapt to poor noise environments. In the case that the SNR of signal is higher than -10dB, it could get the right result. At the same time as the SNR in the simulation experiments decreases, it is easy to find that the effect of decoding has a downward trend.

Tab.1. Results of decoding

SNR	result	evaluation
5dB	HLG HLG HLG	good
2dB	HLG HLG HLG	good
−1dB	HLG HLG HLG	good
−4dB	HLG HLG HLG	good
−7dB	HLG HLG HLG	good
−10dB	EIE L G H L G H L G	medium
−13dB	IELG EIL6 EILG	poor

Using the above algorithm automatically decoding an actually collected Morse signal influenced by noise interference. Fig.2 shows the waveform of Morse signal in time-domain. After STFT and gray mapping, we get the time-frequency image of the signal (Fig.3a). You can see the interference noise in the time-frequency image.

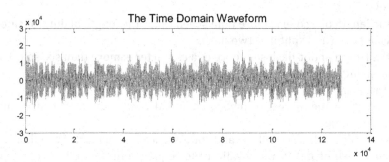

Fig.2. The time domain waveform.

Through the image enhancement, the contrast of the image has significantly improved (Fig.3.b). The calculated value of low is 0.663. Using the method of image binaryzation and morphological processing method to remove noise and leaving only the target area. As shown in Fig.3.c, only leaves rectangular target areas.

We get the length values of 36 codes and 35 intervals in the stage of parameter extraction. The maximum length value of code calculated is 15 and the maximum length value of interval is 25. The two maximums are used to normalize all the length values.

a.Time-Frequency Image

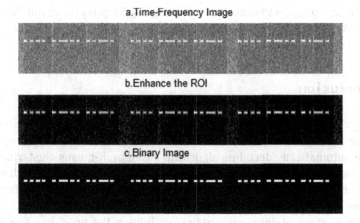

b.Enhance the ROI

c.Binary Image

Fig.3. a. Time-frequency image; b. Enhance the ROI; c. Binary Image.

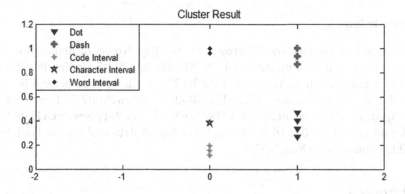

Fig.4. Cluster result.

Cluster dots, dashes and intervals, as shown in the Fig.4, five kinds of elements are correctly classified. Recognition result of dot-dash is obtained by clustering, and group dot-dash correctly according to recognition result of intervals.

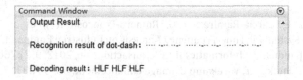

Fig.5. Output the result of decoding.

Trans the sequence of dot and dash into corresponding code text by lookup the code table and outputs the result of decoding. Fig.5 shows the result displayed on the output window, including recognition result of intervals and the corresponding code text.

5 Conclusion

By analyzing the technical problems shortwave Morse telegraph automatic decoding, mainly the impact of noise on the signal and unstable typing of Morse code, an automatically decoding algorithm based on K-means clustering method is proposed which based on the existing Morse code decoding method. In the experimental section of decoding, simulation experiments under different SNR conditions verified the suitability of the proposed method. And decoding experiments of actual signal verified the feasibility of the algorithm. The algorithm proposed has some practical value. How to optimize the clustering algorithm to improve the robustness of clustering is the next step of the work content.

Acknowledgments

This paper is supported by the Project for the Key Project of Beijing Municipal Education Commission under Grant No. KZ201610005007, Beijing Postdoctoral Research Foundation under Grant No.2015ZZ-23, China Postdoctoral Research Foundation under Grant No.2015M580029, 2016T90022, Computational Intelligence and Intelligent System of Beijing Key Laboratory Research Foundation under Grant No.002000546615004, and The National Natural Science Foundation of China under Grant No.61672064.

References

1. Zahradnik P, Simak B. Implementation of Morse decoder on the TMS320C6748 DSP development kit[C]. European Embedded Design in Education and Research Conference. 2014,128-131.
2. Yang C H, Jin L C, Chuang L Y. Fuzzy support vector machines for adaptive Morse code recognition [J]. Medical Engineering & Physics, 2006, 28(9):925-31.
3. Qian S, Chen D. Joint time-frequency analysis[J]. IEEE Signal Processing Magazine, 1999, 16(2):52-67.
4. Cabal-Yepez E, Garcia-Ramirez A G, Romero-Troncoso R J, et al. Reconfigurable Monitoring System for Time-Frequency Analysis on Industrial Equipment Through STFT and DWT[J]. Industrial Informatics IEEE Transactions on, 2012, 9(2):760-771.
5. Gangwal O P, Funke E P, Varekamp C. Image enhancement: US, US 8331711 B2[P]. 2012.
6. Gonzalez R C, Woods R E. Digital Image Processing[J]. Prentice Hall International, 2002, 28(4):484 - 486.
7. Hartigan J A, Wong M A. A K-means clustering algorithm.[J]. Applied Statistics, 2013, 28(1):100-108.

A Novel Digital Rights Management Mechanism on Peer-to-Peer Streaming System

Jin Zhang, Jinwen Cai, Zhengshun Zhang *

Computer Network Information Center & Chinese Academy of Science,
Guangzhou Sub-Center, Guangzhou, China
{zhangjin,caijinwen,zhangyihang}@cnicg.cn
http://www.cnicg.cn/

Abstract. Digitalization of content is both a blessing and a curse, it makes the transmission of the digital content efficiently in the peer-to-peer streaming (P2PS) system. However, the digital content suffers the threats from the Internet and may be pirated easily. Such situations call for a new Digital Rights Management (DRM) method which is not only able to protect the copyright of content owners, but also to facilitate the consumption of digital content easily and safely. In this paper, we proposed a novel P2P streaming based DRM mechanism, which can satisfy the requirements to avoid the digital content being used illegally. Our DRM mechanism implements the authentication between the server and the peer, tracking the unfair use of the digital content and deleting the illegal peer from the P2PS system.

Keywords: digital rights management, peer-to-peer streaming, security

1 Introduction

Digital right was first proposed in 1998 [1], which was defined as a broad range of technologies to control and protect the digital medias. It is the prototype of digital rights management (DRM) in the early stage. Due to development of the Internet technologies, a variety of digital content (e.g., digital medias, softwares and so on), are in need of controlling and managing in the process of use. Therefore, DRM service changes into a general means of technologies provided by the service providers (SPs) to control and manage the use of digital content, which includes the provision, safekeeping, license phrasing and offer creation, distribution, booking, payment, authorization, and consumption of digital content. It is quite different from copy protection because the copy protection does not need to manage the digital content and devices. DRM has been well-studied in the last two decades. Many researchers have ever proposed definitions for DRM service [2, 3], however, it has been found that their definitions of DRM service were too unilateral to describe DRM service as a whole.

In the peer-to-peer streaming (P2PS) system, the illegal use of the digital content is tremendously serious [4]. First, the digital content can be reproduced exactly. After someone get one copy of the digital content, he will reproduced the digital content

---

* Zhengshun Zhang is the corresponding author.

© Springer International Publishing AG 2017   243
J.-S. Pan et al. (eds.), *Advances in Intelligent Information Hiding and Multimedia*
Signal Processing, Smart Innovation, Systems and Technologies 63,
DOI 10.1007/978-3-319-50209-0_30

and distribute the digital content to others [5]. second, P2PS system provides cheap, reliable and very fast distribution channels. P2PS systems are self-organized and decentralized [6]. In such systems every participant contributes its upstream bandwidth, processing ability and storage capacity in a collaborative manner. It greatly helps to establish multimedia communication networks and enables efficient distribution of digital content at a large scale [7]. However, such system will help the distribution of the illegal digital content very easily. From the discussion above, it is clearly that the P2PS system lacks the ability to maintain digital content protection and access control towards copyright material, hence leading to piracy commitment, copyright infringement, and even revenue damnification [8]. Therefore, it is of paramount importance to build an integrated platform which accomplishes both efficient delivery and legal consumption for digital content.

The conventional methods to protect the digital rights of the digital content are the copyright law or software, but they cannot work effectively. On one hand, copyright law allows for a number of instances that allow users to make use of a work that would otherwise be a violation of copyright. These exceptions are known as fair use [9]. However, fair use is really much too vague, and be described as a 'feature for lawyers', applications should be argued in court on an individual basis [10]. On the other hand, DRM software seems to be promoted as the solution to protecting copyright of digital content. However, such software theoretically controls and tracks where, when and how a digital content is used; together with who uses the digital content.

The rest of this paper is structured as follows: In the next section, we discuss the motivation and design goal of our mechanism and proposed a novel P2PS-based DRM system. At last, we give the conclusion of this paper.

2   P2PS-based DRM mechanism

In this section, first we will discuss the motivation and design goals, then propose a novel P2PS-based DRM system according to the design goals.

2.1   Motivation and design goals

The goal of our mechanism is to demonstrate a successful convergence of DRM and P2PS system. The root of this project lies in our search for a DRM solution tailored for P2PS system that enables efficient content distribution. While there are well-established DRM mechanism, applying to P2PS system presents complex functional and security requirements. In our P2PS-based DRM mechanism, any peer can be both a distributor and a receiver, and the content is not accessed through any particular product, the server in the P2PS system is not a complete content distributor, most of the distribution assignments are completed by the peers in the system and the server could monitor the availability of the content. It is challenging to devise a reasonably secure architecture tailored for P2PS system where important security functions must be decentralized.

We aim to look for a solution that would accomplish multiple design goals that we established for our system. These goals are enumerated below. Each design goal has been tagged with an identifier (DG1, DG2, etc.). First, we consider usability-related design goals, i.e., design goals that are not directly related to security.

DG **1** : The roles of the peers and the server should be well-defined in the P2PS system.

DG **2** : The server is the content owner and seller, it is able to manage the use and consumption of the digital content.

DG **3** : The server makes a profit through selling digital content to the peers once it receives the peers' requests.

DG **4** : Essential functions (e.g., distributing digital content) are decentralized.

DG **5** : DRM mechanism should be independent from underlying P2P network.

DG **6** : Minimal overheads for end users and convenient payments are able to be provided.

DG **7** : The server only supports the owner list of a segment if the heat degree of that segment is over a threshold.

Next, we consider the desired security features (within the inherent limitations of DRM).

DG **8** : Secure the transaction process of the digital content.

DG **9** : Secure authentication and authorization with minimal overhead.

DG **10** : Secure license distribution.

DG **11** : Track the peers owning illegal digital content and delete it from the owner list.

DG **12** : Break once, break everywhere (BOBE) resistance, which implies that an approach used to compromise one piece of content cannot be immediately applied to compromise another piece of content.

2.2 Functional architecture

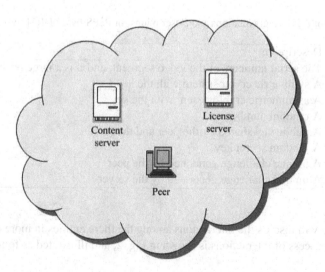

Fig. 1. Entities in P2PS-Based DRM System

Here we proposed a new P2PS-based DRM system, which is composed of three main entities: peers, content server, license server (as shown in Fig. 1). The roles of each entity are as follows:

– Peer - Exchanging the data with server and other peers.
– Content server - Providing the owner list of segment, checking the watermark, finding out and deleting the illegal peers.
– License server - Authenticating the peers, billing the payment when a peer requests a segment, and punishing the illegal peers.

Our DRM system could satisfy all the design goals above, and we hope to constitute a safer, faster and more effective system with less overhead cost. In the P2PS system, the digital goods (the video) is split in many segments for the sake of effective transmission. Before we describe the new DRM system, there are there preconditions we need in the DRM system.

1. It is well known that the peers downloads the segment of the video in the P2PS system, and then play this segment. The segment encrypted in the process of transmission, and there also is a corresponding decrypted key to decrypt the segment which is transmitted separating from the encrypted segment.
2. We will embed a watermark in the encrypted video segment in order to check the integrality of the video segment.
3. The solution we proposed here is a semi-distributed P2PS-based DRM system, but in this paper we only consider the situation that the video segment transmission between the peers.

Table 1. Useful Notations and Descriptions for P2PS-based DRM System

Notation	Description
(α,p)	The serial numbers of the video segment, and p is a large prime number
$SE_k(\bullet)$	A symmetric cryptosystem with the key k
$AE_k(\bullet)$	A asymmetric cryptosystem with the key k
pk	A random number
ps	A public key shared by the peer and the server
k	A random secret key
n_p	A unique challenge generated by the peer
n_s	A unique challenge generated by the server

Next we will discuss the interactions among the there entities in more detail. Every step in the process of interactions is shown in Fig. 2, and illustrated as follows.

step 1) Authentication between the server and peers
   The server must authenticate the identity of a peer first when the peer enters the P2PS system, here we choose the encrypted key exchange protocols proposed by Bellovin and Merritt. The step can be divided in the following sub steps(Fig. 3):
      – The peer generates a random pk, and sends its certificate including its digital signature and $SE_{ps}(pk)$ to the license server;

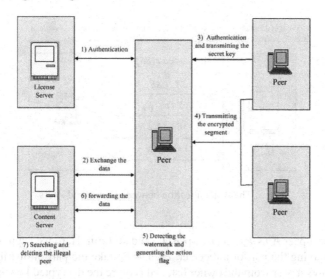

Fig. 2. The Functional Architecture of P2PS-Based DRM System

- Sharing the public key ps, the license server checks the digital signature and decrypts $SE_{ps}(pk)$ to obtain pk. Then it generates a random secret key k, and encrypts it with key pk and ps to produce $SE_{ps}(AE_{pk}(k))$. The license server sends $SE_{ps}(AE_{pk}(k))$ to the license server.
- The peer decrypts $SE_{ps}(AE_{pk}(k))$ to obtain k, generates a unique challenge n_p, and encrypts it with k to produce $SE_k(n_p)$, then the peer sends $SE_k(n_p)$ to the license server.
- The license server decrypts $SE_k(n_p)$ to acquire n_p, generates a unique challenge n_s, and encrypts these two challenges with the key k to produce $SE_k(n_p, n_s)$. The license server sends $SE_k(n_p, n_s)$ to the peer.
- The peer decrypts $SE_k(n_p, n_s)$ to acquire (n_p, n_s), and compares the former against its earlier challenge. If it matches, it encrypted n_s with k to obtain $SE_k(n_s)$, P sends $SE_k(n_s)$ to the license server.
- The license server decrypts $SE_k(n_s)$ to acquire n_s, and compares it against its earlier challenge. If it matches, the authentication is successful.

Then the peer could use the service of content server, and process the safe communication with the content server by using the secret key k.

step 2) Exchange the data between the peers and content server

After the peer confirms that the authentication is successful, it forwards local segment list, which shows the local segments in its cache, to the content server, and the content server will update the owner list of corresponding segments which the peer holds in its cache. Then the peer advances a request for the digital content it wants and pays for the digital content in order to obtain the owner list from the content server. Further more, the peer also gets the serial number (α, p) which is the label of the first segment.

step 3) The authentication between peers and decryption key exchange

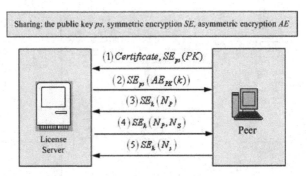

Fig. 3. The Authentication between Server and Peers

We denote peer A as P_A, peer B, peer C, and so forth. Here we suppose that P_A is a peer playing the part of a receiver, P_B and P_C take the part of distributors. Then P_A chooses a peer from the owner list, and acquire the decrypted key to decrypt the corresponding segments. This process can be divided into the following sub steps (seen in Fig. 4):

- P_A generates an independent random number γ_A which is chosen from the set of integers $[1, p - 2] = \{1, 2, , p - 2\}$, such keeps γ_A secret, but place

$$S_A = \alpha^{\gamma_A} mod p \tag{1}$$

P_A sends S_A to P_B.
- P_B generates an independent random number γ_B chosen from the set of integers 1,2,,p-2, such keeps γ_B secret, but place

$$S_B = \alpha^{\gamma_B} mod p \tag{2}$$

P_B sends S_B to P_A.
- P_A obtains K_A by obtaining S_B from P_B and letting

$$K_A = S_B^{\gamma_A} mod p \tag{3}$$

- P_B obtains K_B by obtaining S_A from P_A and letting

$$K_B = S_A^{\gamma_B} mod p \tag{4}$$

- P_B encrypts the decrypted key of the segment by K_B, and send the result to P_A.
- P_A obtains the result and decrypt it with K_A.

Here, we notice that

$$K_A = S_B^{\gamma_A} mod p = (\alpha^{\gamma_B} mod p)^{\gamma_A} mod p$$
$$= \alpha^{\gamma_B \gamma_A} mod p = S_A^{\gamma_B} mod p$$
$$= K_B$$

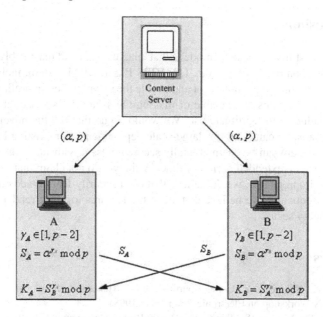

Fig. 4. The Authentication between Peers

step 4) The encrypted segment acquisition

After P_A acquires the decrypted key of the segment, it needs the encrypted content of that segment, it still selects several distributors from the owner list. P_A downloads the encrypted segment and maintains all the distributors's ID (denoted as ID_{list}) locally.

step 5) Watermark detection and action flag generation

P_A decrypts the encrypted segment with p and decrypted key obtaining from P_B, then detects the watermark WM in the segment and generates an action flag, AF_A ('0' represents failure, and '1' represents success).

step 6) Forwarding data to the server

P_A sends $(WM, ID_A, Tstamp_A, AF_A, \alpha)$ to the content server and starts to play the segment it downloads, where ID_A is the ID of P_A.

step 7) Searching and deleting the illegal peers

The server checks WM to confirm the legality and integrity of the segment played by P_A. If the server considers the segment is legal and integrated, the server will add the ID of P_A (ID_A) to the owner list of that segment. If the content server considers the segment is illegal and not integrated, it will request P_A to forward the owner list of the peers (ID_{list}) which it obtains the encrypted segment from, here we call this ID_{list} as 'suspicious illegal owner list' and the server would delete the illegal peers from the owner list of that segment according to the following principle: N peers forward suspicious illegal owner list separately where the peers own the same segment, if a peer's ID emerges on the suspicious illegal owner list more than three times, the server would judge that the peer is an illegal peer and delete it from the P2PS system.

3   Conclusion

We have successfully proposed a model a that enables legal and reasonably secure content distribution on the P2PS system. Our P2PS-Based DRM system, includes various features to deter content infringement and enable the server to obtain profits. We believe that these novel features are capable of making the system less susceptible to attacks and more conducive to legitimate use. We would argue that the the novel mechanism would be a plausible candidate for large-scale deployment over existing P2PS system. However, no system can be deemed totally secure unless it withstands the threats from the hackers over an extended period of time. While we do not claim that the security of our system is impregnable, we do believe that each security feature adds to the overall security. Consequently, we believe that all of the features together render P2PS-based DRM reasonably secure.

References

1. Fujimura, K., Nakajima, Y.: General-purpose Digital Ticket Framework. In: 3rd Conference on USENIX Workshop on Electronic Commerce (1998)
2. Rosenblatt, W., Mooney, S., Trippe, W.: Digital Rights Management: Business and Technology. John Wiley & Sons, Inc. (2001)
3. Jamkhedkar, P.A., Heileman, G.L.: Digital Rights Management Architectures. Computers & Electrical Engineering (2009)
4. Von Lohmann, F.: Peer-to-peer File Sharing and Copyright Law: A Primer for Developers. In: 2nd International Workshop on Peer-to-Peer Systems (2003)
5. Brahmbhatt, D., Stamp, M.: Digital Rights Management for Streaming Media. Handbook of Research on Secure Multimedia Distribution (2009)
6. Chu, Y., Rao, S. G., Seshan, S., Zhang, H.: A case for end system multicast. IEEE Journal on selected areas in communications (2002)
7. Zhang, X., Liu, J., Li, B., Yum, T.: Grid Information Services for Distributed Resource Sharing. In: 24th IEEE International Conference on Computer Communications (2005)
8. Iwata, T., Abe, T., Ueda, K., Sunaga, H.: A DRM system Suitable for P2P Content Delivery and the Study on its Implementation. In: 9th Asia-Pacific Conference on Communications (2003)
9. Arnab, A., Hutchison, A.: Fairer usage contracts for DRM. In: 5th ACM workshop on Digital Rights Management (2005)
10. Felten, E.W.: A Skeptical View of DRM and Fair Use. Communications of the ACM (2003)

Robust Transmission of Compressed Sensing Signals with Error Correction Codes

Hsiang-Cheh Huang [1], Feng-Cheng Chang [2], Yueh-Hong Chen [3], and Po-Liang Chen [1]

[1] Department of Electrical Engineering, National University of Kaohsiung, Taiwan, R.O.C.
hch.nuk@gmail.com
[1] Department of Innovative Information Technology, Tamkang University, Taiwan, R.O.C.
135170@mail.tku.edu.tw
[2] Dept of Computer Science and Information Engineering, Far East University, Taiwan, R.O.C.
yuehhong@gmail.com

Abstract. Compressed sensing is one of the developing techniques with superior compression performances over existing schemes in the field of lossy data compression. Due to the observations that compressed signals are vulnerable to channel errors during transmission, data protection techniques may also be applied to the compressed signals at the encoder before transmitting to the decoder over lossy channels. We apply the error correction codes for the protection of compressed sensing signals of digital images to reach robust transmission in this paper. With the error correction codes, some redundancies would be introduced to protect compressed signals. By doing so, reconstructed image qualities would be improved; however, this may cause the reduction of compression performances in compressed sensing. And there comes the trade-off between data protection capability and compression performance for robust transmission. We propose to use different levels of redundancies for the protection of compressed sensing signals. Simulation results have presented the effectiveness of applying error correction codes and majority voting to protect compressed signals. For different lossy rates, reconstructed image qualities would be improved with error correction codes. Induced errors from lossy channels during transmission can be alleviated, which leads to the effective protection of compressed sensing signals.

Keywords: compressed sensing, error correction codes, transmission

1   Introduction

Data compression forms an important part in our daily lives. Nowadays, lots of multimedia contents, including images, video, or livestreams, are broadcast or delivered along with social media such as Facebook. Due to the ease of using smartphones or tablets, after capturing the multimedia contents, they can be delivered over wireless networks. Consequently, two things may be observed. First, the number

© Springer International Publishing AG 2017   251
J.-S. Pan et al. (eds.), *Advances in Intelligent Information Hiding and Multimedia
Signal Processing*, Smart Innovation, Systems and Technologies 63,
DOI 10.1007/978-3-319-50209-0_31

of multimedia files gets increased rapidly, and the size of each file is large. The enormous amounts of multimedia contents need to be compressed before saving and transmission. Second, for data transmission over wireless networks, channel errors or data lost would be inevitable during the delivery. Even worse, if we combine the two items altogether, we may observe that error propagation may cause serious effects to compressed signals. Therefore, we may need to look for ways to efficiently perform data compression, in addition to the robust transmission of multimedia contents, for practical applications.

Unlike employing compression techniques such as transform coding and quantization in international standards, we employ the recently developed technique named compressed sensing [1,2,3] in this paper. Most papers in compressed sensing focus on the compression performances. Besides the compression performances, we also look for the robust transmission of compressively sensed signals over lossy channels [4,5]. Compressed signals are vulnerable to channel errors or data losses, which may lead to the difficulty for the decoder to correctly reconstruct the decoded images. In order to alleviate the degradation caused by channel errors, we integrate error control codes into compressed sensing for data protection, and to look for the enhancements of reconstructed images.

This paper is organized as follows. We briefly describe the concepts and notations of compressed sensing in Sec. 2. We then present the use of error correction codes for protecting the transmission of compressed sensing signals over lossy channels in Sec. 3. We demonstrate the simulation results with different levels of protections in Sec. 4. And finally, we address the conclusions in Sec. 5.

2   Brief Descriptions of Compressed Sensing

We briefly describe compressed sensing based on the representations in [1,2]. Compressed sensing looks for the sampling rate, which is far below the Nyquist rate. It is composed of the *sparsity* principle, and the *incoherence* principle.

> For the *sparsity* principle, it implies the information rate in data compression. In compressed sensing, a much smaller information rate than the sampling rate is required, and ir can be represented with the proper basis Ψ, $\Psi \in C^{N \times N}$, and C means the complex number. More specifically, Ψ is the basis to reach sparsity with a k-sparse coefficient vector $\mathbf{X}$, $\mathbf{X} \in C^{N \times 1}$, with the condition that
> $$\mathbf{f} = \Psi \mathbf{X}. \tag{1}$$
> Here, $\mathbf{f}$ denotes the reconstruction corresponding to the original signal.

> For the *incoherence* principle, it extends the duality between time and frequency. The measurement basis Φ, $\Phi \in C^{m \times N}$, which acts like noiselet, is employed for sensing the signal $\mathbf{f}$, with the condition that
> $$\mathbf{Y} = \Phi \mathbf{f}. \tag{2}$$
> Here, $\mathbf{Y}$ denotes the measurement vector. We note that Eq. (2) is an underdetermined system.

Decoding serves as another important topic in compressed sensing [6,7,8]. With the parameter settings in [1,2], we choose K_1 and K_2 coefficients for sparsity and incoherence, respectively in the following simulations.

3　Robust Transmission with Error Correction Codes

With the experiences in practical applications, compressed signals, including those generated with international standards such as JPEG, are vulnerable to channel errors during transmission. Thus, we expect that compressed sensing signals may also experience channel errors or data loss for transmitting over lossy channels, and hence data protection techniques [9,10,11] may need to be provided in advance to reach robust transmission.

One intuitive way for robust transmission is to use multiple transmissions, i.e., the same chunk of compressed data is delivered multiple times, often odd-numbered times. At the decoder, by use of majority voting rule, the lost or erroneous data can be restored or corrected. Besides, error correction codes are effective means for robust transmission, which may provide data protection capability to some extent. With the reasonable amount of redundancy introduced for data protection, protected data can be correctly decodable at the decoder.

Multiple transmissions are famous for the ease of implementation. However, the effective transmission rate gets decreased rapidly. If the same chunk of compressed data is transmitted three times, the effective transmission rate becomes one third of its original counterpart. It also implies the redundancy of 200%. Even though the decoder can simply apply the majority voting rule for data correction, it would be desirable to reduce the redundancy if some other means other than multiple transmissions can be explored.

We choose the BCH (Bose - Chaudhuri - Hocquenghem) Codes for performing data protection due to its widely use for practical applications [12]. BCH codes form a large class of multiple random error-correcting codes. Suppose the bitstreams are divided into chunks, with k coefficients in each chunk. After the encoding with BCH codes, one chunk would contain n coefficients, called the code length. Consequently, the newly produced $(n-k)$ coefficients would refer to the redundancy after encoding with BCH code encoder. The redundancy can be represented by $\left(\frac{n-k}{k}\right) \times 100\%$. These notations can be represented by the $\mathrm{BCH}(n, k)$ code. For the integer $m \geq 3$, BCH codes follow the conditions below:

> $t < 2^{m-1}$;
> $n = 2^m - 1$;
> $n - k \leq mt$.

Here, the parameter t refers to the error correcting capability.

With the presentations above, we plan to integrate both the multiple transmission and BCH codes for the protection of compressed sensing signals. For multiple transmissions, because the minimal odd-number is three, it leads to the smallest redundancy of 200%. For comparisons, we would like to choose the redundancy of less than 200% in BCH codes with the proper selection of the coefficient of k.

4   Simulation Results

With the proposed integration for data protection of compressed sensing, we choose the color image bird, taken by the authors with the size of 1024×1024, for simulations in Fig. 1. We first perform the calculations of compressed sensing for the red, green, and blue planes, respectively. And then we set the block size of $n = 511$ in BCH codes. In order to make comparisons with the three-time multiple transmission for data protection, which leads to the capacity of 200%, k should be larger than 171. Thus, we choose two sets of parameters for simulations with the BCH codes, i.e., $\mathrm{BCH}(511, 211)$ with error correcting capability $t = 41$ and the redundancy of 142.1%, and $\mathrm{BCH}(511, 385)$ with $t = 14$ and the redundancy of 32.7%.

Original image, Size: 1024×1024

Fig. 1. Original color image bird, taken by ourselves, with the size of 1024×1024.

With compressed sensing, considering the settings in [1], we choose $K_1 = 48000$ for sparsity, and $K_2 = 960000$ for incoherence, respectively. After completing compressed sensing calculations, these coefficients are transmitted over the lossy channel with the lossy rate of 5%. The red, green, and blue planes in the image are transmitted and then composed.

Figure 2 depicts the reconstructed image without error protection after at the decoder. In Fig. 2(a), it presents the result without error protection. It comes from the compressed sensing reconstruction of the red, green, and blue planes, depicted in Fig. 2(b), Fig. 2(c), and Fig. 2(d), respectively. Degraded qualities in the three planes can be easily observed, and leading to the reconstruction quality of 29.18 dB in Fig. 2(a).

(a) Without error protection
PSNR: 29.18 dB

(b) Without error protection
Red plane

(c) Without error protection
Green plane

(d) Without error protection
Blue plane

Fig. 2. Simulations under 5% lossy rate without error correction for color image `bird`.

By applying the $BCH(511, 385)$ protection, reconstructed results are demonstrated in Fig. 3. In Fig. 3(a), it shows the reconstruction from the three color planes in Fig. 3(b), Fig. 3(c), and Fig. 3(d), respectively. Comparing the reconstructions between Fig. 3 and Fig. 2, degraded qualities from channel errors seem to be improved to some degree. Besides, the reconstructed image in PSNR has also been improved in Fig. 3(a) subjectively. We can also observe the improvements by comparing the counterparts in Fig. 3 and Fig. 2 objectively.

When we increase the redundancy rate, we expect to provide the better protection of the reconstructed image. By applying the $BCH(511, 211)$ protection, reconstructed results are demonstrated in Fig. 4. In Fig. 4(a), it depicts the reconstruction after deco-

(a) BCH(511, 385) protection
PSNR: 29.96 dB

(b) BCH(511, 385) protection
Red plane

(c) BCH(511, 385) protection
Green plane

(d) BCH(511, 385) protection
Blue plane

Fig. 3. Simulations under 5% lossy rate with BCH(511, 385) protection for color image bird.

ding with $BCH(511, 211)$. Also, Fig. 4(b), Fig. 4(c), and Fig. 4(d) are the reconstructions of red, green, and blue planes after protecting with $BCH(511, 211)$ code. Taking the red plane for instance, we can easily observe that the quality in Fig. 4(b) presents much better than its counterparts in Fig. 3(b) and Fig. 2(b). With the better qualities of the three planes in Fig. 4(b), Fig. 4(c), and Fig. 4(d), color image of the reconstruction, demonstrated in Fig. 4(a), presents much better than its counterparts in Fig. 3(a) and Fig. 2(a) both subjectively and objectively. When we evaluate these results from another perspective, we can easily find that the enhancements of image quality are gained with the sacrifice of the increased redundancy. Therefore, how to look for the tradeoff between these metrics may need further studies.

(a) BCH(511, 211) protection
PSNR: 35.30 dB

(b) BCH(511, 211) protection
Red plane

(c) BCH(511, 211) protection
Green plane

(d) BCH(511, 211) protection
Blue plane

Fig. 4. Simulations under 5% lossy rate with BCH (511, 211) protection for color image bird.

5   Conclusions

In this paper, we have employed the integration of error correction codes for the transmission of compressed sensing images. By use of the BCH codes, we can easily observe the capability for error correction to cope with data loss of compressed sensing signals during transmission.

In compressed sensing, selection of parameters directly influences the compression capability. In order to alleviate the effects caused by channel error or data loss during transmission, some redundancy should be added, including multiple transmission or protection with error control codes, for protection. However, these protection schemes may lead to the reduction of overall compression ratio. Therefore, how to effectively choose the balance of parameters between compression and protection would be further explored.

In our simulations, we perform comparisons based on the easily implemented multiple transmission and the BCH codes for the protection of compressed sensing signals. With the increase of redundancy in BCH codes, which implies the better capability for error correction, reconstructed image qualities in the three color planes, get enhanced. Even though the enhancements can be observed, the percentage in the redundancy may need to be decreased. How to find the best combination of parameters can be studied in the future.

Acknowledgement

The authors would like to thank Ministry of Science and Technology, Taiwan, R.O.C., for supporting this paper under Grants No. MOST 104-2221-E-390-012 and MOST 105-2221-E-390-022.

References

1. Romberg, J.: Imaging via Compressive Sampling. IEEE Signal Processing Magazine, 25, 14–20 (2008)
2. Candes, E.J., Wakin, M.B.: An Introduction to Compressive sampling. IEEE Signal Processing Magazine, 25, 21–30 (2008)
3. Arildsen, T., Larsen, T.: Compressed Sensing with Linear Correlation between Signal and Measurement Noise. Signal Processing, 98, 275–283 (2014)
4. Huang, H.C., Chang, F.C.: Error Resilience for Compressed Sensing with Multiple-Channel Transmission. Journal of Information Hiding and Multimedia Signal Processing, 6, 847–856 (2015)
5. Huang, H.C., Chang, F.C.: Robust Image Watermarking Based on Compressed Sensing Techniques. Journal of Information Hiding and Multimedia Signal Processing, 5, 275–285 (2014)
6. Chang, F.C., Huang, H.C.: The Analysis of Reconstruction Efficiency with Compressive Sensing in Different K-Spaces. Proc. Int'l Conf. on Robot, Vision and Signal Processing, 67–70 (2015)
7. Candes, E.J., Tao, T.: Decoding by Linear Programming. IEEE Trans. Inform. Theory, 51, 4203–4215 (2005)
8. Chambolle, A.: An Algorithm for Total Variation Minimization and Applications. Journal of Mathematical Imaging and Vision, 20, 89–97 (2004)
9. Wang, Y., Orchard, M.T., et al.: Multiple Description Coding Using Pairwise Correlating Transforms. IEEE Trans. Image Processing, 10, 351–366 (2001)
10. Goyal, V.K., Kovacevic, J.: Generalized Multiple Description Coding with Correlating Transforms. IEEE Trans. Inform. Theory, 47, 2199–2224 (2001)
11. Huang, H.C., Chen, P.L., Chang, F.C.: Error Resilient Transmission for Compressed Sensing of Color Images with Multiple Description Coding. Proc. Int'l Conf. on Robot, Vision and Signal Processing, 63–66 (2015)
12. Lin, S. and Costello, D.J.: Error Control Coding, 2nd Ed. Pearson Prentice Hall (2004)

Comparison of IPv4-over-IPv6 (4over6) and Dual Stack Technologies in Dynamic Configuration for IPv4/IPv6 Address

aTa Te Lu, aCheng Yen Wu, *bWen Yen Lin,
aHsin Pei Chen, bKuang Po Hsueh

aChien Hsin University of Science and Technology
aInstitute of Computer Science and Engineering
bVanung University
bDigital Multimedia Technology

attlu@uch.edu.tw, ajohnny.wu520@gmail.com,
*bqqnice@mail.vnu.edu.tw, ahpchen@uch.edu.tw
bkphsueh@gmail.com

Abstract. IPv4 and IPv6 technologies face many challenges on LAN or the Internet such as transition mechanisms. The Internet Engineering Task Force (IETF) proposed a variety of solutions, e.g. Dual stack, IPv6-in-IPv4 (IPv4-IPv6), and IPv4-over-IPv6 (4over6). Clients use manual, stateful, or stateless auto configuration to get IPv6 addresses. Auto configuration with Stateful Dynamic Host Configuration Protocol for IPv6 (DHCPv6) protocol automatically configures IPv6 address through the router into the LAN, which is the best allocation choice. The DHCPv4 and DHCPv6 (DHCP 4o6) server used 4over6 and is considered to generally be the best plan.

In this paper, we set up a network communications platform and performed an effectiveness analysis comparing DHCPv4 and DHCPv6 (DHCP 4o6) servers individually with Dual Stack and 4over6 tunnel mechanisms. Experimental routing performances show that the routing path from 7 nodes with Dual Stack was better than 4over6 by 17.329%.

Key words: IPv6 transition; DHCPv6; IPv4-over-IPv6; Dual Stack

1 Introduction

In the future of network environments, IPv4-to-IPv6 or IPv6-to-IPv4 communications is an important step for Internet development. IPv6 and IPv4 environments are different and independent, and how compatible they are will influence how much time they occupy during the transition period. Presently, IP-based solutions do not deal with all these self-management demands, which leads to different approaches being adopted by different standardization bodies and even different approaches within the same

© Springer International Publishing AG 2017   259
J.-S. Pan et al. (eds.), *Advances in Intelligent Information Hiding and Multimedia Signal Processing*, Smart Innovation, Systems and Technologies 63,
DOI 10.1007/978-3-319-50209-0_32

standardization bodies as systems evolve and new network architectures are developed [1]-[2]. IETF has further developed different types of IPv6 conversion techniques, including dual stack, translation, and tunneling for different IPv6 conversion scenarios. We consider IP-based communications quickly changing topologies from dynamic topological addressing auto-configuration mechanisms to be a major concern [3]. DHCP provides dynamical network configuration to hosts in an IPv4 environment. DHCP broadcasts router solicitation message through LAN to provide the address of the default router to hosts. 4over6 tunneling and dual stack technology have become a major transition scenario particularly since more and more IPv6 access networks are being deployed, while IPv4 users need to communicate with the IPv4 Internet through these IPv6 networks as shown in figure 1.

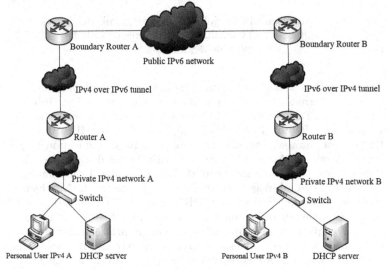

Fig. 1. IPv4-over-IPv6 transition architecture

Stateless address auto-configuration is a standard characteristic of Internet Control Message Protocol version 6 (ICMPv6) [4], permitting a router to advertise an IPv6 prefix to provide clients an individual IPv6 address. On the other hand, stateful address-auto-configuration supplies clients with IPv6 addresses from a central authority through the use of DHCPv6 (RFC 3315). DHCPv6 provides DHCPv6 DNS options when answering to requests or information-request messages. Therefore, IPv6 clients can use DHCPv6 for address distribution or stateless DHCPv6 to receive other configuration information for choosing DNS options. DHCPv6 supplies a mechanism for reconfiguring and propagating new configuration information to DHCPv6 clients when DNS information is transformed or updated. DHCPv6 allows administrators more control over address distribution than Stateless Address Auto Configuration (SLAAC) [5]. When IPv6 clients use the default DCHPv6 and as IPv4 clients use the default DHCP, the management and configuration options provided by the protocol can be useful in extensive, complex networks. Different realms can be established to configure different addresses for different parts of the networks, as shown in figure 2.

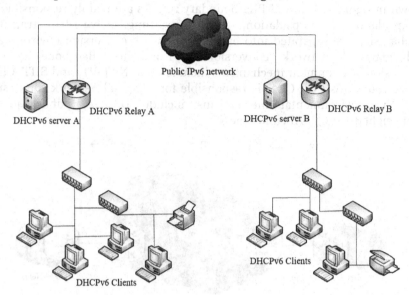

Fig. 2. DHCPv6 Architecture

DHCPv6 involves the DNS recursive name server option and the domain search list option. The DNS servers and domain names are listed in these two options in order of preference for use by the DNS resolver on the client. Therefore, the IPv6 client can choose DNS options using DHCPv6 for address distribution or stateless DHCPv6 to gain other configuration information.

2   Research Background

A Basic Bridging Broad Band (B4) element is only configured from the service provider with IPv6. As such, it can only learn the address of a DNS recursive server through DHCPv6 (or other similar methods over IPv6). As DHCPv6 only defines an option to get the IPv6 address of such a DNS recursive server, the B4 element cannot easily discover the IPv4 address of such a recursive DNS server, and as such will have to perform all DNS resolution over IPv6. The B4 element can pass this IPv6 address to downstream IPv6 nodes, but not to downstream IPv4 nodes. (Dual-Stack Lite Broadband Deployments Following IPv4 Exhaustion) [6] The DNS Recursive Name Server option provides a list of one or more IPv6 addresses of DNS recursive name servers to which a client's DNS resolver may send DNS queries. DNS servers are listed in the order of preference for use by the client resolver.

The current literature on 4over6 technology aspects, such as Lin Liu proposed

"The Research of 4over6 Transition System Deployment for IPv6 Backbone" [7], is based on the above 4over6 tunneling technology, with Provider Edge equipment (PE) being responsible for the conversion of marginal networks. As shown in figure 3, PE1 and PE2 boundary routers are mainly responsible for encapsulation, de-capsulation, and forwarding. IPv4-IPv6 transition mechanisms are separated into backbone network conversion environments and marginal network conversion environments. Backbone network conversion environment mechanisms include 6to4, NAT-PT, and SIIT. Client device router (as CE1, CE2) is responsible for marginal network conversion, and the environmental mechanisms include DS-Lite, public 4over6, lightweight 4over6, MAP, and others.

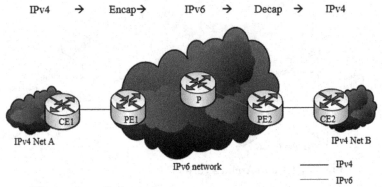

Fig. 3. 4over6 data plane

Dr. Zilong Liu proposed [8] using the lwB4 customer premises equipment (CPE) and lwAFTR the boarder router device (BR) to form a lightweight 4over6 tunnel network CERNET2 mechanism deployed in an IPv6 network environment. The new DHCP4o6 server respectively uses lwB4 and lwAFTR (BR) to does lease query on servers in figure 4.

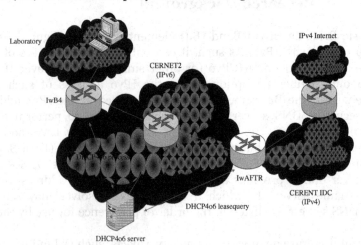

Fig. 4. Deployment in Lightweight 4over6 network at Tsinghua University

3 Tunneling and Dual Stack Routing Introduction

A. IPv4-over-IPv6 tunneling:

An IPv4-based network infrastructure will link to an IPv6 internet network in the future. IPv4 terminals connect to an IPv6 Internet network through IPv4-over-IPv6 tunneling conversion as shown in Figure 5. IPv4 packets are encapsulated into IPv6 packets in R1, and transmitted through IPv6 network links. IPv4 packets are de-capsulated from IPv6 packets in R2. Generic routing encapsulation (GRE) IPv4 tunnel supports IPv6 traffic—IPv6 traffic to carry IPv6 packets in IPv4-based network. The standard GRE tunneling technique is designed to provide secure communication between two boundary devices to implement point-to-point encapsulation scheme. C1 (e3/0 192.168.6.2/30)and R5(e3/1 192.168.6.1/30), R5(e3/0 192.168.3.2/30) and R3(e3/1 192.168.3.1/30), R3(e0/0 192.168.1.1/30) and R1(e0/0 192.168.1.2/30, tunnel 0 192.168.9.1), R2(e0/0 192.168.5.1/30, tunnel 0 192.168.9.2) and R4(e0/0 192.168.5.2/30), R4(e3/1 192.168.4.1/30) and R6(e3/0 192.168.4.2/30), R6(e3/1 192.168.7.1/30) and C2(e3/0 192.168.7.2/30) have six different local area network segment, they set routing protocol to OSPF for IPv4. R1(e0/1 2001:DB8:2:2::1/64) and R2(e0/1 2001:DB8:2:4::2/64) are boundary routers on WAN, with routing protocol set to OSPFv3 for IPv6. R1's tunnel IP is 192.168.9.1/24, the tunnel source is 2001:DB8:2:2::1, and the tunnel destination is 2001:DB8:2:4::2. R2's tunnel IP is 192.168.9.2/24, the tunnel source is 2001:db8:2:4::2, and the tunnel destination is 2001:DB8:2:2::1. When C1 transports data to C2, it must be through R5, R3, R1 by OSPF in the local area network. R1 and R2 does so by OSPFv3 [9] and 4over6 tunneling in the internet. Finally, this can be done through R2, R4, R6 by OSPF in the local area network to C2.

B. Dual stack:

Dual stack means that devices are able to run IPv4 and IPv6 in parallel in a network, so Dual stack offers a very flexible coexistence strategy. Setting Dual stack protocol in the router mode for IPv4 and IPv6, involves running IPv4 and IPv6 at the same time in Figure 6. End nodes (T1 and T2) and routers (R1, R2, R3, R4, R5, R6) run both IPv4/IPv6 protocols, and if IPv6 communication is possible, that protocol is preferred. C1 (e1/0 10.0.5.2/30, 2001:16::2/64) and R5 (e1/1 10.0.5.1/30, 2001:16::1/64), R5 (e1/0 10.0.3.2/30, 2001:14::2/64) and R3 (e1/1 10.0.3.1/30, 2001:14::1/64), R3 (e1/0 10.0.1.2/30,2001:11::2/64) and R1 (e1/0 10.0.1.1/30, 2001:11::1/64), R2 (e1/0 10.0.2.1/30, 2001:13::1/64) and R4 (e1/0 10.0.2.2/30, 2001:13::2/64), R4 (e1/1 10.0.4.1/30, 2001:15::1/64) and R6 (e1/0 10.0.4.2/30,2001:15::2/64), R6 (e1/1 10.0.6.1/30,2001:17::1/64) and C2 (e1/0 10.0.6.2/30,2001:17::2/64) have six different local area network segments, and they set routing protocol to OSPF for IPv4 and OSPFv3 for IPv6. R1 (s2/0 10.10.10.1/30, 2001:12::1/64) and R2 (s2/0 10.10.10.2/30, 2001:12::2/64) are boundary routers on WAN, and they set routing protocol to static routing for IPv4 and

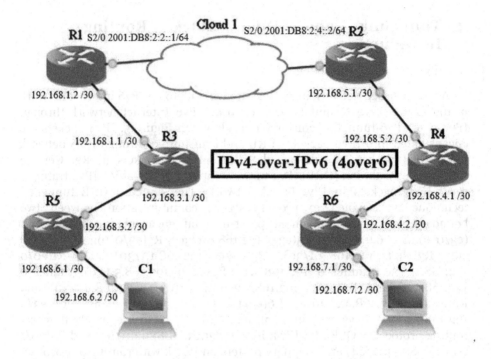

Fig. 5. IPv4 for routing communications through IPv4-over-IPv6 (4over6) architecture.

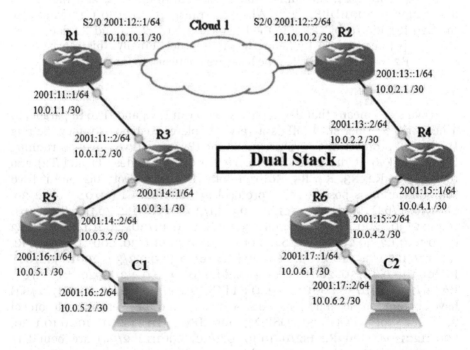

Fig. 6. IPv4 for routing communications through Dual Stack architecture.

IPv6. When C1 transports data to C2, it must be through R5, R3, R1 by OSPF for IPv4 or OSPFv3 for IPv6 in the local area network. R1, R2 does so by static routing for IPv4 and IPv6 on WAN. Finally, this can be done through R2, R4, R6 by OSPF for IPv4 or OSPFv3 for IPv6 in the local area network to C2.

Liu et al. [8] proposed a DHCPv4 server to deliver IPv4 to terminal equipment. After obtaining IPv4 address, if it needs external communication, IPv4 address will encapsulate through 4over6 tunneling server, and then afterwards convert to IPv6 external communication, as shown in figure 7.

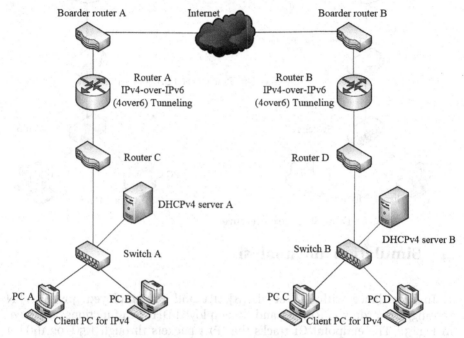

Fig. 7. IPv4-over-IPv6 tunneling architecture

This paper will present two improvement programs:

1. Remove DHCPv4 server A and DHCPv4 server B, as shown in figure 7. Router C and Router D will replace DHCPv4, as figure 8.

    It will have the following three benefits:

    A. Reduced cost: If a router is set to increase the DHCP function, enterprise

        companies do not need to provide additional servers and operating systems.

    B. Reduced information security risks: Stable routing platform that can avoid the risk of hackers attacking a DHCP server.

    C. Reduced administrator workload: Routers use a simple command-line interface and associated command set description, and engineers can reduce the workload of maintaining a DHCP server.

2. Dual stack replaces 4over6 mechanism, as shown in figure 8.

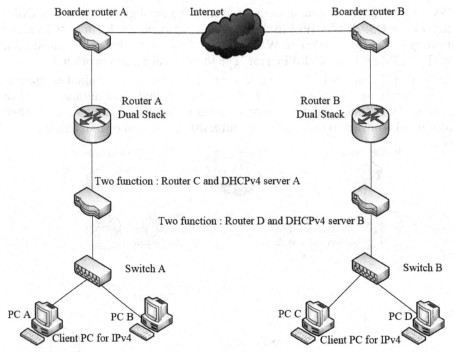

Fig. 8. Dual Stack architecture

4 Simulation and analysis

In accordance with Liu et al. [8], R1 and R2 employed 4over6 VPN routing/forwarding (VRF), R5 and R6 employed DHCPv4 functions, as shown in Fig. 5. The endpoint C1 tracks the IPv4 packets through R5 (192.168.6.1 DHCPv4 router), R3 (192.168.3.1), boundary router R1 (192.168.1.2), 4over6 tunnel from R1 to R2, boundary router R2 (192.168.9.2), R4 (192.168.5.2), R6 (192.168.4.2, DHCPv4 router), to arrive endpoint C2 (192.168.7.2). We tested tracking the route from C1 to C2 (192.168.7.2) fifty times and got an average result as shown in figures 9 and 11.

```
Cl#traceroute 192.168.7.2
Type escape sequence to abort.
Tracing the route to 192.168.7.2
VRF info: (vrf in name/id, vrf out name/id)
 1 192.168.6.1 4 msec 32 msec 28 msec
 2 192.168.3.1 44 msec 72 msec 48 msec
 3 192.168.1.2 104 msec 80 msec 88 msec
 4 192.168.9.2 124 msec 120 msec 120 msec
 5 192.168.5.2 128 msec 144 msec 152 msec
 6 192.168.4.2 160 msec 144 msec 184 msec
 7 192.168.7.2 224 msec 172 msec 184 msec
```

Fig. 9. IPv4-over-IPv6 (4over6) tracking results

R1 to R6 employed Dual Stack architecture in Fig. 6. Client C1 tracks the IPv4 packets through R5 (10.0.5.1 DHCPv4 router), R3 (10.0.3.1), boundary router R1 (10.0.1.1), boundary router R2 (10.10.10.2), R4 (10.0.2.2), and R6 (10.0.4.2, DHCPv4 router) to arrive client C2 (10.0.6.2). We tracking the route from C1 (10.0.5.2) to C2 (10.0.6.2) fifty times and got an average result as shown in figure 10 and 11.

```
Cl#traceroute 10.0.6.2

Type escape sequence to abort.
Tracing the route to 10.0.6.2

 1 10.0.5.1 40 msec 28 msec 4 msec
 2 10.0.3.1 76 msec 44 msec 32 msec
 3 10.0.1.1 68 msec 60 msec 64 msec
 4 10.10.10.2 92 msec 96 msec 104 msec
 5 10.0.2.2 124 msec 124 msec 120 msec
 6 10.0.4.2 188 msec 152 msec 108 msec
 7 10.0.6.2 172 msec 168 msec 144 msec
```

Fig. 10. Dual Stack tracking results

Figure 11 compares the results of Dual Stack and IPv4overIPv6 mechanisms tracked through seven nodes (from C1 to C2) of the average time for Dual Stack and IPv4overIPv6 was 95.615 msec and 112.185 msec, respectively.

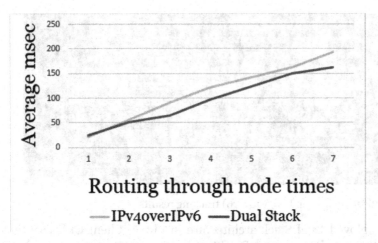

Fig. 11. Dual Stack and IPv4overIPv6 comparison of results

5   Conclusion

IPv6 and IPv4 environments are different, and are compatible for a long time during the transition period. IPv6 to IPv4 or IPv4 to IPv6 routing environments will be seen very frequently in the future.

In this paper, the performances of IPv4overIPv6 and Dual Stack tracking results are compared. The average time for Dual Stack and IPv4overIPv6 was 95.615 msec and 112.185 msec, respectively. Routing performance of Dual Stack is better than IPv4overIPv6 17.329%.

6   References

1. Cui, Y., Wu, J., Wu, P.: Public IPv4-over-IPv6 Access Network. IETF RFC 7040 (2013)
                    2.   Bound, J., Packard, H., Ericsson, B., Lemon, V., Nominum, T., Perkins, C., Carney, M., Microsystems, S.: Dynamic Host
3. Configuration Protocol for IPv6 (DHCPv6). IETF RFC 3315 (2003)
4. Imadali, S., Vèque, V., Petrescu, A.: Analyzing Dynamic
5. IPv6 Address Auto-configuration Techniques for Group IP-Based Vehicular
6. Communications. 39th IEEE Conference on Local Computer Networks
7. Workshops (LCN Workshops), pp. 722 – 729, France (2014)
8. Conta, A., Deering, S., Gupta, M.: Internet Control Message Protocol (ICMPv6)
9. for the Internet Protocol Version 6 (IPv6) Specification. IETF RFC 4443 (2006)
10. Thomson, S., Narten, T., Jinmei, T.: IPv6 Stateless Address

Autoconfiguration.

11. IETF RFC 4862 (2007)
12. Woodyatt, J., Lee, Y., Durand, A., Droms, R.: RFC 6333 Dual-Stack Lite
12. Broadband Deployments Following IPv4 Exhaustion. IETF RFC 6333 (2011)
13. Lin, L., Cui, Y,. Sun, J., Sun, Q.: The research of 4over6 transition system deployment for IPv6 backbone. 2nd International Conference on Computer
14. Science and Network Technology (ICCSNT), pp. 912 –915, Beijing, China (2012)
15. Zilong, L., Jiang, D., Yong, C., Chaokun, Z.: Dynamic Configuration for IPv4/IPv6 Address Mapping in 4over6 Technology. 9th IEEE International Conference on Anti-counterfeiting, Security, and Identification (ASID), pp. 132 –136, Beijing, China (2015)
16. Coltun, R., Ferguson, D., Moy, J., Lindem, A.: OSPF for IPv6. IETF RFC 5340 (2008)

A Framework for Supporting Application Level Interoperability between IPv4 and IPv6

Yeong-Sheng Chen and Shang-Yi Liao

Department of Computer Science, National Taipei University of Education, Taipei, Taiwan

{yschen, syliao80}@tea.ntue.edu.tw

Abstract. IPv6 has many advantages over IPv4. In the long run, IPv4 will be gradually updated and transferred to IPv6. However, IPv6 has not been fully deployed. Therefore, how to achieve interoperability of application services in an IPv4/IPv6 coexistent environment is an important issue. This study proposed a proxy based approach with virtualization technology for this purpose. A proxy server was built so that all the application services were interoperable in an IPv4/IPv6 coexistent environment without any code modifications or adjustments on the settings of the web servers. The proposed proxy based approach could effectively reduce the IPv4/IPv6 transfer risks, and hardware and software costs for network construction and maintenance without increasing architectural complexity of the networks. All the application services in the networks could be properly provided without being interrupted, and hence high service availability and connection quality were achieved in the IPv4/IPv6 coexistent environment. Practical implementation results demonstrated the effectiveness of the proposed design.

Keywords: IPv6-IPv4 Coexistence, Proxy, Virtualization, High Availability.

1 Introduction

Internet Protocol (IP) is a fundamental protocol for packet transmission between the source and destination on the Internet. It defines the addressing methods and the data packet structure [1]. The IP address of a data packet can be seen as the address the data packet on the internet. Currently, the widely used version of the Internet Protocol is Internet Protocol version 4 (IPv4). The address length of IPv4 is 32 bits, giving an address of 2^{30}. This address space has been gradually insufficient to meet the increasing demand from various services and applications on the Internet and the rapid increasing population of mobile device users. Thus, the depletion of IPv4 addresses is expected and will cause limitations on the expansion of Internet applications. To solve the dilemma, Internet Engineering Task Force (IETF) presented development plans for new generation Internet protocol, and published the standard specification of Internet Protocol version 6 (IPv6) in December 1998 [2].

© Springer International Publishing AG 2017

J.-S. Pan et al. (eds.), *Advances in Intelligent Information Hiding and Multimedia Signal Processing*, Smart Innovation, Systems and Technologies 63, DOI 10.1007/978-3-319-50209-0_33

IPv6 has 128 bit address length and hence provides a vastly large address space. It also allows improved route aggregation across the Internet for more efficient packet transmission. Compared with IPv4, IPv6 has a larger address space, more efficient packet routing mechanism and more flexible Quality of Services (QoS) mechanism. Therefore, Internet Protocol will be gradually upgraded to IPv6 through the time.

1.1   Depletion of IPv4 and growth of IPv6

On 3 February 2011, Internet Assigned Numbers Authority (IANA) announced that the last IPv4 address block had been allocated [3]. Then, on 15 April 2011, APNIC was the first regional Internet Registry who run out of allocated IPv4 addresses [3]. That is, from then on, not everyone who needed an IPv4 address could be allocated one. This exhaustion implies that direct end-to-end connectivity will not be universally available on the Internet before IPv6 is fully implemented. Shortages and depletion of IPv4 addresses had sparked worldwide attention. On 8 June 2011, major Internet service providers around the world, including Google, Facebook, Yahoo and more than 1000 other well-known websites participated in the "World IPv6 Day" activity for a global-scale trial of IPv6 [4]. That successful trial demonstrated that major websites around the world had been well-positioned for moving to a global IPv6-enabled Internet. Following that event, organized by the Internet Society, World IPv6 Launch began on 6 June 2012 [5]. It was intended to motivate organizations including Internet service providers, hardware makers, and web companies to prepare for and permanently enable IPv6 on their products and services.

Due to the explosive growth of Internet, and additional factors, such as the popularity of mobile devices and the always-on connection technologies, which increase the address demand and aggravate the shortage of IPv4 addresses, global IPv6 traffics have grown a lot in these few years. According to the statistics from Google about IPv6 adoption on the Internet, the ratio of users that access Google over IPv6 has been growing up rapidly to nearly 14% as on 10 September 2016 [6].

1.2   Transition from IPv4 to IPv6

The difference between IPv4 and IPv6 is not just on the version, but also they can be identified to be two different protocols. For instance both Chinese and English are languages for communication, however, their text composition and grammars are very different. Hosts using IPv4 cannot directly communicate with those using IPv6.

Currently, most of the Internet applications and services are still provided through IPv4 networks. How to provide IPv6 users to access them is an important issue. In the interim period before IPv6 completely replaces IPv4, it is necessary to develop mechanisms for the interoperability and communication between the IPv4 networks and the IPv6 networks.

The IPv6 protocol and the IPv4 protocol are incompatible and hence IPv6 users cannot read the contents of the website or use web applications in the IPv4 networks. However, deploying the new IPv6 protocol does not mean to tear down the IPv4 network and replace it with an IPv6 enabled network. Instead, we want to achieve seamless and painless upgrades from IPv4 to IPv6. There are some options for migration to IPv6 from the existing IPv4 network infrastructure: dual-stack network, tunneling, and translation [7] [8] [9]. However, with these solutions, the hardware and software may need to be upgraded, or the existing network may need to be redesigned. In many cases, the current network may not be ready for hardware and software upgrades. Modifications for the upgrade may incur additional costs for network construction and maintenance. In addition, it may cause the risks that stable operations in the original IPv4 networks become unstable. Furthermore, in some cases, the existing web server software and operating systems may not support IPv6.

To avoid the above mentioned problems, this study is aimed to develop mechanisms for interoperable application services in IPv4/IPv6 coexistence environments with the following conditions: (1) no modifications on the source codes of the applications, (2) no adjustments on the settings of the application hosts and network devices, and (3) no re-design or re-development of the applications.

Thus, this study proposed a proxy based approach with virtualization technology to achieve this purpose. A proxy server is built so that all the web application services are interoperable in an IPv4/IPv6 coexistent environment without any code modifications or adjustments on the settings of the web servers while maintaining high availability, service quality, and reducing construction and maintenance costs and infrastructure complexity.

The rest of this paper is organized as follows. Section 2 describes the background and related work. Section 3 describes the framework proposed in this study. In Section 4, implementation of a prototype system based on the proposed scheme is described. Conclusions and future work are drawn in Section 5.

2 Background and related work

IPv4 protocol has been used for more than 30 years since it was released. IPv6 is the successor technology to IPv4 and has many advantages over IPv4. The prefix aggregation in IPv6 address is flexible and its simplified protocol header improves the forwarding efficiency. In addition, IPv6 protocol provides flow label based QoS support and has better mobility and security supports than IPv4. In general, IPv6 is a redesign of IPv4 and is believed to be a mature and feasible solution for the next-generation Internet. Although the continuous demands for new IP addresses and better protocol efficiency are driving IPv6 towards a wide deployment, yet currently, services and applications still remain in IPv4 networks. The main difficulty is essentially caused by the cost of transition from IPv6 to IPv4. Therefore, IPv6 and IPv4 will coexist for a long period and how to achieve interoperability of web application services in an IPv4/IPv6 coexistent environment is an important issue.

Many researches have been dedicated to developing transition mechanisms for interoperability of web application services in the IPv4/IPv6 coexistent environment. A number of transition mechanisms have been proposed by the Internet Engineering Task Force (IETF). However, most of them consider initiating sessions only from the IPv6 networks to the IPv4 networks, but not in the reverse direction [7]. Park *et al.*

propose an IPv4-to-IPv6 dual stack transition mechanism which can operate in the case that hosts in the IPv4 network initiate connections within hosts in the IPv6 network [7]. Narayan and Tauch [8] empirically evaluate the performance of two transition mechanisms, configured tunnel and 6to4 transition mechanism, on two different Linux distributions. In [9], the authors propose a flexible transition scheme called Bi-Directional Mapping System (BDMS) to deal with the IPv4/IPv6 address mapping transition which converts the received packets to their destination environment. The proposed method reduces the packet size compared with tunneling mechanisms and also reduces the costs compared with the dual stack mechanisms.

In general, the options for the transition from IPv4 to IPv6 networks include: dual-stack network, tunneling, and translation. However, to implement these solutions, the hardware and software may need to be upgraded, or the existing network may need to be redesigned. This may incur additional costs for network construction and maintenance, and cause the risks that stable operations in the original IPv4 networks become unstable. To cope with this problem, the proxy-based solution, which is easy to implement and of little cost and does not need to redesign the network architecture, is considered to be a good choice [10] [11] [12].

The basic function of a proxy is acting as an intermediary between a client and a web server across the networks. A Proxy server that receives a request from a client host in the IPv6 network to the destination server residing in the IPv4 network can forward the request to the destination server and then send back the acquired data to the client. There are three main types of proxy servers: Regular caching proxy servers, transparent proxy servers, and reverse proxy servers. Regular caching proxy servers and transparent proxies are both intended to speed up performance by prefetching and holding on the requested data. This process can result in data being delivered more quickly if it is requested repeatedly. Reverse proxies direct client requests to the appropriate backend server and typically sits behind the firewall in a private network [12]. A reverse proxy can distribute the load from incoming requests to several servers, with each server serving its own application area. In this study, we utilize the reverse proxy server to implement the proposed framework for supporting application level interoperability between IPv4 and IPv6.

3   Operations of the Proposed Framework

As discussed in Section 1, mechanisms for supporting interoperability of application services between IPv4 and IPv6 networks are imperative. The proposed framework for this purpose is aimed to keep existing web applications and services unchanged, without modifying the network settings and the codes of applications. Thus, the IPv4/IPv6 transfer risks can be effectively reduced and all the application services in the networks work properly without being interrupted so as to achieve high service availability and connection quality in an IPv4/IPv6 coexistent environment.

As shown in Figure 1, it is assumed that the web application servers are located in an IPv4 network, and users that want to access these servers are located in an IPv4, IPv6, or IPv4/IPv6 coexistent network. A proxy server is built, and the records in the DNS server are modified so that the connection request from an IPv6 user will be directed to the proxy server. The mechanisms are described as follows. When a user wants to connect a web application server, the DNS server will return the IPv6 address of the proxy server and the IPv4 address of the web application server to the user. Thus, if the connection request is from an IPv6 user, the request will be sent to the proxy server. Then, the proxy server will connect the web application server, retrieve the requested information and return it to the user. However, for an IPv4 user, since he or she is not able to use the IPv6 address (of the proxy server), the connection is sent to the web application server instead of being directed to the proxy server.

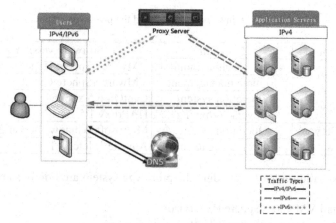

Fig. 1. System architecture

Besides, in order to effectively utilize the computing power of servers, virtualization technology is applied [13]. As shown in Figure 2, the proxy server is built on the virtualization platform Hypervisor. The operating system is CentOS, which is a free Linux distribution. The proxy server can be easily built by installing the HAProxy [14] with proper firewall policy and IP settings. HAProxy is an open source software offering high availability, load balancing, and proxying for TCP and HTTP-based applications [14].

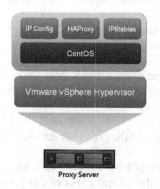

Fig. 2. Implementation of the Proxy server in this study

4 Implementations and Experiments

A prototype system of the proposed framework was implemented in our campus network (domain: ntue.edu.tw), which is an IPv4/IPv6 dual stack environment. Brief implementation details are described and some results of our preliminary experiments are presented in this section.

The main software tools used to build the prototype system are listed in Table 1. All the software tools are popular and hence it is easy to implement the proposed framework with them.

Table 1. Main software tools used the prototype system.

Category	Function	Software version
Virtualization	Virtual machine monitor	VMware vSphere Hypervisor 6.0
	Remote management	VMware vSphere Client
Proxy	Operation system	CentOS 6.5 x86_64
	Reverse proxying	HAProxy 1.5.18
DNS Server	Operation system	Microsoft Windows Server 2012
Simulation	Performance testing	Apache JMeter 2.13

Implementation steps for building the prototype system are briefly summarized as follows.

(1) Install VMware vSphere Hypervisor.
(2) Install VMware vSphere Client.
(3) Create a virtual machine on vSphere Hypervisor.
(4) Install CentOS on the virtual machine.
   ● Add "NETWORKING_IPv6=yes" into the file "/etc/sysconfig/network"
   ● Add IPv4 address and IPv6 address into the file "/etc/sysconfig/network-scripts/ifcfg-eth0"
   ● Verify the IP settings using the command "/sbin/ifconfig"
(5) Install EPEL (Extra Packages for Enterprise Linux).
   ● Check the version of CentOS using the command "uname –a".
   ● Download EPEL according to the version of CentOS.
(6) Install HAProxy.
   ● After installation, restart HAProxy using the command "/etc/init.d/haproxy restart".
(7) Set the firewall policies
   ● Edit the file "/etc/sysconfig/ip6tables"
   ● Make new policies effective using the command "/etc/init.d/ip6tables restart"
(8) Add new DNS records
   ● Add IPv6 address of the proxy server (AAAA record).

The IPv6 Lab [15] developed by Taiwan Network Information Center (TWNIC) was used to test and verify the effectiveness of our prototype system. It can provide different type of connections (IPv4 or IPv6) to access the web servers in our campus network. The results of experiments showed that our prototype system can effectively provide web application services for both IPv4 users and IPv6 users without any code modifications or adjustments on the settings of the web servers.

Besides, we use Apache JMeter [16], which is open source software mainly designed for testing web applications, to evaluate the performance of our prototype system. With Apache JMeter, different numbers (500, 1000, 1500, and 2000) of connection requests to a particular web server in our campus are generated. The results of experiments are shown in Table 2. From Table 2, we can see that in our proxy-based framework, IPv6 connections always have better performance than IPv4 connections in term of average connection response time, connections per second, and transmission bandwidth. This should be because that in our campus environment, there were many existing IPv4 traffics which interfered with the IPv4 connections in our experiments.

Table 2. Performance comparisons of the experimental results.

Number of connections	Average Connection Response Time (Seconds)		Connections per Second		Transmission Bandwidth (KB/s)	
	IPv6	IPv4	IPv6	IPv4	IPv6	IPv4
500	2.746	2.775	46.985	45.843	611	597
1000	4.564	5.608	50.446	49.510	656	645
1500	7.799	8.803	52.398	45.892	681	598
2000	10.991	12.447	50.662	46.264	642	602

5 Concluding Remarks

IPv6 is a redesign of IPv4 and is believed to be a mature and feasible solution for the next-generation Internet. Although currently services and applications still remain in IPv4 networks, in the long run, the Internet Protocol IPv4 will be gradually updated and transferred to IPv6. However, during the transition period, how to achieve interoperability of web application services in an IPv4/IPv6 coexistent environment is a very important issue.

This study proposed a proxy based approach, which cost little, to achieving seamless and painless upgrades from IPv4 to IPv6. A proxy server was built so that all the web application services are interoperable in an IPv4/IPv6 coexistent environment without any code modifications or adjustments on the settings of network devices and the application servers. We practically designed a prototype system in a campus network environment. A test environment was built without affecting the operations of the existing services in our campus network. In the test environment, the system architecture is simulated and verified to be a feasible solution. In addition, with the benefits of virtualization, the proposed system can be quickly deployed for online operation. Practical implementation of a prototype system demonstrated that the proposed proxy based approach can effectively reduce the IPv4/IPv6 transfer risks, and hardware and software costs for network construction and maintenance without increasing architecture complexity. It keeps all the application services in the networks work properly without being interrupted, and hence achieves high service availability and connection quality.

References

1. J. Postel, "Internet Protocol", RFC 791, September 1981.
2. S. Deering and R. Hinden, "Internet Protocol, Version 6 (IPv6) Specification", RFC 2460, December 1998.
3. Geoff Huston, IPv4 Address Report, Retrieved 10 September 2016, from http://www.potaroo.net/tools/ipv4/
4. 2011 World IPV6 Day, Retrieved 10 September 2016, from http://www.internetsociety.org/ipv6/archive-2011-world-ipv6-day
5. World IPv6 Launch, Retrieved 10 September 2016, from http://www.worldipv6launch.org/
6. Google IPv6 Statistics, Retrieved 10 September 2016, from http://www.google.com/ipv6/statistics.html
7. E. Park, J. Lee, and B. Choe, "An IPv4-to-IPv6 dual stack transition mechanism supporting transparent connections between IPv6 hosts and IPv4 hosts in integrated IPv6/IPv4 network," 2004 IEEE International Conference on Communications, Vol. 2, pp. 1024-1027.
8. S. Narayan, and S. Tauch, "IPv4-v6 configured tunnel and 6to4 transition mechanisms network performance evaluation on Linux operating systems," 2nd International Conference on Signal Processing Systems (ICSPS), 5-7 July 2010.
9. R. AlJa'afreh, J. Mellor, M. Kamala, and B. Kasasbeh, "Bi-directional Mapping System as a New IPv4/IPv6 Translation Mechanism," Tenth International Conference on Computer Modeling and Simulation, UKSim 2008, 1-3 April 2008
10. W. Yuan, H. Sun, X. Wang, and X. Liu, "Towards efficient deployment of cloud applications through dynamic reverse proxy optimization," 2013 IEEE International Conference on High Performance Computing and Communications, and 2013 IEEE International Conference on Embedded and Ubiquitous Computing (HPCC_EUC), 13-15 Nov. 2013, pp. 651-658.
11. Q. Liao, and Z. Tian, "Design and Realization of the IPv6 Web Proxy," Intelligent Computer and Applications, Vol. 4, p. 17, 2013.
12. E. Kawai, A. Shirahase, K. Tsukada, and S. Yamaguchi, "Practical migration strategy to IPv6 for enterprise Web services," 11th International World Wide Web Conference, May 2002
13. VMware Compatibility Guide, Retrieved 10 September 2016, from http://www.vmware.com/
14. HAProxy, Retrieved 10 September 2016, from http://haproxy.1wt.eu/
15. TWNIC IPv6 Lab, Retrieved 10 September 2016, from http://img.twnic.net.tw/ipv6launch1/ipv6lab/intro.html
16. Apache JMeter, Retrieved 10 September 2016, from http://jmeter.apache.org/

Design and Implementation of an IPv4/IPv6 Dual-Stack Automatic Service Discovery Mechanism for an Access Control System

Hui-Kai Su[1]*, Chih-Hsueh Lin[1], Jia-Long Hu[1],
Chun Liang Chang[1] and Wen Yen Lin[2]

[1] Department of Electrical Engineering,
National Formosa University
*hksu@nfu.edu.tw
[2]Department of Digital Multimedia Technology,
Vanung University

Abstract. With the development of the Internet network technologies, a vast number of things are connecting to the Internet. The available IPv4 addresses are used up almost. IPv6 and IPv4 will coexist for many years. Therefore, the Internet capabilities will be a significant issue. Moreover, we let the home monitoring system as a use case. This paper proposes an IPv4/IPv6 Dual-Stack Automatic Service Discovery Mechanism for an Access Control System. Our system architecture is divided into two parts: Door Access Controller Module and Door Access Control Server. The Door Access Control Server is implemented with an IPv4/IPv6 Dual-Stack network and Automatic Service Discovery on an Android-based IPTV Box. The functionalities of system configuration, access configuration and access monitoring are designed in the IPTV Box. Additionally, the Door Access Controller Module is designed by an embedded system. The modules of RFID reader and Camera can hot plug in anytime. In our implementation result, the IPv4/IPv6 Dual-Stack Automatic Service Discovery Mechanism works well. The test results of different service discovery methods with IPv4/IPv6 dual stack are presented in this paper. In the result, we can observe that the polling method is inefficiency. Especially with IPv6, it is unworkable due to the host ID space with 264 addresses. The performance of service discovery with multicast is a good solution in our system.

Keywords: IPv6, Dual Stack, Multicast, Service Discovery, Home Network, Internet of Things

1   Introduction

In recent years, the technology of communication and network is improved rapidly. The internet service is expanded anywhere in the world. The new term of IoT (Internet of Things) is proposed. Every object in our life and in our environment would be connected together and linked to the internet [1][2][3]. To the huge amount of objects, how to configure and manage efficiently is a significant issue.

© Springer International Publishing AG 2017   279
J.-S. Pan et al. (eds.), *Advances in Intelligent Information Hiding and Multimedia
Signal Processing*, Smart Innovation, Systems and Technologies 63,
DOI 10.1007/978-3-319-50209-0_34

With the development of the Internet network technologies, a vast number of things are connecting to the Internet. The available IPv4 addresses are used up almost. IPv6 and IPv4 will coexist for many years. Therefore, the Internet capabilities will be a significant issue. Moreover, we let the home monitoring system as a use case.

In this paper, the existing standard IPv6 network protocol, internet transition, routing schemes are applied to door access management and monitor APP into IPTV box. Moreover, the service discovery mechanism with IPv6 multicast is proposed. We build an access control monitoring system situation, then implement an access control system with IPv4/IPv6 dual-stack automatic service discovery and modularization. In our system, the IPTV box can discovery the door access control nodes in the home network automatically and quickly.

The remaining part of this paper is organized as follows. In Section 2, we introduce the background. In Section 3, the architectures of our system and modules are presented. Our experiment and evaluation is shown in Section 4. Finally, Section 5 concludes this paper.

2   Background

The IPv4 cannot support the requirements of huge IoT devices. The IPv6 is the alternative in the next generation of internet. However, to huge IoT nodes, how to provide an efficient service discovery mechanism for simple configuration and management is very significant[4].

The IPv6 provide unicast, multicast and anycast transmissions[5][6][7].

- The unicast transmission is to send data to a single network destination that is identified by a unique address.
- The multicast is a group communication. The group is identified by a group address. In the multicast domain, clients can join and leave the group anytime. When the source sends a packet to the group address. The participants in the group will received the same packet.
- Anycast is a network addressing and routing method. If the network destination is an anycast address, the packet will be delivered to the nearest node of the anycast address.

In a general case, the IPv6 prefix length of a LAN (Local Area Network) is 64. The maximum capacity is 2^{64} host addresses in this LAN. Using the pooling service discovery and scanning from host address 0 to 2^{64} is impossible. Additionally, the anycast is using in the inter network environment generally. Thus, multicast may be a good solution for service discovery.

In this paper, we design and Implement an IPv4/IPv6 dual-stack automatic service discovery mechanism for an access control system. The performance of service discovery with IPv4 unicast, IPv4 broadcast, IPv6 unicast and IPv6 multicast is shown below.

3   System Architecture

In order to respond to changing things in the Internet, IPv4 and IPv6 protocol is bound to coexist for a long time in the state, so this paper combined IPv4/IPv6 dual stack network, automatic service discovery feature to integrate RFID technology as identity identification, embedded system implementation by modular access control system. This chapter describes and explains the system architecture, function and operation of the process, the following discussion are divided into three parts:

- Part I: Introduction to support IPv4/IPv6 dual-stack architecture automatic service discovery and access control system of modular.
- Part II: Introduction to system architecture and operation of the access control process.
- Part III: The access control system architecture and operational processes of the server.

3.1   System Architecture

According to Figure 1, the system architecture divided into two parts: Door Access Controller Module and Door Access Control Server. We used embedded system board as Door Access Controller Module's control core, and external a Web Cam return image data to the Door Access Control Server. In Door Access Control Server is to use TV Box based on Android operating system. When Door Access Controller Module get IP address, user can use TV Box to do search Door Access Controller Module.

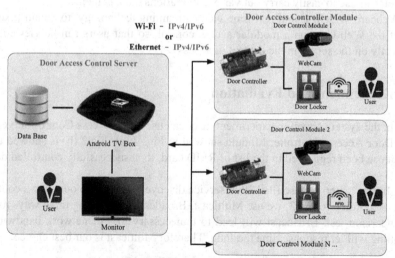

Fig 1. The system architecture.

3.2   Access Controller of Module

Access control architecture is using the Beagle Bone Black core of embedded system development platform, combined with RFID identification devices, door locks, Web Cam and internal Data Base of the composition. Access Controller with IPv4/IPv6 dual stack mechanism, when obtained automatically assigned by the router down after IPv4 or IPv6 addresses will be automatically added to Multicast Group waiting to receive a message, the internal Data Base will be used to store the user can enter the door white list , so users can use RFID sensor card authenticating operation when the recognition result is determined in line with the user identity, the door will open, and vice versa kept locked state, the user can also be an arbitrary access control server whitelist for each modular access control operation of content to add, modify and remove. Web Cam detect whether the device when the device is detected, the user can watch the real-time monitoring of access control on the server screen.

3.3   Access Control of Server

The TV Box is running Android operating system. The built-in SQLite Data Base and our designed access control App are performed. Functions are as follows:
- Access Controller Search: Search by Multicast all the modular access control within the LAN, and establish a connection.
- User list management: the user lists all modular controllers synchronize access to the inside of the SQLite Data Base, and allows users to modify data on users.
- Graphical user interface: To make the operation more humane, to reduce the cognitive burden on the user, it provides simple graphical user interface, so that you can use to easily carry out various operations management.
- Webcam Watch: Using real-time video streaming technology, to obtain images of the Web Cam on a modular access control, so that users can be viewed directly on the server in the control of access.

4   Experiment and Evaluation

In the system situation experiment, user can use Door Access Control Server to search Door Access Controller Module showed in Fig. 2 (a). Fig.2 (b) is showed that user can use been registered in the list of RFID card, then successfully controlled door locker.

In this paper, we use IPv4/IPv6 service discovery applied to our access control system. According to Fig.3, we use Multicast, Broadcast and Polling three ways to do the comparison test. Broadcast will lead to unnecessary loss of network bandwidth, and Polling will let response time too long. Therefor Multicast is our best choice.

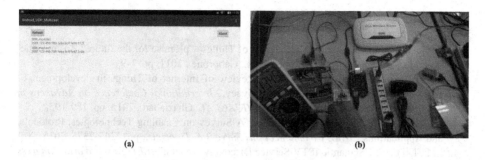

(a)   (b)

Fig. 2 The experiment of access control system.

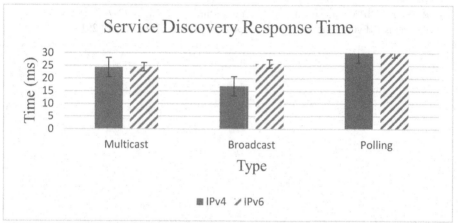

Fig. 3 The response time of service discovery.

5   Conclusion

The paper proposes an IPv4/IPv6 dual-stack automatic service discovery mechanism for an access control system. The service discovery mechanism is based on
IPv4/IPv6 multicast. In the IPv4/IPv6 Dual-Stack network, the devices obtain both of
IPv4 and IPv6 address. The system architecture is divided into two parts: Door Access Controller Module and Door Access Control Server. Users can use Door Access
Control Server to search Door Access Controller Module, and use RFID card to identify the users to control the door locker. In the implementation results, the system
works well, and the IPv4/IPv6 multicast is a good solution for service discover mechanism. In the future, the service discovery mechanism would be applied to more different situation.

Reference

1. L. Coetzee and J. Eksteen, "The Internet of Things – promise for the future? An introduction," *IST-Africa Conference Proceedings.*, Gaborone., 2011, pp. 1-9.
2. A. W. Burange and H. D. Misalkar, "Review of Internet of Things in development of smart cities with data management & privacy," *International Conference on Advances in Computer Engineering and Applications (ICACEA).*, Ghaziabad., 2015, pp. 189-195.
3. A. Al-Fuqaha et al., "Internet of Things: A Survey on Enabling Technologies, Protocols, and Applications," *IEEE Communications Surveys & Tutorials.*, pp. 2347-2376, 2015.
4. F. L. Lai et al., "Semantic IPTV Service Discovery System," *IEEE Eighth World Congress on Services.*, Honolulu., 2012, pp. 353-360.
5. J. Chen et al., "A New Design of Embedded IPv4/IPv6 Dual-Stack Protocol," *International Conference on Network Computing and Information Security (NCIS).*, Guilin., 2011, pp. 163-167.
6. R. Hinden, "IP Version 6 Addressing Architecture," *IETF RFC 4291*, February 2006.
7. R. Droms, "IPv6 Multicast Address Scopes," *IETF RFC 7346*, August 2014.

Defending Cloud Computing Environment Against the Challenge of DDoS Attacks Based on Software Defined Network

Shuen-Chih Tsai, I-Hsien Liu, Chien-Tung Lu, Chan-Hua Chang, and
Jung-Shian Li

Department of Electrical Engineering,
Institute of Computer and Communication Engineering,
National Cheng Kung University, Tainan, Taiwan, R.O.C.
{sctsai,dannyliu,chientung,frankchang}@hsnet.ee.ncku.edu.tw
jsli@mail.ncku.edu.tw

Abstract. With the explosive growth of cloud computing, virtualization technology has become more and more mature. However, it also increases the complexity of the network topology and causes many new important issues. One of the important issues is the security problem. It is hard to directly monitor the network traffic between Virtual Machines (VMs) through the external network devices, which make VMs more vulnerable in virtual environments. This research focuses on how to efficiently and rapidly protect VMs from malicious attacks without consuming its resources. We combine virtualization platform with the concept of Defense in Depth based on Software Defined Network (SDN), and implement a real-time detection and defense system for DDoS attacks. Moreover, we propose an enhanced entropy-based DDoS detection method to improve its detection accuracy, and we deploy it in SDN architecture.

Keywords: Cloud Computing, SDN, Defense in Depth, DDoS, KVM, Virtual switch, NIDS

1 Introduction

With the explosive growth of the Internet, the growing number of network devices will increase the complexity of network architecture. Conventional network will face the following three problems: Manageability, Innovation and Scalability. Unlike conventional network has the above disadvantages, cloud computing and Software Defined Network (SDN) have attracted more and more attention recently.

Network brought a lot of convenience to our lives, but it also take some information security threats to us, people usually being attacked or infected with malware without their knowledge. According to IBM Research [1], Distributed Denial of Service (DDoS) attack is one of the most common attack types in 2014, which primarily goal is to make an online service unavailable.

© Springer International Publishing AG 2017
J.-S. Pan et al. (eds.), *Advances in Intelligent Information Hiding and Multimedia
Signal Processing*, Smart Innovation, Systems and Technologies 63,
DOI 10.1007/978-3-319-50209-0_35

This paper is primarily concerned with how to protect virtual machine and SDN controller from DDoS attacks; therefore, we deploy a lightweight DDoS attacks detection and mitigation system in cloud computing environment based on SDN architecture, which can effectively detect the attacks at its early stage and do not consume virtual machines resources.

1.1   Related Work

The problem of DDoS attacks has attracted much research recently; however, most of them are based on conventional networks. In [2], although long-term entropy detection method has higher accuracy, it needs to observe a large number of packets that has the defects in detection at early stage. However, since it uses a fixed threshold for detection, it might have the problem of high false positive rate. ZHANG Jie and QIN Zheng [3] proposed an entropy-based DDoS attacks detection method, which can be used to detect low rate or high rate DDoS attacks.

Since SDN is new network architecture and not fully developed, there are only a few studies discussing the topic of DDoS attacks on it. Most of them are applying the existing conventional solutions to it. The authors in [4] proposed a moving target defense (MTD) technique on SDN called OpenFlow Random Host Mutation (OF-RHM). This method can effectively defend against malicious traffic, but does not work when the attackers know the internal virtual IP address. Xing et al. [6] proposed a SDN-based Introduction Detection and Prevention System (IDPS) solution, which deployed an IDPS on a physical host (Domain 0) to protect virtual machines. This method is concerned with improving the feasibility and efficiency in virtual network; however, when the network environments become more complex, those SDNIPS will hard to control. In [7], since a conventional firewall usually deploys on the border between the public Internet and the private network, the internal traffic cannot be examined by the firewall. As a result, the authors introduced an SDN-based firewall application called FLOWGUARD, which was designed to exclude the security policy violations in OpenFlow [5] networks, but it did not mention about the details of detection method.

2   System Architecture

Our primary purpose is to protect the internal network devices and the controller against malicious attacks in its early stage. In our system, after DDoS attacks arise, the threat will be automatically detected and excluded in a short time with the following steps. First, SDN controller application calculates the entropy value of the incoming packets and monitors its distribution with consecutive intervals. Then, when DDoS attacks arise, it will appear a large gap and the controller will achieve the fast mitigation strategy to block the specific port. Besides, it will also forward the suspicious flows to NIDS for future examination. Finally, NIDS will find out the malicious traffic and stop it.

2.1 System Design Overview

We use Ryu controller which is a Python-based controller as our SDN controller. Besides, Ryu controller can not only support multiple southbound protocols for managing network devices but also be integrated with OpenStack, which let network resource become more flexible. Software-based Open vSwitch (OVS) is chosen as our SDN switch that has better flexibility in handling packets. We use Kernel-based virtual machine (KVM) as our virtualization platform. KVM is a full virtualization technology for Linux, and it turns the Linux kernel into a hypervisor. Snort is an open source network intrusion detection system (NIDS) capable of providing a comprehensive real-time monitor, and it has three main benefits including protocol, signature, and anomaly-based inspection methods.

2.2 Scenario Description

There are several attack vectors on SDN, which containing attacks at the data plane layer and the controller plane layer shown on Fig. 1. Orange, yellow and red lines are represented as external attack, insider attack and the attacks that toward the SDN controller respectively.

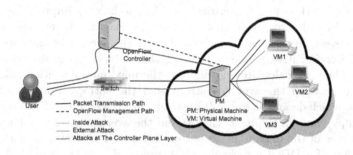

Fig. 1. Attacks

2.3 Proposed Method

Module 1: Dynamic entropy detection module. In information theory, entropy is used to measure of randomness. Our system is based on SDN; we propose an enhanced entropy-based DDoS attacks detection mechanism on the controller application, which can dynamically change its threshold with a classified algorithm. We assume the incoming packets are random uncertainty in general case, so the entropy distribution is close to a flat. By contrast, we can find a gap when DDoS attacks arise. Let S denote a set with all the protected hosts IP addresses in Eq. (1), and R is the hash table containing destination IP addresses x and the number of their occurrence y in a window size z in Eq. (2). The probability of occurs in a window size z is defined in Eq. (3).

$$S = x_1, x_2, x_3, ..., x_k, \tag{1}$$

$$R = (x_1, y_1), (x_2, y_2), (x_3, y_3), ..., (x_k, y_k), \tag{2}$$

$$P_i = \frac{y_i}{z}. \tag{3}$$

According to Oshima's research [2] that 50 packets were the optimal size to effectively detect DDoS attacks, which can identify the attacks earlier than other size; and the other reason is that each host can only accept a finite number of new incoming connections in the network. Besides, the reason why we choose destination IP address as our basis for entropy calculation is that our system is based on SDN which follows the OpenFlow protocol; hence the different source IP address is useless in our system [8]. After getting the probability of all hosts in a window, we can sum up them and calculate the entropy value that defined in (4).

$$H = -\sum_{i=1}^{k} p_i \log_2 (p_i). \tag{4}$$

Therefore, knowing the entropy distribution can help us to find out DDoS attacks easily. However, the critical problem is how to improve its detection accuracy. In order to tackle this problem, we classify the entropy value into different levels. In the beginning, we calculate the average of the entropy value within 10 consecutive windows before DDoS attacks arise, and denote as in Eq. (5). Then, we calculate and find out the maximum deviation, which is the difference between the average entropy and the entropy value in the next 5 windows, which denote as in Eq. (6). Finally, in Eq. (7), if the difference between the entropy value in the next window and the average entropy value is greater than n times of the maximum deviation, it denote DDoS attacks has arisen [3]. In addition, if a DDoS attacks does not appear during the period of time, the average entropy value and the maximum deviation value will be recalculated in next period.

$$H_{\text{avg}} = \frac{1}{m} \sum_{a=1}^{m} H_a, m \in R^+. \tag{5}$$

$$d_{\max} = \max_{1 \le w \le u} (|H_w - H_{\text{avg}}|). \tag{6}$$

$$|H_w - H_{\text{avg}}| > n d_{\max}. \tag{7}$$

Module 2: Snort integration module. The major objective of this module is to construct a real-time detection system for malicious attacks. We use Snort to monitor network traffic and deploy it on our core switch, which can instantly find out the malicious packets and generate alert messages when an attack arises. Our detection system is under the SDN architecture, we combine Snort alert message with SDN controller for further processing. However, since Snort requires a lot of resources to analyze packets, we let SDN controller and Snort be installed on different hosts, which can avoid consuming the controllers resources and affecting its performance.

Module 3: Defend in Depth with OpenFlow-based firewall module. The purpose of this module is to provide a comprehensive defense system for DDoS attacks. Although Snort can be configured as a firewall system by using iptable to filter and block malicious traffic, it has several major disadvantages, which includes the problems of the latency, complexity, resource consumption and inflexible network reconfigurations [6]. To overcome the above-mentioned drawbacks, we construct a new form of firewall in SDN switches.

3 Implementation and Simulation Result

3.1 Environment

Fig. 2 shows the experimental environment of our system, which consists of one Ryu controller, one software-based OepnFlow switch as OOF-firewall, a physical machine with OpenFlow switch as IOF-firewall, a detection machine with Snort, one client and one attacker. In addition, KVM is installed on the physical machine, which has 3 virtual machines on it.

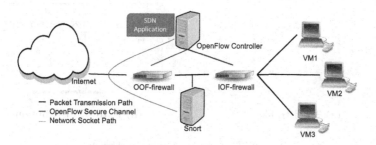

Fig. 2. System environment

3.2 Traffic Generation

In our experiment, we use an open-source Scapy as our packet generator, which is a powerful interactive packet manipulation tool based on Python programing language for forging and decoding with different protocols, and it also provides the basic functions to create each layers packets of the OSI model and send them over the wire.

3.3 Experimental Result and Analysis

The experiment demonstrates the effectiveness of our system with different SDN attack vectors, which including attacks at data plane layer and controller plane layer. All the experiments consist of normal and attack traffic. In order to simulate DDoS attacks on our experiments, we assume that normal traffic is randomly run to all VMs, and attack traffic is specifically targeted to a particular VM that manually launched after one third of total packets.

External Attack. We assume that attacker is outside of the protected network, who wants to make an online service VM2 unavailable by overwhelming its bandwidth or system resource so that user cannot access it anymore. In Fig. 3, we can see the network traffic and its randomness among three VMs via the entropy distribution. In normal traffic, the traffic is about 40 Mbps and its entropy value is between 1.4 and 1.6. After attackers launch attacks and target to a specific VM, the traffic will increase rapidly to around 100Mbps and the randomness will decrease so that we can easily see its entropy value drop. However, as soon as the controller identify a VM under DDoS attacks, it will instantly add the deny policy to OOF-firewall for blocking the malicious traffic. As we can see in these figures, the attack will be mitigated within a short time that is around 250 packets. At the same time, the controller will also add the allow policies to the firewall that are used to forward the original known traffic to Snort for detailed inspection.

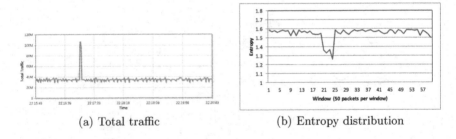

(a) Total traffic   (b) Entropy distribution

Fig. 3. The results of external attack.

Insider Attack. We assume that the attacker is inside of the protected network. In this experiment, attacker is VM1, which is unfortunately infected with a virus, and it wants to make an online service VM2 unavailable by running out its bandwidth or system resource. In Fig. 4, the protected network traffic is about 30 Mbps and its entropy value is between 1.4 and 1.6 at the beginning. After the attacker launches attacks and target to VM2, the traffic will go up substantially to around 100Mbps and the entropy value will drop dramatically. The reason why the entropy value will decrease significantly is that we increase attack rate to prove the effectiveness of our enhanced DDoS detection method; therefore, the entropy value has great difference with the normal condition and the malicious attacks will be detected in seconds that is around 150 packets. Then, the controller will immediately add the deny policy to the IOF-firewall for blocking the ingress port that connected with attacker; as a result, the traffic can rapidly return to the normal state, which is around 20Mbps and the entropy value is between 0.8 to 1.

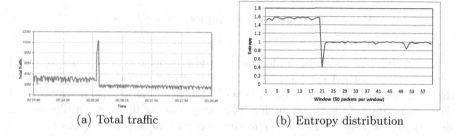

(a) Total traffic   (b) Entropy distribution

Fig. 4. The results of insider attack.

Attacks at the controller layer. The last experiment assumes that the attackers target is the controller. The normal traffic is generated by clients, and the attacker wants to overwhelm the controller's bandwidth or system resource so that it can make the connection between the controller and the switch failure. Fig. 5 shows the traffic and the entropy value of the protected network. In the beginning, the normal traffic is around 30 Mbps and its entropy value is between 1.4 and 1.6. After the attackers launch attacks and target to the controller, the traffic will rise dramatically to around 110Mbps, and the entropy value will also go up to about 1.85. The reason why the entropy value will increase is that the entropy detection module find out the destination IP address is the controller itself, so it will add the address to the protected set. As we can see in these figures, the attacks will be immediately eliminated with adding the deny policy in a short time, which is about 250 packets.

4   Conclusion

In this paper, we deal with the problem of DDoS attacks without consuming VMs resources in virtual environments. We first propose an enhanced entropy-based DDoS detection method to improve its detection accuracy and implement it on SDN controller in Module 1. Then, in Module 2, we integrate a NIDS with SDN for deep packet inspection. Finally, Module 3 achieves fast mitigation strategy through the concept of Defend in Depth. Besides, the network host can arbitrarily increase or decrease and our system will automatically adjust

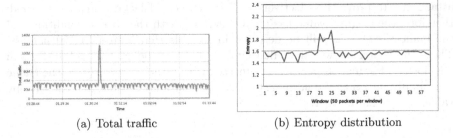

(a) Total traffic   (b) Entropy distribution

Fig. 5. The results of DDoS attacks on the controller.

based on different network configurations. The experiments demonstrate that our detection method is effective in different scenarios. Moreover, we compare the CPU utilization of the controller with different methods, and we also test the performance of the switch under different number of flow entries. The results show that it is a lightweight and effective detection and defense system.

References

1. IBM X-Force Threat Intelligence Quarterly,1Q 2015, Retrieved 2015/06/02 from http://public.dhe.ibm.com/common/ssi/ecm/wg/en/wgl03073usen/WGL03073USEN.PDF
2. Shunsuke Oshima, Takuo Nakashima, and Toshinori Sueyoshi.: Early DoS/DDOS detection method using short-term statistics. IEEE International Conference on Complex, Intelligent and Software Intensive Systems (CISIS), 2010, pp. 168-173.
3. Jie Zhang, Zheng Qin, Lu Ou, Pei Jiang, JianRong Liu and Alex X. Liu.: An Advanced Entropy-Based DDOS Detection Scheme. IEEE International Conference on Information, Networking and Automation (ICINA), 2010, pp. 67-71.
4. Jafar Haadi Jafarian, Ehab Al-Shaer, and Qi Duan.: Openflow random host mutation: transparent moving target defense using software defined networking. Proceedings of the first ACM workshop on Hot topics in software defined networks, 2012, pp. 127-132.
5. Nick McKeown, Tom Anderson, Hari Balakrishnan, Guru Parulkar, Larry Peterson, Jennifer Rexford, Scott Shenker and Jonathan Turner.: OpenFlow: Enabling Innovation in Campus Networks. ACM SIGCOMM Computer Communication Review, vol. 38, no. 2, 2008, pp. 69-74.
6. Tianyi Xing, Zhengyang Xiong, Dijiang Huang and Deep Medhi.: SDNIPS: Enabling Software-Defined Networking Based Intrusion Prevention System in Clouds. IEEE 10th International Conference on Network and Service Management (CNSM), 2014, pp. 308-311.
7. Hongxin Hu, Wonkyu Han, Gail-Joon Ahn and Ziming Zhao.: FLOWGUARD: building robust firewalls for software-defined networks. Proceedings of the third ACM workshop on Hot topics in software defined networking, 2014, pp. 97-102.
8. Seyed Mohammad Mousavi and Marc St-Hilaire.: Early Detection of DDoS Attacks against SDN Controllers. IEEE International Conference on Computing, Networking and Communications (ICNC), 2015, pp. 77-81.

Decision Support for IPv6 Migration Planning

Shian-Shyong Tseng[1], Ai-Chin Lu[2], Wen-Yen Lin[3*], Ching-Heng Ku[2],
Geng-Da Tsai[2]

[1]Dept. of Applied Informatics and Multimedia, Asia University, Taiwan
sstseng@asia.edu.tw
[2]Taiwan Network Information Center, Taiwan
{aclu,chku,dar}@twnic.net.tw
[3]Dept. of Digital Multimedia Technology, Vanung University, Taiwan
qqnice@gmail.com

Abstract. While confronting the global IPv4 address exhaustion, it is important and crucial for the entire Internet environment to smoothly upgrade to the next generation Internet Protocol, IPv6. In previous years, all the external network services of Taiwan governmental organizations have been upgraded to be IPv6 capable. In this paper, we are concerned about the planning of the IPv6 upgradation for the internal services. Due to a variety of the internal network services, the planning must take account of the network management policy, legacy equipment, monitoring and tracking requirement, Wi-Fi service requirement, the user accountability, the budget, and the scope. Therefore, EMCUD is applied to find the decision rules to help users to develop their IPv6 migration plan.

Keywords: IPv6 upgradation, Repertory Grid-based KA, EMCUD, IPv6 IPAM, knowledge acquisition, network management

1 Introduction

While confronting the global IPv4 address exhaustion, it is important and crucial for the entire Internet environment to smoothly upgrade to the next generation Internet Protocol, IPv6. In the last four years, all the external network services of Taiwan governmental organizations have been upgraded to be IPv6 capable, and we have obtained lots of experiences and knowledge about the IPv6 upgrade. In this paper, we are concerned about the planning of the IPv6 upgradation for the internal services. Due to a variety of the internal network services, the planning must take account of important attributes.

Because Repertory grid-based KA is a good approach for the classification problem, it will be used to help the network managers conquer the IPv6 migration planning problem. Up to now, we have obtained eight objects and eight objects. Moreover, EMCUD is then applied to find the decision rules to help users to develop their IPv6 migration plan.

© Springer International Publishing AG 2017   293
J.-S. Pan et al. (eds.), *Advances in Intelligent Information Hiding and Multimedia
Signal Processing*, Smart Innovation, Systems and Technologies 63,
DOI 10.1007/978-3-319-50209-0_36

2   Related work

In this section, the Repertory Grid, based on Kelly's The Psychology of Personal Constructs [1] will be introduced. It reports how people make sense of the world, could be used as an efficient KA technique in identifying different objects and distinguishing these objects in a domain. As Repertory Grid technique has been widely used by researchers, EMCUD, a Repertory Grid-based KA method, was proposed to elicit the embedded meanings of knowledge from the existing hierarchical grids given by experts [2-3]. The embedded meanings referred here represents the information that domain experts take for granted but are implicit to the people who are not familiar with the application domain. In this paper, EMCUD is applied to facilitate the acquisition of new inference rules for the planning of the IPv6 upgradation for the internal services. Furthermore, a prototype is implemented to evaluate the effectiveness of our system.

EMCUD [4-5] is used to elicit the embedded meanings of IPv6 migration knowledge (embedded rules bearing on IPv6 address allocation cases and some management attributes). Besides, the attribute ordering table (AOT) records the relative importance of each attribute to each object in EMCUD to capture the embedded meanings with acceptable certainty factor value by relaxing or ignoring some minor attributes. It can guide the experts to determine the certainty degree of each embedded rule, in order to extend the coverage of original rules generated by traditional repertory grid (acquisition table).

3   IPv6 migration Planning for internal network services

As we know, there are a couple of methods to perform address configuration on IPv6, including static addressing, static addressing with DHCPv6 (stateless), dynamic addressing via DHCPv6 (Stateful), SLAAC alone, and SLAAC with DHCPv6 (Stateless). Besides, IPv6 static addressing works exactly the same as IPv4 static addressing does, and needs to be manually configured [12].

As to auto-configure IPv6, there are two different ways, DHCPv6 and SLAAC. According to the functionality of the IPv6 DHCP server, the stateful DHCPv6 tracks the status of the assigned addresses, while the stateless DHCPv6 does not. Instead, Stateless DHCPv6 provides the DNS server information and then does the IPv6 address assignment by another mechanism.

In the Stateless address auto configuration (SLAAC), each client device can get the IPv6 address automatically, where nodes listen for ICMPv6 Router Advertisements (RA) messages periodically sent out by routers on the local link or requested by the node using an RA solicitation message. Therefore, it provides the ability to address a host based on a network prefix that is advertised from a local network router via Router Advertisements (RA). RA messages (which are sent by default by most IPv6 routers) are sent out periodically by the router, and include one or more IPv6 prefixes (Link-local scope), default router to use, and its lifetime. So, the client device can auto configure IPv6 address by RA.

In the last four years, we have successfully updated all the public network services of Taiwan Government Network to be IPv6 capable. According to the discussion with the domain experts, there are eight different approaches (objects) about the IPv6 address management as shown in Table 1.

- **Object1 (RA+SLAAC without EUI64): it can be** easily configured but can't work in any Native IPv6 environment.

- **Object2 (RA+SLAAC with EUI64)**: it can be easily configured by using the EUI64 to be the host address but can't work in Native IPv6 environment, where the Extended Unique Identifier (EUI) allows a host to assign itself a unique 64-Bit IP Version 6 interface identifier (EUI-64) and the IPv6 EUI-64 format address is obtained through the 48-bit MAC address.

- **Object3 (Static** IPv6 Address): it works the same as IPv4, and users can apply the address by themselves.

- **Object4 (RA+Stateless DHCPv6 without EUI64)**: it is similar to Object1 except that it can get the DNS server setting from Stateless DHCPv6 server, where the Stateless DHCPv6 server is not required to store the dynamic state information of any individual clients. For the large networks which has huge number of end points attached to it, the stateless DHCPv6 approach will highly reduce the number of DHCPv6 messages that are needed for address state refreshment.

- **Object5 (RA+Stateless DHCPv6 with EUI64)**: it is similar to the Object2 except that it can get the DNS server setting from Stateless DHCPv6 server.

- **Object6 (RA+Stateful DHCPv6 by Prefix):** only the Gateway address can be received from Router Advertisement(RA), and the functions are exactly the same as IPv4 DHCP in which hosts receive both their IPv6 address and additional parameters from the DHCP server. So, the components of a DHCPv6 infrastructure consist of DHCPv6 clients that request configuration and DHCPv6 servers that provide configuration. The Host IP address is randomly created from Prefix of RA and the Network address is created from the Prefix.

- **Object7 (RA+Stateful DHCPv6 From IP pool): it is similar to** the Object6 and the Gateway address is received from Router Advertisement (RA), but the host IP address is received from the IP pool of DHCPv6.

- **Object8 (RA+Stateful DHCPv6 with DUID): it is similar to** the Object6, but the host address is created from the DHCP Unique Identifier (DUID). Each DHCP client or server has a DUID. DHCP servers use DUIDs to identify clients for the selection of configuration parameters, and DHCP clients use DUIDs to identify the server in a message while the server needs to be identified.

Table 1. Eight objects for IPv6 migration

Objects	Gateway	Prefix	Host ID	DNS
Object1 (RA+SLAAC without EUI64)	by Router advertisement	by Router advertisement	Random	using the IPv4 DNS setting
Object2 (RA+SLAAC with EUI64)	by Router advertisement	by Router advertisement	EUI64 From MAC address	using the IPv4 DNS setting
Object3 (Static Address)	Manually Configured	Manually Configured	Manually Configured	Manually Configured
Object4 (RA+Stateless DHCPv6 without EUI64)	by Router advertisement	by Router advertisement	Random	by DHCPv6 server
Object5 (RA+Stateless DHCPv6 with EUI64)	by Router advertisement	by Router advertisement	EUI64 From MAC address	by DHCPv6 server
Object6 (RA+Stateful DHCPv6 by Prefix)	by Router advertisement	by DHCPv6 server	Random from DHCPv6 server	by DHCPv6 server
Object7 (RA+Stateful DHCPv6 From IP pool)	by Router advertisement	by DHCPv6 server	From the Pool of DHCPv6 server	by DHCPv6 server
Object8 (RA+Stateful DHCPv6 with DUID	by Router advertisement	by DHCPv6 server	Using DUID	by DHCPv6 server

4   Concerned Attributes

Based on the EMCUD method for the selection of IPv6 migration approaches, the following 8 attributes are proposed as shown in Table 2.

Attributes	Data Type	Attribute Values	Definition
A1- Difficulty of IP configuration	Multiple values	5~1	The difficulty level to configure the Client device
A2-Monitoring capability	Multiple values	5~1	Monitoring and Tracing the IP address
A3- Difficulty of setting access control rules for firewall	Multiple values	5~1	The difficulty level of setting the access control rules for firewall
A4-Friendlyness of configuring the Wi-Fi Network IP	Multiple values	5~1	Value 5 means more friendly to configure the Wi-Fi device
A5-Required skill level of network manager	Multiple values	5~1	The required skill level of network manager
A6- workload of network manager	Multiple values	5~1	The workload of network manager
A7- Budget of migration	Multiple values	5~1	The Budget of IPv6 migration
A8- Difficulty of Identity Hiding	Multiple values	5~1	The difficulty of Identity hiding, where Value 5 means hard to identity hiding

Table 3. The acquisition Repertory Grid of IPv6 migration Object

Objects / Attributes	Object1 (RA+SLAAC without EUI64)	Object2 (RA+SLAAC with EUI64)	Object3 (Static Address)	Object4 (RA+Stateless DHCPv6 without EUI64)	Object5 (RA+Stateless DHCPv6 with EUI64)	Object6 (RA+Stateful DHCPv6 by Prefix)	Object7 (RA+Stateful DHCPv6 From IP pool)	Object8 (RA+Stateful DHCPv6 with DUID)
A1- Difficulty of IP configuration	5	4	2	4	3	2	2	1
A2-Monitoring capability	1	5	5	1	5	2	4	5
A3- Easiness of access control rule setting for firewall	1	3	5	1	3	1	4	5
A4-Friendlyness of IP configuration of Wi-Fi Network	5	5	2	5	5	5	5	1
A5-Required skill level of network manager	2	3	1	2	3	3	4	5
A6- workload of network manager	1	2	4	1	2	3	4	5
A7- Budget of migration	1	1	1	2	2	3	4	5
A8- Difficulty of Identity Hiding	1	5	3	1	5	1	2	3

After constructing the IPv6 Migration AT, we then construct the AOT table as shown in Table 4.

Table 4. The Attribute Ordering Table (AOT) of IPv6 Migration Planning

Objects / Attributes	Object1 (RA+SLAAC without EUI64)	Object2 (RA+SLAAC with EUI64)	Object3 (Static Address)	Object4 (RA+Stateless DHCPv6 without EUI64)	Object5 (RA+Stateless DHCPv6 with EUI64)	Object6 (RA+Stateful DHCPv6 by Prefix)	Object7 (RA+Stateful DHCPv6 From IP pool)	Object8 (RA+Stateful DHCPv6 with DUID)
A1- Difficulty of IP configuration	D	2	x	1	1	x	x	x
A2-Monitoring capability	x	D	D	x	D	x	1	1
A3- Difficulty of access control rule setting for firewall	x	3	1	x	3	x	x	2
A4-Friendlyness of IP configuration of Wi-Fi Network	1	1	x	D	x	D	D	x
A5-Required skill level of network manager	2	x	x	x	2	1	2	D
A6- workload of network manager	x	x	x	x	x	x	x	x
A7- Budget of migration	x	x	2	3	x	3	x	3
A8- Difficulty of Identity Hiding	3	x	3	2	x	2	3	x

With both acquisition table and AOT, the EMCUD method can be applied. The embedded rules can then be generated and some of them have low CF value.

5   Conclusion

Due to a variety of the internal network services, the planning must take account of the network management policy, legacy equipment, monitoring and tracking requirement, Wi-Fi service requirement, the user accountability, the budget, and the scope. Therefore, we have found eight objects and eight attributes, and then the EMCUD is applied to find the decision rules to help users to develop their IPv6 migration plan. Based upon our methods, we can help the network managers to choose the appropriate approach for the upgrade planning. In the near future, we are going to refine the rules by relaxing or ignoring some minor attributes.

Acknowledgement

This paper is partially sponsored by Ministry of Transportation and Communications, Republic of China, under Grant 1050219coco, and Taiwan Network Information Center (TWNIC).

Reference

1. Kelly, G., *The Psychology of Personal Constructs* New York: Norton. Reprinted by Routledge (London), 1955.
2. Hwang, G.J. and S.S. Tseng. *Building a multi-purpose medical diagnostic system under uncertain and incomplete environment*. in *Computer-Based Medical Systems, 1990., Proceedings of Third Annual IEEE Symposium on*. 1990.
3. Hwang, G.J. and S.S. Tseng. *Eliciting embedded meanings in building expert systems under uncertainty*. in *Computers and Communications, 1990. Conference Proceedings., Ninth Annual International Phoenix Conference on*. 1990.
4. Hwang, G.J. and S.S. Tseng, *EMCUD: A knowledge acquisition method whichcaptures embedded meanings under uncertainty*. International Journal of Man-Machine Studies, 1990. 33(4): p. 431-451.
5. Lin, S.-C., S.-S. Tseng, and C.-W. Teng, *Dynamic EMCUD for knowledge acquisition*. Expert Systems with Applications, 2008. 34(2): p. 833-844.
6. Tseng, S.-S. and S.-C. Lin, *VODKA: Variant objects discovering knowledge acquisition*. Expert Systems with Applications, 2009. 36(2): p. 2433-2450.
7. Zhao, Q. and Y. Ma, *An object-oriented model of IPv4/IPv6 network management*. The Journal of China Universities of Posts and Telecommunications, 2010. 17, Supplement 2: p. 89-92.
8. Nowicki, K., et al. *Extension Management of a Knowledge Base Migration Process to IPv6*. in *Applications and the Internet (SAINT), 2011 IEEE/IPSJ 11th International Symposium on*. 2011.
9. Chandra, D.G., M. Kathing, and D.P. Kumar. *A Comparative Study on IPv4 and IPv6*. in *Communication Systems and Network Technologies (CSNT), 2013 International Conference on*. 2013.

10. Tseng, S.S., et al. *Building a Self-Organizing Phishing Model Based upon Dynamic EMCUD.* in *Intelligent Information Hiding and Multimedia Signal Processing, 2013 Ninth International Conference on.* 2013.

11. Tseng, S.S., C.H. Ku, and A.C. Lu. *Building an IPv6 upgrade model based upon cost-effective strategies.* in *TENCON 2014 - 2014 IEEE Region 10 Conference.* 2014.

12. Yechiel, N. *IPv6 address assignment – stateless, stateful, DHCP.* 2014 [cited 2014 July,2]; Available from: https://thenetworkway.wordpress.com/2014/07/02/ipv6-address-assignment-stateless-stateful-dhcp-oh-my/.

13. Aravind, S. and G. Padmavathi. *Migration to Ipv6 from IPV4 by dual stack and tunneling techniques.* in *Smart Technologies and Management for Computing, Communication, Controls, Energy and Materials (ICSTM), 2015 International Conference on.* 2015.

14. Terli, V.K.K., et al. *Software implementation of IPv4 to IPv6 migration.* in *2016 IEEE Long Island Systems, Applications and Technology Conference (LISAT).* 2016.

Part IV
Encryption and Authentication Methods

A Secure Authentication Scheme for Telecare Medical Information Systems

Chin-Chen Chang[1], Jung-San Lee[1], Yu-Ya Lo[2], and Yanjun Liu[1,*]

[1]Department of Information Engineering and Computer Science,
Feng Chia University, Taichung, Taiwan, R.O.C.
alan3c@gmail.com, leejs@fcu.edu.tw, yjliu104@gmail.com
[2]Department of Computer Science and Information Engineering,
National Chung Cheng University, Chiayi, Taiwan, R.O.C.
yy04180526@gmail.com

Abstract. In 2012, Chen et al. proposed an ID based authentication scheme for Telecare Medical Information Systems. However it has some security weaknesses. Later Xie et al. have proposed a new scheme to improve the security so that various attacks can be resisted, such as off-line password guessing attacks, user anonymity attacks, impersonation attacks, and perfect forward secrecy. However, Xie et al.'s scheme is vulnerable to the denial of service (Dos) attack when a patient submits a login request to the server. We proposed a secure authentication scheme for telecare medical information systems in out method, we use visual secret sharing which was proposed by Shamir to generate one-time password to overcome the denial of service attack problem.

Keywords: Timestamp; TMIS; one-way collision-free hash function

1 Introduction

The Telecare Medical Information System (TMIS) is a good way to bring telemedicine directly into patients' homes. The medical server protects various private data and information of registered users, such as their names and electronic medical records (EMRs). The main problem that researchers face when deploying a TMIS is ensuring the security and privacy of important data and information.

In 2010, Wu et al. [1] claimed that their TMIS was more efficient than previous TMISs because they added a pre-computation step. However, He et al. [2] pointed out that Wu [1] et al.'s scheme could not resist insider attacks and impersonation attacks. Then, Wei et al. [3] showed that He et al.'s [2] protocol was vulnerable because it could not resist off-line password guessing attacks, and they also proposed an improved scheme. However, Zhu [4] pointed out that Wei et al.'s [3] scheme also was vulnerable to the same security attacks as He et al.'s [3] scheme. In 2012, Zhu [4] proposed an improved scheme to solve the problems mentioned above. Muhaya [5] proved that Zhu's [4] protocol was insecure. Islam and Biswas [6] found that Wu et al.'s scheme was still vulnerable to the privileged insider attack, off-line password guessing, and ephemeral secret leakage.

© Springer International Publishing AG 2017
J.-S. Pan et al. (eds.), *Advances in Intelligent Information Hiding and Multimedia Signal Processing*, Smart Innovation, Systems and Technologies 63,
DOI 10.1007/978-3-319-50209-0_37

Jiang et al. [7] showed that Chen et al.'s [8] scheme cannot resist various attacks, such as impersonation attacks, off-line password guessing attacks, and denial of service attacks. In 2014, Islam and Khan [9] presented some improved authentication and key agreement protocols for TMISs.' However, Zhang and Zhou [10] pointed out that Islam and Khan's [9] protocols also were subject to various kinds of attacks. Liu et al. [11] showed that Zhang and Zhou's [10] approach was vulnerable to lost/stolen smartcard attacks and off-line password guessing attack. Then, Liu et al. proposed an improved protocol using biometric keys to resolve the security problems.

In our paper, we used JAVA code to determine the number of times that login requests were made. We determined that the system would deny service if patient U submitted 3,850 login requests to the server. Therefore, Server S will deny to offer service and the system will fail. Xie et al.[12] proposed a new scheme to eliminate the weaknesses of Chen et al.'s scheme. However, we determined that Xie et al.'s[12] scheme was still subject to Dos attacks. Therefore, we proposed a new scheme using visual secret sharing and a one-time password system to better resist Dos attacks.

When we made more than 3,850 requests for login to the server, the terminal displayed the message 'java.net.SocketException.' Thus, we thought that the server had denied service after we had submitted 3,850 login requests.

2   Review of Xie et al.'s scheme

In this section, we focus on the Login and Verification phases of Xie et al.'s [12] scheme. The notations are shown below:

U_i: name of the ith patient

S: name of the server

ID_i: identity of ith patient

PW_i: password of ith patient

α: a random number generated by the patient

$h(\cdot)$: one-way, collision-resistant, cryptographic hash function

p, q: two large prime numbers

$n : n = p \times q$

X: medical server S's long-term secret key

e: medical server S's public key

$E_X()$: symmetric encryption function with key X

N: registration time

x: master key

2.1   Registration phase

When the patient wants to obtain medical service from medical servers, he/she may transfer the registration request to the medical servers and obtain a smart card to log in to the server by the following steps, which are shown in Figure 1.

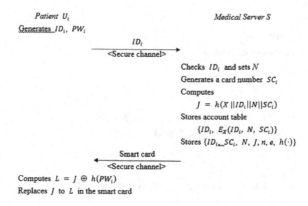

Fig. 1. Registration phase of Xie et al.'s scheme

Step 1 *Patient $U_i \to$ Medical Server $S : \{ID_i\}$*
The patient selects his/her identity and transfers it to the server.

Step 2 *Medical Server $S \to$ Patient $U_i :$ {Smart card}*
If U_i is a new user, sets $N = 0$. Otherwise, U_i is registering in the system, server S sets $N = N + 1$. S continues to generate a card number SC_i and computes $J = h(X||ID_i||N||SC_i)$, and maintains an account table for the registration patient as $\{ID_i, E_X(ID_i, N, SC_i)\}$, which records the identity ID_i, registration time N, and card number SC_i. Otherwise, terminates it. Server S stores $\{ID_i, SC_i, N, J, n, e, h(\cdot)\}$ into a smart card and issues the smart card to U_i through the secure channel.

Step 3 *Patient $U_i :$*
Patient U_i computes $L = J \oplus h(PW_i)$ and replaces J with L in the smart card. Therefore, the smart card contains $\{ID_i, SC_i, N, L, n, e, h(\cdot)\}$.

2.2 Login phase

After a patient obtains a smart card from server S, he/she can login to the medical server to obtain service. The login procedure follows the steps shown in Figure 2.

Patient inputs the identity ID_i and password PW_i first. Then, the smart card generates a random number α and computes $J = L \oplus h(PW_i), A = J^\alpha \bmod n, C_1 = h(T_i||J||A) = h(T_i||h(X||ID_i||N||SC_i)||A)$ and $AID_i = (ID_i||N||SC_i|| C_1||A)^e \bmod n$ where T_i is the current timestamp. After that, The smart card sends AID_i, T_i to S.

2.3 Verification phase

Before patient U_i obtains service from medical server S, both the server and the patient must authenticate each other to confirm that a legitimate patient is requesting service. The verification procedure consists of the steps shown in Figure 3.

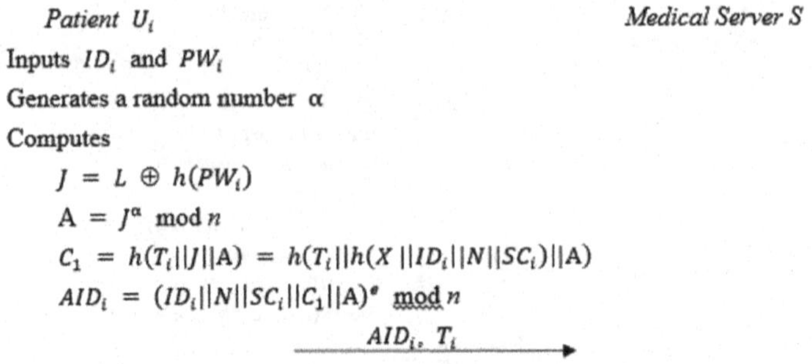

Patient U_i   *Medical Server S*

Inputs ID_i and PW_i

Generates a random number α

Computes

$$J = L \oplus h(PW_i)$$
$$A = J^\alpha \bmod n$$
$$C_1 = h(T_i||J||A) = h(T_i||h(X||ID_i||N||SC_i)||A)$$
$$AID_i = (ID_i||N||SC_i||C_1||A)^e \bmod n$$

$$\xrightarrow{\quad AID_i,\ T_i \quad}$$

Fig. 2. Login phase of Xie et al.'s scheme

Step 1. *Patient $U_i \to$ Medical Server $S : \{AID_i, T_i\}$*
The patient selects his/her identity and transfers it to the server. Upon receiving $\{AID_i, T_i\}$ from patient at time T', S checks whether $T_i - T' < \Delta T$ or not, where ΔT is a valid interval. If not, the session is terminated. Otherwise, S computes $(AID_i)^X \bmod n = \{ID_i||N||SC_i||C_1||A\}, J = h(X||ID_i||N||SC_i)$,
$C_1' = h(T_i||J||A)$ and decrypts $E_X(ID_i, N, SC_i)$ with key X to obtain ID_i, N, SC_i, then checks the validity of SC_i, C_1', ID_i. If so, U_i is authenticated. Otherwise, S terminates it. After that, S generates a random number b and computes $B = J^b \bmod n, C = A^b \bmod n, sk = h(T_i||C||T_s), C_2 = h(C_1|(|C|)|T_s||B)$. Then, S send $\{C_2, T_s, B\}$ to U_i, where T_s is the current timestamp.
Step 2. *Medical Server $S \to$ Patient $U_i : \{C_2, T_s, B\}$.*
When the smart card receives $\{C_2, T_s, B\}$ at time T'', it checks whether $T_s - T'' < \Delta T$ or not. If not, terminates it. Otherwise, the smart card computes $C = B^\alpha = J^{ab} \bmod n, C_2' = h(C_1|(|C|)|T_s||B)$ and checks whether $C_2' \stackrel{?}{=} C_2$ or not. If not, terminates it. Otherwise, S is authenticated. U_i computes the session key $sk = h(T_i||C||T_s)$.

2.4 Password change phase

If patients wish to improve the security of their cards, they can get a new password to replace their old password. The procedure for updating a password is shown in Figure 4.

If U_i wants to improve the security of smart card and changes password of smart card with new password. He/She can insert their smart card into card reader and inputs their old PW_i and new PW_i^{new}, then the smart card computes $L^{new} = L \oplus h(PW_i) \oplus h(PW_i^{new})$ and replaces L with L^{new}.

3   Weaknesses of Xie et al.'s scheme

From Subsection 2.2 and Subsection 2.3, we saw that Patient U_i submitted a login request AID_i and T_1 to medical server S. Server S will check whether the

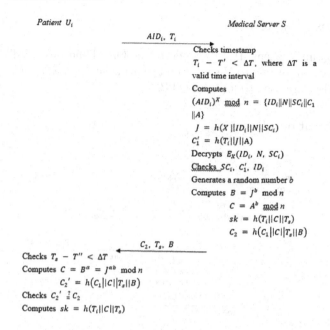

Fig. 3. Verification phase of Xie et al.'s scheme

Patient U_i   **Smart card**

Inserts his/her smart card into a device

Inputs old password PW_i and new password PW_i^{new}

$$\text{Compute } L^{new} = L \oplus h(PW_i)$$
$$h(PW_i^{new})$$

$$\text{Replaces } L \text{ with } L^{new}$$

Fig. 4. Password change phase of Xie et al.'s scheme

timestamp $T_i - T' < \Delta T$ and computes $(AID_i)^X \bmod n = \{ID_i||N||SC_i||C_1||A\}$, $J = h(X||ID_i||N||SC_i)$ and $C_1' = h(T_i||J||A)$. We consider that there is denial of service attack when the server computes $C_1' = h(T_i||J||A)$. Since many patients log in to the medical server, they may transfer many login requests, AID_i and T_i, to the server. If transferring timestamps to the server is made 3,850 times, the server will deny the request for service after computing $C_1' = h(T_i||J||A)$. Therefore, the server may deny offering service to the patient, thereby causing a denial of service attack.

Using JAVA cryptography programming, we calculated the number of login requests that will make the server deny service. Transferring login request over 3850 times will occur denial of service attack since server has to compute $C_1' = h(T_i||J||A)$ during each login request is submitted .

4   Proposed scheme

We used Shamir's scheme [13] to generate the One Time Password in our proposed scheme. According to Visual Secret Sharing scheme, a picture base and a picture shadow are used to form an OTP.

4.1   Registration phase

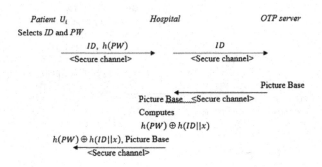

Fig. 5. Registration phase of the proposed scheme

Step 1. $U_i \rightarrow$ Hospital:$\{ID, h(PW)\}$
  The user selects his/her identity ID and password PW and then computes and transfers $ID, h(PW)$ to the hospital server.
Step 2. Hospital $\rightarrow OTP\ server : \{ID\}$
  The hospital server obtains the identity ID and transfers the identity ID to $OTP\ server$ in order to obtain picture base. The hospital server will retransfer picture base to patient U_i.
Step 3. $OTP\ server \rightarrow$ Hospital:$\{$Picture Base$\}$
  According to the identity ID received from the hospital server, the $OTPserver$ can generate the picture base, which combines the OTP with the shadow.
Step 4. Hospital $\rightarrow U_i : \{h(PW) \oplus h(ID||x),$ Picture Base$\}$
  The hospital computes $h(ID||x)$ with master key x and calculates $h(PW) \oplus h(ID||x)$ with $h(PW)$. After computing the value $h(PW) \oplus h(ID||x)$, the hospital transfers $h(PW) \oplus h(ID||x)$ and Picture Base to $PatientU_i$. $PatientU_i$ stores $h(PW) \oplus h(ID||x)$, Picture Base in his/her mobile phone's smartcard.

4.2   Login and authentication phase

Step 1. $U_i \rightarrow OTP\ server : \{ID\}$
  After $OTPserver$ obtains identity ID, the $OTPserver$ encrypts ID and corresponding OTP with key K_{OH}. The $OTPserver$ generates picture shadow SH by corresponding picture base and OTP.

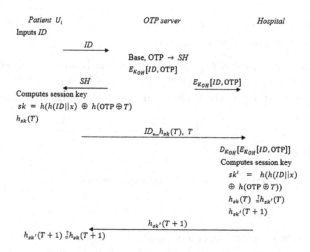

Fig. 6. Login and authentication phase of the proposed scheme

Step 2. $OTP\ server \to U_i : \{SH\}$

After the $OTPserver$ obtains the identity from the patient U_i, it can find the corresponding OTP and the picture base from the obtained ID. According to the corresponding OTP and picture base, the $OTPserver$ can determine the corresponding picture shadow SH and transfer it to patient U_i.

Step 3. $OTP\ server \to$ Hospital:$\{E_{K_{OH}}[ID,\text{OTP}]\}$

After the $OTPserver$ obtains the identity ID, the $OTPserver$ encrypts ID and corresponding OTP with secret key K_{OH} and delivers $E_{K_{OH}}[ID,\text{OTP}]$ to the hospital server. After that, the hospital server can obtain the identity of the patient and the corresponding OTP by secret key K_{OH} and can compute a session key with secret key K_{OH}.

Step 4. $U_i \to$ Hospital:$\{ID, h_{sk}(T), T\}$

Patient U_i computes session key $sk = h(h(\text{ID}||x) \oplus h(\text{OTP} \oplus T))$, which includes $h(\text{ID}||x)$ and $h(\text{OTP} \oplus T)$. According to $h(PW) \oplus h(\text{ID}||x), U_i$ can generate $h(\text{ID}||x)$ by his/her own password $h(PW)$, and U_i can compute $h(\text{OTP} \oplus T)$ by the generated timestamp T and then combines the obtained picture base in the registration phase and picture shadow SH in the login and authentication phase to generate OTP. Therefore, Patient U_i can compute $sk = h(h(\text{ID}||x) \oplus h(\text{OTP} \oplus T))$ and $h_sk(T)$.

Step 5. Hospital $\to U_i : \{h_{sk'}(T+1)\}$

The hospital computes session key sk' by the obtained identity ID and timestamp T. The hospital calculates $h(\text{ID}||x)$ from the obtained ID and master key x, which is generated by the hospital server, and the hospital server calculates $h(\text{OTP} \oplus T)$ by the timestamp T obtained from Patient U_i and decrypts $E_{K_{OH}}[ID,\text{OTP}]$ with key K_{OH} to obtain the OTP. Therefore, the hospital server can compute $h_{sk'}(T)$ and compare $h_{sk'}(T)$ to the obtained $h_{sk}(T)$. If $h_{sk'}(T)$ is equal to the obtained $h_{sk}(T)$, the hospital authenticates patient U_i, computes $h_{sk'}(T+1)$, and transfers $h_{sk'}(T+1)$ to patient U_i. Otherwise, the session is terminated.

After patient U_i obtains $h_{sk'}(T+1)$ from the hospital's server, patient U_i computes $h_{sk}(T+1)$ and compares $h_{sk}(T+1)$ to $h_{sk'}(T+1)$. If $h_sk(T+1) = h_{sk'}(T+1)$, patient U_i authenticates the hospital. Otherwise, the session is terminated.

5 Security analyses

In this section, we make some analyses on our protocol.

5.1 Replay attack

According to our proposed protocol, using the current timestamp to compute the session key $sk = h(h(\text{ID}||x) \oplus h(\text{OTP} \oplus \text{T}))$, we must obtain the current timestamp to calculate session key sk and h_{sk} (T). Suppose the attacker has current timestamp T' to compute session key sk, if and only if $\text{T}' = \text{T}$, then $h(h(\text{ID}||x) \oplus h(\text{OTP} \oplus \text{T}')) = h(h(\text{ID}||x) \oplus h(\text{OTP} \oplus \text{T}))$. However, it is not feasible to make this computation to obtain $\text{T}' = \text{T}$ during a different time slot. The attacker must obtain $\text{T}' = \text{T}$ during the same time slot.

Therefore, the attacker performs the computations infeasibly to launch a replay attack by letting $T_1, T_2, T_3, \ldots, T_n$ be equal to T. Thus, our proposed protocol can resist replay attacks.

The OTP will be different for every transaction. If an attacker launches a replay attack, the session key, which includes the OTP, is different in the two transactions. Therefore, attackers are unable to compute or guess the correct OTP when they launch replay attacks in our proposed scheme.

5.2 Denial of service

In our proposed scheme, we allow many users to log in to the OTP server many times. The OTP generator will send different OTPs to each user when many users want to obtain hospital service. In our authentication phase, the hospital calculates the session key and transfers $h_{sk}(T+1)$ to log in the patient who wants to authenticate the hospital's server. The patient only has to check whether $h'_{sk}(T+1) \overset{?}{=} h_{sk}(T+1)$ is equal or not. If the equation is equal, the patient authenticates the server, and the server authenticates the patient. The patient only has to compute $h'_{sk}(T+1)$ when the hospital's server wants to authenticate the patient. This is different from Xie et al.'s scheme, in which the medical server has to calculate $C'_1 = h(T_i||J||A)$ each time if many patients login at the same time. That will cause a large burden for the medical server since it must calculate C'_1 each time a patient logs in. However, in our proposed protocol, the server does not have to calculate timestamp each time when patients want to log in to the server. The hospital's server only has to calculate the session key in each session. Therefore, we consider our proposed protocol to be resistant to denial of service attacks.

5.3 Online and offline password attack

(1) Since we use a one-way, collision-free hash function $h()$, if one wishes to obtain $h(PW)$, it can be done if and only if the patient has obtained the correct PW. If the patient does not have PW and has PW' instead, he/she can obtain $h(PW') = h(PW)$ if, and only if, $PW' = PW$. Therefore, an attacker must guess the correct PW to compute $h(PW)$. However, there is a master key x to compute the hash function $h(PW) \oplus h(\text{ID}||x)$.

(2) However, suppose that the attacker has the password to derive $h(\text{ID}||x)$, and the attacker wants to guess session key sk to obtain service from the server. According to session key $sk = h(h(\text{ID}||x) \oplus h(\text{OTP} \oplus \text{T}))$, the attacker must obtain the current timestamp to compute session key sk to compute $h(\text{OTP} \oplus \text{T}))$. It is not possible to obtain $sk = h(h(\text{ID}||x) \oplus h(\text{OTP} \oplus \text{T}))$ without using the current timestamp T.

5.4 Server spoofing attack

Suppose an attacker spoofs an OTP server or hospital server and issues ID and OTP to the hospital's server. The server will compute a session key and transfer it to the patient in order to authenticate the patient. The attacker must know the current timestamp T and corresponding master key x to make $h'_{sk}(T+1)$ equal to $h_{sk}(T+1)$, which is computationally infeasible. Therefore, our proposed protocol is resistant to server spoofing attacks.

5.5 User impersonation attack

An attacker may transfer ID, $h_{sk}(T)$ and T to the hospital to impersonate a legal user. Before transferring the data to the hospital, the attacker must compute session key sk to appear to be a legal user. However, the attacker must have the correct password PW of the user. Because of the properties of the one-way, collision-free hash function, it is not feasible for the attacker to determine the PW of $h(PW)$. If $h(X_1) = h(X_2)$ if, and only if, $X_1 = X_2$. If $X_1 \neq X_2, h(X_1) = h(X_2)$ is computing infeasible. The one-way, collision-free hash function protects PW from $h(PW)$. An attacker cannot derive the session key without the password of a user. Therefore, it resists impersonation attacks.

5.6 Forward secrecy and backward secrecy

Our proposed scheme uses a one-time password as the password; picture base and picture shadow are available to generate a one-time password. The one-time password is a dynamic password that can only be used one time. Therefore, the attacker cannot use the password to derive the following password and the preceding password.

6 Conclusions

We have discussed the weaknesses of Xie et al.'s scheme and proposed our new protocol, which uses Shamir's secret key and one-time password generation. We described the weaknesses of Xie et al.'s scheme, and we found that submitting more than 3,850 login requests to the server will result in the JAVA cryptographic programming making the server deny any further service since the server computes $C_1' = h(T_i||J||A)$. Therefore, our protocol is more secure than Xie et al.'s method.

References

1. Wu, Z. U., Lee, Y. H., Lai, F. P., Lee, H. C., Chung, Y. F.: A secure authentication scheme for telecare medicine information systems. Journal of Medical systems, 36(3),1529–1535 (2012)
2. He, D. B., Chen, J. H., Zhang, R.: A more secure authentication scheme for telecare medicine information systems. Journal of Medical Systems, 36(3), 1989–1995 (2012)
3. Wei, J. H., Hu, X. X., Liu, W. F.: An improved authentication scheme for telecare medicine information systems. Journal of Medical Systems, 36(6), 3597–3604 (2012)
4. Zhu, Z.: An efficient authentication scheme for telecare medicine information systems. Journal of Medical Systems, 36(6), 3833–3838 (2012)
5. Muhaya, F. T.: Cryptanalysis and security enhancement of Zhu's authentication scheme for Telecare medicine information system. Security and Communication Networks, 8(2), 149–158 (2015)
6. Islam, S. K., Biswas, G. P.: Cryptanalysis and improvement of a password-based user authentication scheme for the integrated EPR information system. Journal of King Saud University - Computer and Information Sciences, 27(2), 211–221 (2015)
7. Jiang, Q., Ma, J. F., Ma, Z., Li, G. S.: A privacy enhanced authentication scheme for telecare medical information systems. Journal of Medical Systems, 37(1), 1–8 (2013)
8. Chen, H. M., Lo, J. W., Yeh, C. K.: An efficient and secure dynamic ID-based authentication scheme for telecare medical information systems. Journal of Medical Systems, 36(6), 3907–3915 (2012)
9. Islam, S. K., Khan, M. K.: Cryptanalysis and improvement of authentication and key agreement protocols for telecare medicine information systems. Journal of Medical Systems, 38(10), 1–16 (2014)
10. Zhang, L. P., Zhu, S. H.: Robust ECC-based authenticated key agreement scheme with privacy protection for telecare medicine information systems. Journal of Medical Systems, 39(5), 1–11 (2015)
11. Liu, W. H., Xie, Q., Wang, S. B., Hu, B.: An improved authenticated key agreement protocol for telecare medicine information system. SpringerPlus, 5(1), 1–16 (2016)
12. Xie, Q., Zhang, J., Dong, N.: Robust anonymous authentication scheme for telecare medical information systems. Journal of Medical Systems, 37(2), 1–8 (2013)
13. Shamir, A.: How to share a secret. Communications of the ACM, 22(11), 612–613 (1979)

Reversible authentication scheme for demosaicked images without false detection

Xiao-Long Liu, Chia-Chen Lin*, Cheng Han Lin, Li Juan Lin, and Bo Jun Qiu

College of Computer and Information Sciences,
Fujian Agriculture and Forestry University
Fuzhou 350002, China
Department of Computer Science and Information Management,
Providence University,
Taichung, Taiwan
shallen548@gmail.com, {mhlin3,s1012565,s1032809,s1032842}@pu.edu.tw

Abstract. The integrity of an image transmitted over the Internet is easily breached, and as such, research into image authentication has become increasingly urgent. In this paper, a new ecient fragile watermark-based authentication scheme for demosaicked images is proposed. In the proposed scheme, each watermark bit is embedded into two rebuilt components of each demosaicked pixel by using a designed authentication table. The watermarked image can be recovered back to the original demosaicked image if necessary. The experimental results demonstrated that our method preserves very satisfactory image quality and tamper detection accuracy. Most important, there was no false detection in the proposed scheme.

Keywords: Image demosaicking; Image authentication; Tamper detection; Fragile watermarking

1 Introduction

To maintain the full integrity of digital images transmitted over the Internet, there has been considerable research for several years into fragile watermark-based image authentication. Fragile watermark-based image authentication techniques are also known as strict authentication [1] and are designed to make the embedded watermark easily corrupted once the watermarked image is attacked by any kind of manipulation. Instead of resulting in meaningless encrypted data when directly applying traditional cryptography [2] to an original image, a watermark embedding procedure just modifies certain features of the original image and the resulting object remains a meaningful image. Most of the time, the distortion cannot be perceived by the human visual system and therefore it is not easy for malicious attackers to suspect that a watermark is placed in an image. Moreover, if an image is suspected of being manipulated or modified,

---

*

© Springer International Publishing AG 2017   313
J.-S. Pan et al. (eds.), *Advances in Intelligent Information Hiding and Multimedia
Signal Processing*, Smart Innovation, Systems and Technologies 63,
DOI 10.1007/978-3-319-50209-0_38

the authenticity and integrity of that suspicious image can be easily verified by watermarking-based image authentication techniques. Moreover, when the image is declared unauthentic, the tampered areas can be detected or even restored using the detection or recovery functions provided by the authentication techniques.

Digital still cameras are widely used in contemporary daily life to capture digital images of the real world. In general, a digital camera needs three sensors with three color filters to capture the R, B and G component of each color pixel, respectively. However, the hardware cost of having three sensors in a digital camera is quite expensive. To reduce hardware costs, a single sensor with a grid of dierent color sensors, called a color filter array (CFA) [3] is often used. The commonly used pattern for a CFA is the Bayer pattern, where only one component, i.e., red, blue or green, of each color pixel is significantly sampled by the corresponding grid. Inevitably, the use of CFA results in missing the other two components in each color pixel. To reconstruct the missing color components in each color pixel, an interpolation process called image demosaicking [4] is performed on the CFA sampled data, as shown in Fig. 1. Various image demosaicking schemes [5, 6] with varying reconstruction performance have been proposed.

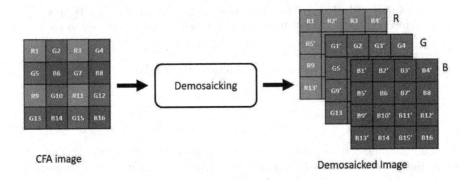

Fig. 1. Example of image demosaicking.

In order to protect the integrity of demosaicked images, a reversible fragile watermark-based image authentication scheme for image demosaicking was proposed by Hu et al. [7] in 2014. Although good tamper detection accuracy can be achieved by using this scheme, the visual quality of the watermarked image is unsatisfactory. The distortion also enlarges when increasing the length of the fragile watermark. Furthermore, the watermarked image cannot be recovered back to the original demosaicked image. To provide reversibility and improve tamper detection accuracy and visual qualities, two reversible image authentication schemes were proposed by Lin et al. [8] and Hu et al. [9] in 2015 and

2016, respectively. However, both of their schemes also introduce false detection, where several unmodified pixels are detected as modified pixels.

This paper proposes a new ecient fragile watermark-based authentication scheme for demosaicked images. In the proposed scheme, each watermark bit is embedded into the two rebuilt components of each demosaicked pixel by using a designed authentication table. If the watermarked image is tampered, the tampered areas of the image are accurately detected, otherwise, the watermarked image can be recovered back to the original CFA form when necessary. The proposed scheme was compared with some related schemes and showed outstanding performance.

2   THE PROPOSED SCHEME

This section presents a detailed exposition of the proposed reversible fragile watermark-based image authentication scheme. After obtaining a sampled CFA image, any potential image demosaicking process can be applied on it to acquire a demosaicked image. The proposed watermark embedding procedure works directly on the two rebuilt components of the demosaicked pixels. Since the original sample components are not modified during the watermark embedding procedure, the watermarked image can be recovered back to the original demosaicked image with CFA sampling and image demosaicking processes. If the watermarked demosaicked image needs authentication, the proposed tamper detection procedure can be directly implemented on the suspicious image.

2.1   Watermark Embedding Procedure

Assuming the demosacked image with H*W pixels is to be watermarked. A fragile watermark with total number of H*W watermark bits is first generated. The watermark is a sequence of random values from 0 to 24, to generate it, the pseudo random number generator (PRNG) with a predefined seed is used to generate H*W random values. Each random value rv is then converted in to the watermark bit wb by Eq. (1):

$$wb = rv \bmod 25. \tag{1}$$

For watermark embedding, a 5*5 authentication table (ACT) with values 0~24 is predesigned, note that any ACT with values 0~24 is suitable in this scheme, an example of it is shown in Fig. 2. Assume CP denotes the original color pixel with RGB components in the demosaicked image, where only one component of each CP is sampled by CFA, we denote it as o, and p and q denote the reconstructed two components of demosaicked pixel. Each watermark bit wb will be embedded into (p, q) by using the designed authentication table ACT.

To embed each watermark bit wb into (p, q), we first compute the coordinate values x and y by Eq. (2):

$$x = p \bmod 5, \; y = q \bmod 5. \tag{2}$$

4	16	23	5	22
10	17	24	6	13
11	18	0	7	14
12	19	1	8	15
3	20	2	9	21

Fig. 2. Example of a 5*5 authentication table (ACT).

where x and y represent the x-coordinate and y-coordinate in authentication table ACT, respectively. In accordance with coordinate x and y, the authentication value $av = ACT[x, y]$ can be found in the authentication table. If av equals wb, no change is made on (p, q). Otherwise, we need to adjust p and q into p' and q'. To adjust p and q, the coordinate value x' and y' of wb in authentication table is found. Then the difference between x and x', y and y' is calculated by Eq. (3):

$$dx = x' - x, \ dy = y - y'. \tag{3}$$

In order to minimize the distortion of modified pixel, dx and dy are then adjusted by Eq. (4):

$$
\begin{aligned}
dx &= dx + 5, if\ dx \leq -2 \\
dx &= dx - 5, if\ dx > 2 \\
dy &= dy + 5, if\ dy \leq -2 \\
dy &= dy - 5, if\ dy > 2.
\end{aligned}
\tag{4}
$$

The results include adjusted values to overcome the overflow and underflow problem, p' and q' will be calculated by Eq. (5):

$$
\begin{aligned}
p' &= dx + p, if\ 1 < p < 254 \\
p' &= dx + p + 5, if\ p \leq 1 \\
p' &= dx + p - 5, if\ p \geq 254 \\
q' &= dx + q, if\ 1 < q < 254 \\
q' &= dx + q + 5, if\ q \leq 1 \\
q' &= dx + q - 5, if\ q \geq 254.
\end{aligned}
\tag{5}
$$

An example of the watermark embedding procedure is described in the following. Assume the watermark bit wb to be embedded is 23 and the authentication table ACT is shown in Fig. 1. Given the color pixel CP= (157 , 82 , 229), where 157 is the sampled component o of CFA, 82 and 229 are the rebuilt components

p and q, respectively. By Eq. (2), x is calculated as 2 and y is calculated as 4. The authentication value av of (p, q) is $ACT[x][y]=ACT[2][4]=14$, which is not equal to wb. Since $wb=23$ is located in ACT[0][2], $x'=0$ and $y'=2$. According to Eq. (3) and Eq. (4), dx is calculated as -2, and dy is calculated as -2. Therefore, with Eq. (5), $p = 82$ and $q = 229$ is modified into $p' = 80$ and $q' = 227$, respectively. As a result, the original color pixel CP= (157 , 82 , 229) will be changed into (157 , 80 , 227).

2.2 Tamper Detection Procedure

The goal of the tamper detection procedure is to detect whether the given image is modified or not, and locate the tampered area if the given image is detected as modified. To detect the H*W sized image, the original fragile watermark with total number of H*W watermark bit wb is first generated by PRNG the same as watermark embedding procedure. Then the extracted watermark with a sequence of H*W authentication values av is generated from the given image by Eq. (2) and looking up the authentication table. After that, we can determine whether each pixel is tampered or not by comparing wb and av. If the watermark bit wb is equal to extracted authentication value av, this pixel is regarded as clear pixels and marked as white. Otherwise, this pixel is regarded as modified pixels and marked as black.

3   Experimental Results and Discussions

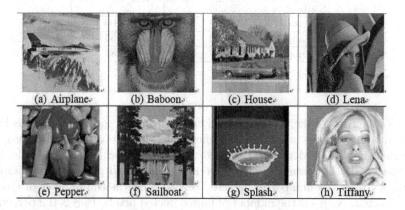

(a) Airplane	(b) Baboon	(c) House	(d) Lena
(e) Pepper	(f) Sailboat	(g) Splash	(h) Tiffany

Fig. 3. Eight test images.

This section presents some performance results of the proposed scheme. Subsequently, a performance comparison is done with other related work to demonstrate the superiority of the proposed scheme. All experiments were performed with eight commonly used full color images: Airplane, Baboon, House, Lena,

Pepper, Sailboat, Splash, and Tiffany, of the same size 512*512, as shown in Fig. 3. During experiments, the CFA images are captured by applying Bayer pattern to sample the original color images. Bilinear interpolation technique is selected to reconstruct the CFA image into demosaicked image.

Table 1 presents the results for peak signal-to-noise ratio (PSNR) values of the demosaicked images reconstructed by bilinear interpolation and the watermarked images of the proposed scheme. The average visual qualities of the demosaicked images and the watermarked image were 29.86 dB and 29.83 dB, respectively. The PSNR values of image Baboon were significantly less than the average value, because it is a complex image, and its adjoining pixels have low correlation. Therefore, the image demosaicking process for Baboon has a significant distortion compared to the other images. However, the characteristics of the images do not influence the proposed watermarking scheme. The PSNR loss from the demosaicked images to the watermarked image was less than 0.1 dB, which indicates that the watermark embedding procedure of the proposed scheme incurs only a very tiny distortion to the demosaicked images.

Table 1. Results of PSNR Values

Images	Bilinear Interpolation	Proposed scheme
Airplane	31.24	31.12
Baboon	23.60	23.58
House	29.33	29.26
Lena	31.98	31.84
Pepper	30.12	29.99
Sailboat	28.36	28.29
Splash	35.39	35.00
Tiffany	29.83	29.56
Average	29.98	29.83

To demonstrate the tamper detection performance, some of the pixels in the water-marked images were first modified, and then detected by the proposed scheme. The tamper detection results of the proposed scheme are listed in Table 2, where N_{MP} represents the total number of modified pixels in the tampered image, N_{DP} is the number of detected pixels, N_{TDP} is the number of true detected pixels, N_{FDP} is the number of false detected pixels, P_{TD} is the percentage of true detection in modified pixels and P_{FD} is the percentage of false detection in modified pixels. The average percentage of true detection P_{TD} of the proposed scheme is nearly 96%, this is because the extracted authentication value av of a modified pixel may be the same as watermark bit wb. The probability is 1/25 according to the designed authentication table. Therefore, only 4% of the modified pixels may miss by the detection. Most important, no false detection is introduced by the proposed scheme during tamper detection.

Table 2. Detection Results of The Proposed Scheme

Images	N_{MP}	N_{DP}	N_{TDP}	P_{TD}	N_{FDP}	P_{FD}
Airplane	29106	27931	27931	95.96%	0	0.00%
Baboon	29106	27992	27992	96.17%	0	0.00%
House	29106	27936	27936	95.98%	0	0.00%
Lena	29106	27974	27974	96.11%	0	0.00%
Pepper	29106	27934	27934	95.97%	0	0.00%
Sailboat	29106	27902	27902	95.86%	0	0.00%
Splash	29106	27923	27923	95.94%	0	0.00%
Tiffany	29106	27894	27894	95.84%	0	0.00%
Average	29106	27935	27935	95.98%	0	0.00%

To further demonstrate the watermarking performance, we compared the proposed scheme with LSB substitution[10], Lin et al.'s scheme[8] and Hu et al.s scheme [9] by using the averaging results of the eight test images. Table 3 shows the performance results in PSNR, true detection rate P_{TD} and false detection rate P_{FD} of each scheme. The proposed scheme and scheme [8] achieve the best PSNR values (29.83dB), followed by LSB scheme (29.81dB) and scheme[9] (29.72dB). The true detection rate P_{TD} of scheme [8], [9] and the proposed scheme are larger than 95%, which indicates that all of the schemes except LSB scheme (87.61%) can almost accurately detect every modified pixels in the tampered image. With respect to the false detection rate P_{FD}, the proposed scheme and LSB scheme achieve the best performance, where no false detection is introduced. However, the P_{FD} of scheme [8] and [9] are 1.21% and 7.10%, which means several clear pixels would be detected as modified pixels in their schemes. Thus, after considering the global performance of different schemes, the proposed image authentication scheme is demonstrably superior.

Table 3. Comparison of The Related Schemes

Schemes	PSNR	PTD	PFD
LSB[10]	29.81	87.61%	0.00%
Lin et al.s[8]	29.83	98.20%	1.21%
Hu et al.s[9]	29.72	99.83%	7.10%
Proposed	29.83	95.98%	0.00%

4   Conclusions

A reversible fragile watermark-based authentication scheme for demosaicked images was proposed in this paper. By utilizing a pre-designed 5*5 authentication table, a watermark bit was embedded into the two rebuilt components of each demosaicked pixel. The tamper detection accurate rate was nearly 96% and there was no false detection. Comparing the proposed scheme with LSB substitution[10], Lin et al.'s scheme[8] and Hu et al.s scheme [9], the experimental results demonstrated that our method preserved the best image quality while maintaining satisfactory tamper detection accuracy. Most importantly, the proposed scheme provided demosaicked image reversibility without any false detection.

Acknowledgments. This work is supported by MOST of Taiwan, No. 103-2632-E-126-001-MY3.

References

1. Haouzia, A. Noumeir, R.: Methods for image authentication: a survey. Multimedia Tools and Applications. 39, 1, 1–46 (2008)
2. Rivest, R. L.: The MD5 message-digest algorithm. Internet Activities Board, Internet Privacy Task Force. (1992)
3. Lukac, R. Plataniotis, K. N.: Color filter arrays: Design and performance analysis. IEEE Transactions on Consumer Electronics. 51, 4, 1260–1267 (2005)
4. Menon, D. Calvagno, G.: Color image demosaicking: an overview. Signal Processing Image. 26, 8, 518–533 (2011)
5. Lu, Y.M. Karzand, M. Vetterli, M.: Demosaicking by alternating projections: theory and fastone-step implementation. IEEE Transactions on Image Processing. 19, 8, 2085–2098 (2010)
6. Ferrandas, S. Bertalmio, Caselles, M. V.: Geometry-based demosaicking. IEEE Transactions on Image Processing. 19, 3, 665–670 (2009).
7. Hu, Y.C. Chen, W.L. Hung, C.H.: A novel image authentication technique for color image demosaicking. in Proceedings of 2014 International Conference on Information Technology and Management Engineering. Hong Kong. (2014)
8. Lin, C.C. Lin, C.H. Liu, X.L. Yuan, S.M.: Fragile watermarking-based authentication scheme for demosaicked images. 2015 International Conference on Intelligent Information Hiding and Multimedia Signal Processing (IIH-MSP). Adelaide, SA. (2015)
9. Hu, Y.C. Lo, C.C. Chen, W.L.: Probability-based reversible image authentication scheme for image demosaicking. Future Generation Computer Systems. 62, 92–103(2016)
10. Wang, R. Z. Lin, C. F. Lin, J. C.: "Image hiding by optimal LSB substitution and genetic algorithm. Pattern recognition. 34, 3, 671–683 (2001)

A New Image Encryption Instant Communication Method Based On Matrix Transformation

Jiancheng Zou[1], Tengfan Weng[1]

[1]Institute of Image Processing and Pattern Recognition, North China University of Technology.,100144,Beijing,China
{Tengfan Weng 948191403@qq.com}

Abstract. With the rapid development of the technology of big data mining and cloud computing. People can easily get images and videos from Internet by mobiles or iPads. How to protect the copyrights of the ownership of these media is a very important problem in information security, especially in the instant communication. In order to improve the security of the transfer processing of images and videos, an improved method of image and video encryption based on Arnold transformation is proposed in this paper. An image is split randomly into several regions to increase the security, each region is then encrypted by Arnold transformation. A security instant communication software is also suggested in this paper. The advantages of our methods include that the periodic recovery of encrypted image is removed and the calculation speed is faster than the traditional Arnold transformation. The methods and the software proposed in this paper are hopefully applied in mobiles or ipads to protect the instant communication of images and videos.

Keywords: Image encryption, Matrix translation, Arnold algorithm, Encrypted communication

1 Introduction

The rapid development of Internet promotes the arrival of the big data era. But the popularity of big data technology has brought new challenges to information security, if the enterprise user's personal information or business information even state secrets is obtained for criminals, it may provide convenience for the fraud and other illegal acts. Digital image is one of the carriers of information in the network, especially with the widely use of the instant communication software, larger numbers of images have been delivered to the internet. It also has the potential risk of information leakage. So it is necessary to encrypt the image before it is released on the network or deliver to

[1] This work was supported by National Science and Technology Major Project (2014ZX025020 03) and The National Natural Science Foundation of China (61170327).

© Springer International Publishing AG 2017
J.-S. Pan et al. (eds.), *Advances in Intelligent Information Hiding and Multimedia Signal Processing*, Smart Innovation, Systems and Technologies 63,
DOI 10.1007/978-3-319-50209-0_39

321

the receiver. In instant communication, not only the security of the encrypted image, but also the calculation speed is needed in order to save time for encryption and ensure it's instant. In 2001, digital image scrambling technology based on Arnold transformation by Ding Wei et al in [1]. An improved Arnold transform is presented in [2], which can enhance the efficiency of scrambling algorithm and watermarking security. An improved digital image scrambling method which can be used for the rectangle image by splitting rectangle image into several square images based on Arnold transform is proposed in [3].Zhou et al employ discrete parametric cosine transform to achieve simultaneous image encryption and compression in [4].An improved digital image scrambling method which can be used for the rectangle image by splitting rectangle image into several square images based on Arnold transform is proposed in [5].

This paper focuses on the practical needs of users for image information security transmission, the traditional Arnold transformation is analyzed and an improved algorithm is proposed [6-7]. Our method is to generate the transform matrix, whose elements are obtained from random number sequence, then the image is split into multiple square regions and each region is encrypted. Based on this method, instant communication software is designed by VS2012 software platform, which can encrypt images and videos instantly during the communication.

2 Arnold Transformation Algorithm

2.1 Arnold Transform of Digital Image

Arnold translation algorithm is proposed by VI Arnold, known as cat mapping, which can be applied to digital image. The definition of Arnold transformation is listed as formulas (1) [8]:

$$\begin{pmatrix} x' \\ y' \end{pmatrix} = \begin{pmatrix} 1 & 1 \\ 1 & 2 \end{pmatrix} \begin{pmatrix} x \\ y \end{pmatrix} mod|N \quad x,y \in \{0,1,\dots,N-1\} \tag{1}$$

The (x,y) is the coordinate of the pixel of the original image, and the (x',y') is the corresponding pixel in the tranformed image, the image must be a square image with a size of N×N. The position of the pixel is scrambled by Arnold translation so the original image is changed to another. We can transform the original image several times to encrypt the image. But Arnold has a cyclical, if we keep repeating the Arnold translation to a certain steps, the encrypted image may be restored to the original image. Such as the image with the size of 256×256, if we keep transforming the image in 192 times, the encrypted image will be the same as the original image[9]. The cyclical is shown in table 1. Figure 1 shows results of the Arnold algorithm on an image with an order of 256.

Table 1. Periods of Arnold transformation with different image sizes

Image sizes	25×25	50×50	100×100	120×120	128×128	256×256	512×512
Periods	50	150	150	60	96	192	384

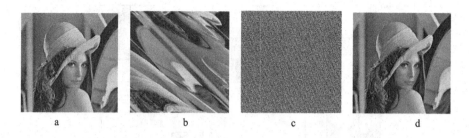

Fig.1. Results of the Arnold algorithm on a image with an size of 256×256. (a) Original image, (b) Image transformed once, c) Image transformed 10 times, (d) Image transformed 192 times

The recovery algorithm of Arnold transform has two methods; one is to use the periodicity of Arnold transform. For an image with an order of N, the period of Arnold transformation in the image is n (The method that calculating the period of an $N \times N$ image is described in [10]); if we have transformed the image in m times, we can rebuild the original image by continue the transform in $n - m$ times. The other method to restore an image is to seek the inverse matrix method to carry out the inverse transformation.

3   An Improved Arnold Transformation and the Software design

The improved Arnold transform can be described as follows [11]:

$$\begin{pmatrix} x' \\ y' \end{pmatrix} = \begin{pmatrix} a & b \\ c & d \end{pmatrix} \begin{pmatrix} x \\ y \end{pmatrix}$$

(2)

The elements of the matrix a, b, c, d are from random sequence, if the value of the $r = (a \times d - b \times c)$ and N(the size of image) are coprime, then the matrix can be used to encrypt the image . Then a square region is chosen on the image randomly and is encrypted with the random matrix (the area of the chosen region should be more than one fourth of the original image). The four vertexes of the image are selected from the image, then the right-bottom vertex and a pixel selected from the chosen region are used to construct a new square region (the size of the square region should be as large as possible), the new region is encrypted by a new random matrix. And then the right-top vertex and a pixel selected from the image are used to construct the next square region to be encrypted. The new algorithm keeps repeating these steps until all the four vertexes are used and all regions of the image are encrypted (as shown in figure 2 and figure 3). After that, the image is split into 5 regions and we have got five transformation matrices for each region. When we encrypt the second image, there is no need to generate 5 new matrices and choose five new regions again. The position and size of the first square region and the transform matrices will be recorded as the key so that we can encrypt the next image and restore the encrypted image in decryption.

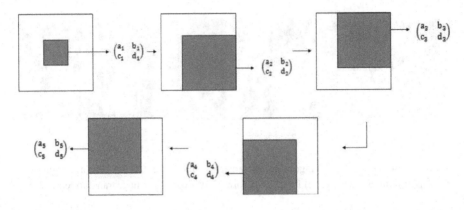

Fig.2. The procedure of encrypting an image

In the encryption procedure of each region, before the transformtaion, the period of the Arnold transformation n will be worked out and the transformation will be executed $n/2$ times to ensure the security. When the image is restored, the transformation will be executed $n/2$ times again for the region to be dercrypted, and each region will be decrypted in the opposite order of the encryption to decrypt the whole image.

The times of transformation can be decided by the size of region, and the size of each region can be concluded from the position and size of the first square region. So the key of encrytion contains the position and size of the first region and the five transform matrices (The key generator is shown in figure 5). The user can generate an appropriate key to encrypt their images.

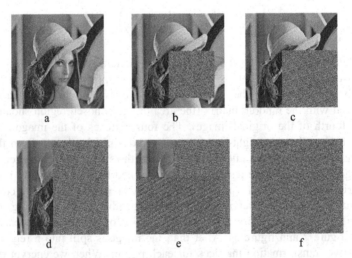

Fig. 3. The result of the encryption. (a)Original image, (b) Encrypt the first region, (c) Encrypt the second region, (d) Encrypt the third region, (e) Encrypt the fifth region, (f) Encrypt the last region

a   b

Fig.4. The encryption of a video (a) The original video, (b) The encrypted video

The new algorithm is implemented by using C++ programming language, and is integrated into an instant communication software designed by Visual Studio2012. The software with a key generator can encrypt the images and videos (a video can be split to images so that it can be encrypted, as shown in figure 4) instantly during the communication. The design of the module of the image encryption in the software is shown as figure 5.

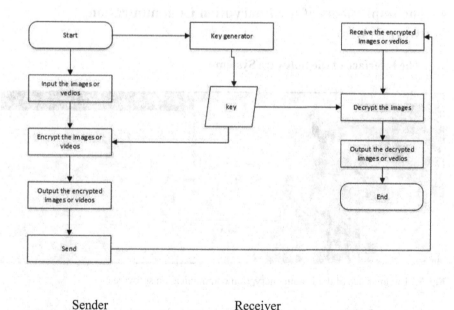

Sender   Receiver

Fig. 5. The design of the image encryption module

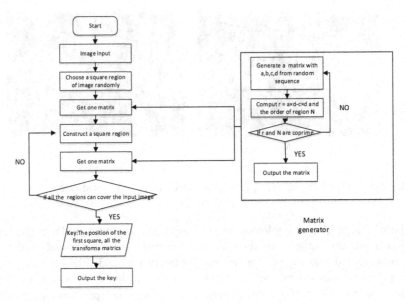

Fig. 6. The design of the key generator

4 The Implements of the Encryption Communication Software System

4.1 The Interface of the Software System

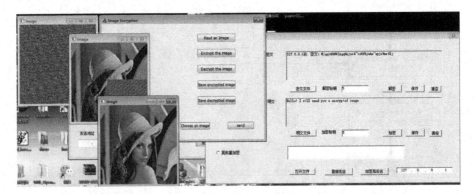

Fig. 7. The interface of the instant encryption communication software

The figure 7 shows the interface of the software, the users can encrypt an image to be sent and decrypt a received image instantly during communication [12]. And vedios can also be encrypted and decrypted effectively by using this software.

Different from the traditional method, our method is splitting the image into five regions and generating five different transform matrices for each region to be encrypted. The encryption procedure no longer has a cyclical[13]. Without the key, the recovery of an encrypted image is extremely difficult.

4.2 Attack Experiments Analysis

In order to verify whether the data is lost during the transmission of the encrypted image, the following two kinds of attack experiments are carried out.

4.2.1 Attack of Cut

The pixel of a part of the encrypted image is set to zero to cut the region with the size of 50×50, as shown in figure 8. The image decrypted by the algorithm is shown in figure 9.

a b

Fig. 8. The results of the attack of cut (a)The encrypted image which is cut ,(b)the image restore from the image which has been cut

Though the encrypted image is cut, after the decryption, the restored image is still close to the original image, with some black noise. The Signal-to-Noise Ratios of the images restored from the encrypted images with differnet size of region which has been cut is shown in table2. The noise will rise with the increasing of the cut region of the images.

Table 2. The PSNR of the restored images from the encrypted images which have been cut

Size of region been cut	30×30	50×50	80×80	100×100
PSNR	26.4537	21.4627	15.5390	13.7257

4.2.2 Attack of Noise

The images which are added Salt and pepper noise and Gaussian noise respectively are shown in figure 9.

a b c

Fig. 9. The original image and the noised images, (a) Original image, (b) Salt and pepper noise, (c) Gaussian noise

Then we add noise on the encrypted images, they are shown in figure 10.The decrypted images are shown in figure 11, and the PSNR are shown in table 3.

We can conclude from the experiment that if a decrypted image is attacked from the noise, the image restored from the decrypted image is almost the same as the original image. It is shown that this algorithm is relatively weak against noise attacks.

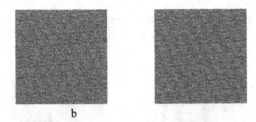

a   b

Fig. 10. The encrypted image after be added noise. (a) Salt and pepper noise, (b) Gaussian noise

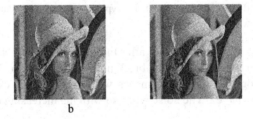

a   b

Fig. 11. The original image and the noised image (a) Salt and pepper noise, (b) Gaussian noise

Table 3. The PSNR of the restored images from the noised images

	Salt and pepper noise	Gaussian noise
PSNR of the original image	18.5555	19.5328
PSNR of the restored image	18.2357	18.9278

5   Conclusion

In this paper, a new image encryption method which can be used in instant communication is improved based on Arnold transformation. And the image encryption communication software is designed based on this method. The calculation speed and the security of encryption are both considered .Our method can encrypt images or videos with a fast speed, good security, and a good result against the attack of noise and cut, which can be well used in instant communication. And the software with a user-friendly interface can execute the encryption and decryption effectively during communication.

5   References

1. Ding Wei, Yan Weiqi, Qi Dongxu, J.:Digital Image Scrambling Technology based on Arnold Transformation.
   Journalof Computer-aided Design& Computer Graphics Vol.13,No.4, pp.338-341(2001)
2. Ding ,M., Jing,F., C.: Digital Image Encryption Algorithm Based on Improved Arnold Transform. International   Forum on Information Technology and Applications, pp.174-176(2010)
3. Yang Yali, Cai Na, Ni Guoqiang, J.: Digital Image Scrambling Technology Based on the Symmetry of Arnold Transform. Journal of Beijing Institute of Technology,15(2):pp.216-220(2006)
4. Zhou ,Y. Panetta, K. and Agaian, S. C.: Image encryption using discrete parametric cosine transform. Rec.43rd Asilomar Conf.Signals,Syst.,Comput.,vol.28,pp.395-399(2009)
5. Yang Yali, Cai Na, Ni Guoqiang, J.:Digital Image Scrambling Technology Based on theSymmetry of Arnold Transform. Journal of Beijing Institute of Technology, 15(2):pp.216-220(2006)
6. Chen,W., Quan ,C., Tay ,C.J., J.: Optical color image encryption based on Arnold transform and interference method. Elsevier Opt. Commun. vol.18, pp. 282-286(2009)
7. Saha,B.J., Pradhan,C.,Kabi,K.K., C.: Robust Watermarking Technique using Arnold's Transformation and RSA in Discrete Wavelets. International Conference on Information Systems and Computer Networks (ISCON), pp.83-87(2014)
8. Wang Xiaofan, C.:Information hiding techniques and applications. Beijing: Mechanical Industry Press, pp.85-109(2001)
9. Li Bing, Xu Jiawei, D.:Arnold transform and its application cycle. Acta Scientiarum Naturalium Universitatis Sunyatseni, 11Supplement (2) (2004)
10. Z hao Hui, J.: N-dimensional Arnold transformation and cyclical. Journal of North China University, (2002,)
11. Qi Dongxu, J.:Matrix Transformation and its Applications to ImageHiding. Journal of North China University of Technology, 11(1), pp. 24-28(1999)
12. David,M., Marie, N.,Bruno M., C.: A study Case on User Experience with IMS using Instant Communication Services. International Conference on IP Multimedia Subsystem Architecture and Applications.pp.1-5(2007)
13. Umamageswari, A., Suresh,G.R., C.: Security in Medical Image Communication with Arnold's Cat map method and Reversible Watermarking. International Conference on Circuits, Power and Computing T echnologies (ICCPCT), pp.1116-1121(2013)

A Three-Party Password Authenticated Key Exchange Protocol Resistant to Stolen Smart Card Attacks

Chien-Ming Chen, Linlin Xu, Weicheng Fang, and Tsu-Yang Wu

School of Computer Science and Technology,
Harbin Institute of Technology Shenzhen Graduate School,
chienming.taiwan@gmail.com, 980742703@qq.com,
980742703@qq.com, wutusyang@gmail.com

Abstract. Authenticated Key Exchange (AKE) is an important cryptographic tool to establish a confidential channel between two or more entities over a public network. Various AKE protocols utilize smart cards to store sensitive contents which are normally used for authentication or session key generation. It assumed that smart cards come with a tamper-resistant property, but sensitive contents stored in it can still be extracted by side channel attacks. It means that if an adversary steals someones smart card, he may have chance to impersonate this victim or further launch another attacks. This kind of attack is called Stolen Smart Card Attack. In this paper, we propose a three-party password authentication key exchange protocol. Our design is secure against the stolen smart card attack. We also provide a security analysis to show our protocol is still secure if sensitive information which is stored in a smart card is extracted by an attacker.

Keywords: authentication, key exchange protocol, stolen smart card attacks

1 Introduction

Secure communication over public network depends on secure encryption and an encryption is secure only if an encryption key is established securely. 2015 Turing awards laureates Diffie and Hellman [1] established a fundamental model for key setup as known as the Diffie-Hellman key exchange which inspires a series of Authenticated Key Exchange (AKE) protocols [2–5]. The purpose of a AKE protocol is to let two communication entities to authenticate each other and establish a common session key which is used for encryption and description the following communication.

Various AKE protocols require participants to keep some information as a secret, for example, a hashed value [6]. Certainly, keeping these secrets in mind may be the most secure way, but remembering it is indeed infeasible for human beings. For this reason, we need a device to store these sensitive information.

© Springer International Publishing AG 2017
J.-S. Pan et al. (eds.), *Advances in Intelligent Information Hiding and Multimedia Signal Processing*, Smart Innovation, Systems and Technologies 63,
DOI 10.1007/978-3-319-50209-0_40

Smart cards seem a reasonable choice because smart cards are portable and inexpensive. Now smart cards are widely used in lots of AKE protocols for different network environments and applications [5–10]. Although smart cards may come with a tamper-resistant property, sensitive contents stored in it can be extracted by some kinds of side channel attacks [10, 11]. More specifically, if a malicious attacker steals a victims smart card and extract the sensitive contents store in it, this attacker may have chance to impersonate this victim or further launch another attacks [12–15]. This kind of attack is called "Stolen Smart Card Attack" [16]. Now, the stolen smart card attack is one of the most important concerns when researchers design their AKE protocols if using smart cards. However, several AKE protocols are still found insecure against the stolen smart card attack [7, 8]. We also found a very recent three-party AKE protocol [6] cannot resistant to this attack [17].

In this paper, we propose a three-party password authenticated key exchange protocol based on Diffie- Hellman key exchange. Our design can be secure against the stolen smart card attacks. More specifically, even a malicious attacker obtains sensitive information stored in a smart card of a legitimate user, he still has no ability to impersonate this legitimate user. We further provide a security analysis to demonstrate our protocol is indeed sure against this kind of attack.

2   The Proposed Protocol

In this section, we propose a three-party password authenticated key agreement scheme using smart card. The proposed scheme has three phases, the initialization phase, the registration phase and the authentication and key agreement phase. Notations used in this section are listed as follows.

- U_i: a legitimate user i.
- S: a trusted server.
- ID_i, ID_S: The identity of user U_i and server S, respectively.
- PW_i: U_i's password.
- $h_1(\cdot)$: a one-way hash function.
- p: a large prime number.
- SK: an established session key

2.1   The Initialization Phase

In this phase, S initializes and chooses the following system parameters:

- A large prime number p.
- A primitive root α of p.
- A one-way hash function $h_1(\cdot)$.

Finally, S publishes the public parameters $\{p, \alpha, h_1(\cdot)\}$

2.2 The Registration Phase

This phase is invoked if a user U_i attempts to register himself to S. The following steps are performed.

Step 1: U_i freely selects his ID_i and PW_i then sends $\{ID_i, PW_i\}$ to S via a secure channel.

Step 2: After receiving $\{ID_i, PW_i\}$, S selects a random number $X_{iS} \in_R$ $[1, p-1]$ and computes $Y_{iS} = \alpha^{X_{iS}} \bmod p$ and $R_{iS} = (Y_{iS} + PW_i) \bmod p$. Then, S stores $\{ID_i, R_{iS}\}$ into a smart card and issues this smart card to U_i. S also stores $\{ID_i, X_{iS}\}$ in its database.

2.3 The Authentication and Key Agreement Phase

Assume that two users U_A and U_B try to establish a common session key, they perform this phase. Figure 1 illustrates our authentication key agreement phase.

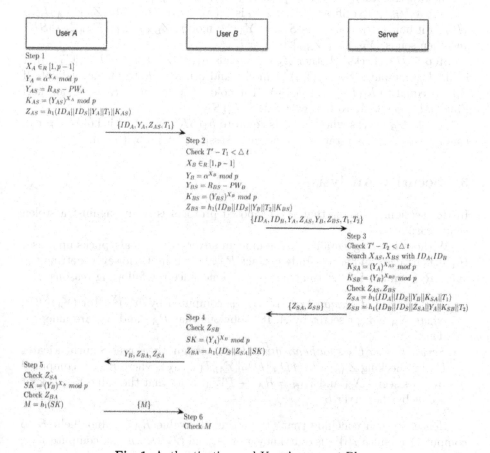

Fig. 1. Authentication and Key Agreement Phase

Step 1: U_A selects a random number $X_A \in_R [1, p+1]$ and computes $Y_A = \alpha^{X_A} \bmod p$, $Y_{AS} = R_{AS} - PW_A$ and $K_{AS} = (Y_{AS})^{X_A} \bmod p$. U_A also utilizes the current time T_1 as a timestamp to compute $Z_{AS} = h_1(ID_A\|ID_S\|Y_A\|T_1\|K_{AS})$. Then U_A sends $\{ID_A, Y_A, Z_{AS}, T_1\}$ to user U_B.

Step 2: Assume U_B receives the above messages at time T', he first verifies the freshness of the messages from U_A by checking if the difference between T' and T_1 is within a valid period Δ_t. Then, U_B selects a random numbers $X_B \in_R [1, p+1]$ and computes $Y_B = \alpha^{X_B} \bmod p$, $Y_{BS} = R_{BS} - PW_B$, and $K_{BS} = (Y_{BS})^{X_B} \bmod p$. Similarly, U_B utilizes the current time T_2 as a timestamp to computes $Z_{BS} = h_1(ID_B\|ID_S\|Y_B\|T_2\|K_{BS})$. After that, U_B sends $\{ID_A, ID_B, Y_A, Z_{AS}, Y_B, Z_{BS}, T_1, T_2\}$ to S.

Step 3: S first validates the freshness of the messages from U_B. Besides, S computes $K_{SA} = (Y_A)^{X_{AS}} \bmod p$ and $K_{SB} = (Y_B)^{X_{BS}} \bmod p$. Then, S verifies whether Z_{AS} is equals to $h_1(ID_A\|ID_S\|Y_A\|T_1\|K_{AS})$ and Z_{BS} is equals to $h_1(ID_B\|ID_S\|Y_B\|T_2\|K_{BS})$ respectively. If both equations hold, S further computes $Z_{SA} = h_1(ID_A\|ID_S\|Y_B\|K_{SA}\|T_1)$ and $Z_{SB} = h_1(ID_B\|ID_S\|Z_{SA}\|Y_A\|K_{SB}\|T_2)$. Eventually, S sends $\{Z_{SA}, Z_{SB}\}$ to user U_B.

Step 4: U_B first verifies whether Z_{SB} is equal to $h_1(ID_B\|ID_S\|Z_{SA}\|Y_A\|K_{SB}\|T_2)$. If it holds, U_B computes $SK = (Y_A)^{X_B} \bmod p$, $Z_{BA} = h_1(ID_S\|Z_{SA}\|SK)$ and then sends $\{Y_B, Z_{BA}, Z_{SA}\}$ to user U_A.

Step 5: U_A checks whether Z_{SA} is equals $h_1(ID_A\|ID_S\|Y_B\|K_{SA}\|T_1)$. If it holds, U_A computes $SK = (Y_B)^{X_A} \bmod p$ and utilizes SK to further validate if Z_{BA} is equal to $h_1(ID_S\|Z_{SA}\|SK)$. If it holds, U_A accepts SK as session key. Finally, U_A sends M to U_B, where $M = h_1(SK)$.

Step 6: U_B checks whether M is equal to $h_1(SK)$ or not. If it holds, U_B can confirm that U_A has established the same session key SK with him.

3   Security Analysis

In this section, we show that the proposed protocol is secure against a stolen smart card attack.

Without loss of generality, we assume an adversary E steals/pucks up a user U_A's smart card and successfully extract R_{AS} stored in it. However, getting the value R_{AS} cannot help E to attack our scheme since the following reasons.

- *The security of session key SK.* SK is computed by $(Y_B)^{X_A}$ (or $(Y_A)^{X_B}$), where X_A is kept secret by A. It is obvious that R_{AS} and X_A are independent.
- *Server to user U_A's authentication.* In our design, the server S authenticates U_A by checking $Z_{AS} = h_1(ID_A\|ID_S\|Y_A\|T_1\|K_{AS})$, where K_{AS} is computed by U_A's secret X_A and $Y_{AS} = R_{AS} - PW_A$. Note that the value Y_{AS} is kept secret by the server S.

Hence, we can conclude that (1) getting the value R_{AS} cannot help E to compute Y_{AS} since PW_A is only known by A, and (2) even E can compute Y_{AS}, E is not capable of computing K_{AS} because X_A is kept secret by A.

4   Conclusion and Future Works

Several AKE protocols have been shown insecure [19–22]. In this paper, we proposed a three-party password authenticated key exchange protocol. According to our security analysis, the proposed protocol is indeed secure against stolen smart card attacks.

Because of the space limitation of this submission, we only provide a simple security analysis to the proposed protocol. We put emphasis on the stolen smart card attack since this attack is our main highlight. In the future, we will demonstrate the proposed protocol is theoretically secure with a security proof using formal proof model by Abdalla and Pointcheval [18] Actually, we have almost finished this part. Besides, we will compare our design with previous related protocols to show our protocol is efficient. We also try to implement this protocol to mobile phones. This part is still a work in progress.

5   Acknowledgements

The work of Chien-Ming Chen was supported in part by the Project NSFC (National Natural Science Foundation of China) under Grant number 61402135 and in part by Shenzhen Technical Project under Grant number JCYJ2015051315170 6574. The work of Tsu-Yang Wu was supported in part by Natural Scientific Research Innovation Foundation in Harbin Institute of Technology under Grant number HIT.NSRIF.2015089.

References

1. Diffie, W., Hellman, M.: New directions in cryptography. IEEE transactions on Information Theory 22(6), 644–654 (1976)
2. Chen, C.M., Wang, K.H., Wu, T.Y., Pan, J.S., Sun, H.M.: A scalable transitive human-verifiable authentication protocol for mobile devices. IEEE Transactions on Information Forensics and Security 8(8), 1318–1330 (2013)
3. Sun, H.M., He, B.Z., Chen, C.M., Wu, T.Y., Lin, C.H., Wang, H.: A provable authenticated group key agreement protocol for mobile environment. Information Sciences 321, 224–237 (2015)
4. Farash, M.S., Attari, M.A.: An enhanced and secure three-party password-based authenticated key exchange protocol without using servers public-keys and symmetric cryptosystems. Information Technology and Control 43(2), 143–150 (2014)
5. Gope, P., Hwang, T.: An efficient mutual authentication and key agreement scheme preserving strong anonymity of the mobile user in global mobility networks. Journal of Network and Computer Applications 62, 1–8 (2016)
6. Li, X., Niu, J., Kumari, S., Khan, M.K., Liao, J., Liang, W.: Design and analysis of a chaotic maps-based three-party authenticated key agreement protocol. Nonlinear Dynamics 80(3), 1209–1220 (2015)
7. Yeh, H.L., Chen, T.H., Shih, W.K.: Robust smart card secured authentication scheme on sip using elliptic curve cryptography. Computer Standards & Interfaces 36(2), 397–402 (2014)

8. Islam, S.H., Khan, M.K.: Cryptanalysis and improvement of authentication and key agreement protocols for telecare medicine information systems. Journal of medical systems 38(10), 1–16 (2014)
9. Xie, Q., Hu, B., Wu, T.: Improvement of a chaotic maps-based three-party password-authenticated key exchange protocol without using servers public key and smart card. Nonlinear Dynamics 79(4), 2345–2358 (2015)
10. Farash, M.S., Kumari, S., Bakhtiari, M.: Cryptanalysis and improvement of a robust smart card secured authentication scheme on sip using elliptic curve cryptography. Multimedia Tools and Applications 75(8), 4485–4504 (2016)
11. Xie, Q.: A new authenticated key agreement for session initiation protocol. International Journal of Communication Systems 25(1), 47–54 (2012)
12. Chaudhry, S.A., Naqvi, H., Shon, T., Sher, M., Farash, M.S.: Cryptanalysis and improvement of an improved two factor authentication protocol for telecare medical information systems. Journal of Medical Systems 39(6), 1–11 (2015)
13. Lai, H., Xiao, J., Li, L., Yang, Y.: Applying semigroup property of enhanced chebyshev polynomials to anonymous authentication protocol. Mathematical Problems in Engineering 2012 (2012)
14. Farash, M.S.: Cryptanalysis and improvement of an improved authentication with key agreement scheme on elliptic curve cryptosystem for global mobility networks. International Journal of Network Management 25(1), 31–51 (2015)
15. Zhang, L., Zhu, S., Tang, S.: Privacy protection for telecare medicine information systems using a chaotic map-based three-factor authenticated key agreement scheme. IEEE Journal of Biomedical and Health Informatics (2016)
16. Zhao, F., Gong, P., Li, S., Li, M., Li, P.: Cryptanalysis and improvement of a three-party key agreement protocol using enhanced chebyshev polynomials. Nonlinear Dynamics 74(1-2), 419–427 (2013)
17. Chen, C.M., Xu, L., Wu, T.Y., Li, C.R.: On the security of a chaotic maps-based three-party authenticated key agreement protocol. Journal of Network Intelligence (2), 61–65 (2016)
18. Abdalla, M., Pointcheval, D.: Interactive diffie-hellman assumptions with applications to password-based authentication. In: International Conference on Financial Cryptography and Data Security. pp. 341–356. Springer (2005)5
19. Chen, C.M., Ku, W.C.: Stolen-Verifier Attack on Two New Strong-Password Authentication Protocols. IEICE Transactions on Communications, vol.E85-B, no.11, pp.2519-2521. (2002)
20. Ku, W.C., Chen, C.M., Lee, H.L.: Cryptanalysis of a Variant of Peyravian-Zunic's Password Authentication Scheme. IEICE Transactions on Communications, vol.E86-B, no.5, pp.1682-1684. (2003)
21. Ku, W.C., Chen, C.M., Lee, H.L.: Weaknesses of Lee-Li-Hwang's Hash-Based Password Authentication Scheme. ACM Operating Systems Review, vol.37, no.4, pp.19-25. (2003)
22. Sun, H.M., Wang, K.H., Chen, C.M.: On the Security of an Efficient Time-Bound Hierarchical Key Management Scheme. IEEE Transactions on Dependable and Secure Computing, vol. 6, no. 2, pp. 159160.(2009)

Printed in the United States
By Bookmasters